28th International Symposium on Space Terahertz Technology (ISSTT 2017)

Cologne, Germany
13-15 March 2017

ISBN: 978-1-5108-5972-2

Printed from e-media with permission by:

Curran Associates, Inc.
57 Morehouse Lane
Red Hook, NY 12571

Some format issues inherent in the e-media version may also appear in this print version.

Printed by Curran Associates, Inc. (2018)

For permission requests, please contact International Symposium on Space Terahertz Technology at the address below.

International Symposium on Space Terahertz Technology
c/o NRAO Headquarters
520 Edgemont Road
Charlottesville, VA 22903-2475
USA

Phone: (434) 296-0254
Fax: (434) 296-0278

mbishop@nrao.edu

Additional copies of this publication are available from:

Curran Associates, Inc.
57 Morehouse Lane
Red Hook, NY 12571 USA
Phone: 845-758-0400
Fax: 845-758-2633
Email: curran@proceedings.com
Web: www.proceedings.com

ISSTT 2017

March 13 – 15, 2017
Cologne | Germany

THE 28TH INTERNATIONAL SYMPOSIUM ON SPACE TERAHERTZ TECHNOLOGY

Hosted by the Kölner Observatorium für Submm-Astronomie (KOSMA),
I. Physikalisches Institut, Universität zu Köln

Program and Abstract Book

ISSTT 2017 in Cologne

group photo

It was an honour to host the 28th International Symposium on Space Terahertz Technology (ISSTT 2017). The I. Physikalisches Institut of the Universität zu Köln (University of Cologne), also known as the Kölner Observatorium für Submm Astronomie (KOSMA), organized the conference that was held from March 13th to 15th 2017.

We want to thank all participants for their contribution.

Local Organizing Committee

Patrick Pütz, Netty Honingh, Jürgen Stutzki, Frank Schlöder,
Florian Blauth, Johanna Böhm, Denis Büchel, Sina Fathi, Sina Widdig

Scientific Organizing Committee

The scientific organizing committee had met for during the lunch break on Monday. For this year we decided to invite all the reviewers, the traditional SOC members and a few more in order to have a more complete representation of the research groups worldwide. In particular, we wanted to express our gratitude towards the reviewers that ensured a fair and unbiased review process of all submissions.

Legend: [x] Reviewer, [+] not attending symposium.

Andrey Baryshev [x]	(University of Groningen, The Netherlands)
Victor Belitsky	(Chalmers University of Technology, Sweden)
Sergey Cherednichenko [x]	(Chalmers University of Technology, Sweden)
Tom Crowe	(Virginia Diodes Inc., USA)
Brian Ellison [x] [+]	(Rutherford Appleton Lab, UK)
Anne-Laure Fontana	(IRAM, France)
Jian-Rong Gao	(TU Delft, SRON)
Gregory Goltsman [x]	(Moscow State Pedagogical University, Russia)
Christopher Groppi [x]	(Arizona State University, USA)
Jeffrey Hesler [x]	(Virginia Diodes Inc., USA)
Netty Honingh [x]	(University of Cologne, Germany)
Heinz-Wilhelm Hübers [x]	(DLR Institute of Planetary Research, Germany)
Boris Karasik	(Jet Propulsion Laboratory, USA)
Valery Koshelets [x]	(Kotelnikov IREE RAS, Moscow, Russia)
Arthur Lichtenberger [x] [+]	(University of Virginia, USA)
Alain Maestrini [x]	(LERMA, CNRS, France)
Imran Mehdi [x]	(Jet Propulsion Laboratory / Caltech, USA)
Patrico Mena	(Universidad de Chile, Chile)
Patrick Pütz [x]	(University of Cologne, Germany)
Christophe Risacher [x]	(Max Planck Institute for Radio Astronomy, Germany)
Yutaro Sekimoto [x] [+]	(National Astronomical Observatory of Japan, Japan)
Sheng-Cai Shi [x]	(Purple Mountain Observatory, China)
Jan Stake [+]	(Chalmers University of Technology, Sweden)
Karl Schuster [x] [+]	(Institut de Radioastronomie Millimétrique, France)
Edward Tong [x]	(Harvard-Smithsonian Center for Astrophysics, USA)
Yoshinori Uzawa	(NICT, Japan)
Christopher Walker [x] [+]	(University of Arizona, USA)
Ghassan Yassin	(Oxford University, UK)

Invited talks

We were happy to present invited contributions from following speakers:

Alfred Krabbe (Deutsches SOFIA Institut, University of Stuttgart)

"SOFIA, the First Three Years of Full Operation"

Teun M. Klapwijk (Kavli Institute of Nanoscience, Delft University of Technology)

"Engineering the physics of superconducting hot-electron bolometer mixers"

Stephan Schlemmer (I. Physikalisches Institut, Universität zu Köln)

"Application of Terahertz Technologies in Laboratory Astrophysics"

Website

For detailed information on the conference and the final version of the abstract book in pdf format, please refer to the website:

http://www.astro.uni-koeln.de/isstt2017

Sponsors

We greatly value the contribution of our sponsors.

Raith GmbH and Radiometer Physics GmbH have a booth in our coffee break room.

Sonderforschungsbereich / Collaborative Research Council 956 funded by the Deutsche Forschungsgemeinschaft DFG

https://www.sfb956.de

Raith GmbH is a leading precision technology solution provider for nanofabrication, electron beam lithography, focused ion beam fabrication, nanoengineering and reverse engineering applications.

https://www.raith.com

Radiometer Physics GmbH is one of the leading suppliers for remote sensing, millimeter wave, sub-mm and THz instrumentation and components.

http://www.radiometer-physics.de

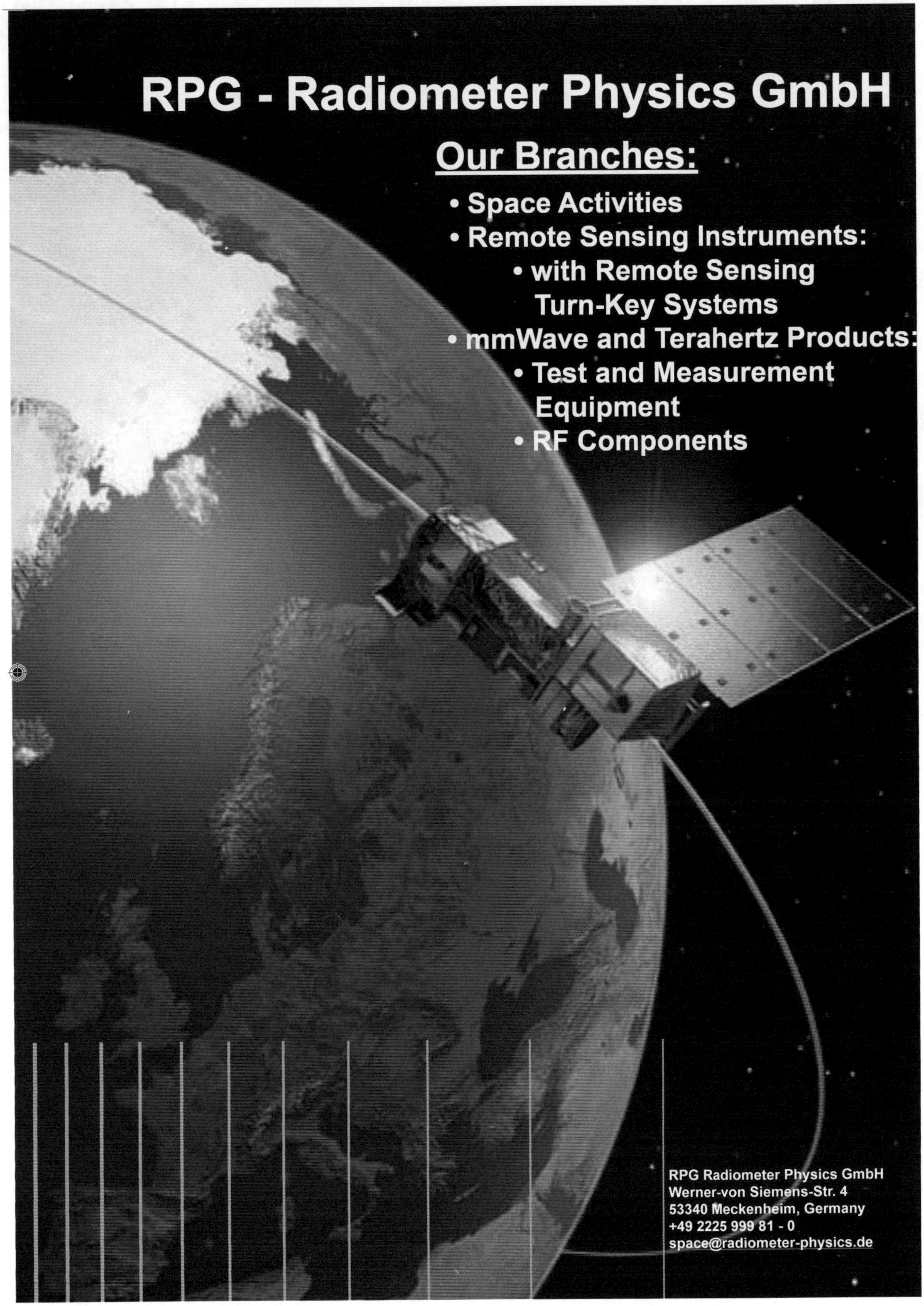

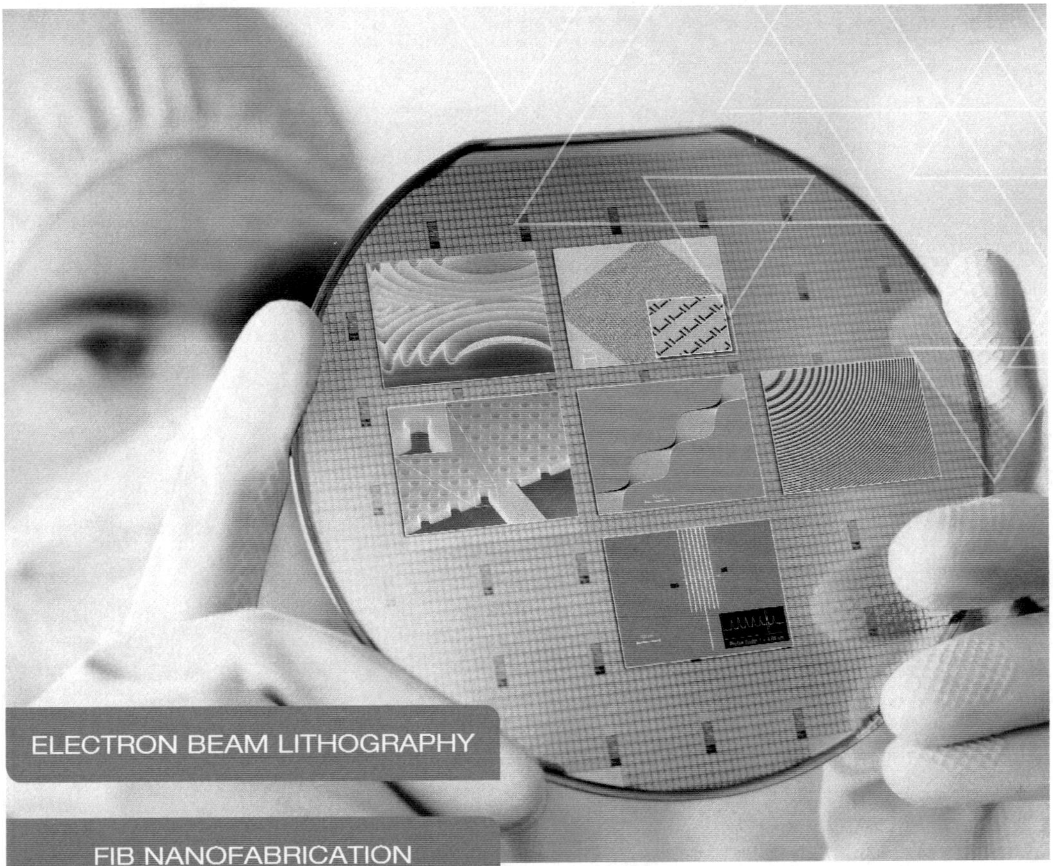

Overview of program

Time	Sunday Mar 12th	Monday Mar 13th	Tuesday Mar 14th	Wednesday Mar 15th	Thursday Mar 16th
8:00 AM		Registration			
8:30 AM					
9:00 AM		Welcome	Inv. talk T. Klapwijk	Inv. talk S. Schlemmer	
9:30 AM		Session M1 Receivers 1 (4 talks)	Session T1 Superconductor Devices (4 talks)	Session W1 MKID (4 talks)	
10:00 AM					
10:30 AM					
10:50 AM		Coffee break	Coffee break	Coffee break	
11:20 AM		Session M2 Mixers & Backends (4 talks)	Session T2 HEB Device Development (4 talks)	Session W2 Optics & Waveguide (4 talks)	
11:50 AM					
12:10 PM					Excursion to the Effelsberg radio telescope (1h bus ride, only 40 seats) departure 12:15 PM (near Physics building) return around 6:00 PM
12:40 PM		Lunch break and SOC lunch meeting	Lunch break	Lunch break	
1:10 PM					
2:00 PM		Inv. talk A. Krabbe	Session T3 Receivers 2 (5 talks)	Session W3 Sources (5 talks)	
2:30 PM		Session M3 Space & Calibration (5 talks)			
3:00 PM					
3:40 PM			Coffee break	Coffee break	
4:10 PM		Coffee break	Session T4 Supra-THz Mixers (5 talks)	Session W4 Semiconductor Devices & Receivers (4 talks)	
4:40 PM		Poster Session (cont'd to Wedn.)			
5:10 PM					
5:50 PM				Wrap-up & Farewell	
6:00 PM	Registration and Welcome Reception at the Wolkenburg	Tour of institute I		Tour of institute II	
6:30 PM					
7:00 PM			Reception for conference dinner (Wolkenburg)		
7:30 PM					
8:00 PM			Conference dinner at the Wolkenburg		
8:30 PM					
9:00 PM					
9:30 PM					
10:00 PM					
10:30 PM					
11:00 PM					
11:30 PM					
12:00 AM					Version V.1.4 Mar 09 2017

Symposium schedule

Sunday, March 12

6:00 PM	**Welcome Reception and Registration**	Wolkenburg, Mauritiussteinweg 59
9:00 PM	End	

Monday, March 13

8:00 AM	**Registration**	seminar rooms in front of lecture hall
9:00 AM	**Welcome**	lecture hall "Großer Hörsaal Botanik" Gyrhofstr. 15

Session M1 Receivers 1 Chair: Edward Tong

9:30 AM The upGREAT THz heterodyne arrays for SOFIA: 1.9 THz and 4.7 THz first results 1
Christophe Risacher — Max-Planck-Institute for Radio Astronomy

9:50 AM Demonstration and stabilization of a 4x2 HEB array receiver at 1.4 THz based on a Fourier phase grating LO 2
José Silva — SRON

10:10 AM The Ice Cloud Imager Front End Receivers onboard MetOp-SG satellite - Preliminary Design and Results 5
Bertrand Thomas — Radiometer Physics GmbH

10:30 AM Stratospheric Terahertz Observatory 2016, Sub-orbital flight from McMurdo, Antarctica 6
Abram Young — University of Arizona

10:50 AM **Morning Coffee Break, 30 min**
seminar rooms in front of lecture hall

Session M2 Mixers & Backends Chair: Christopher Groppi

11:20 AM A High-Performance 650 GHz Sideband-Separating Mixer — Design and Results 7
Ronald Hesper — University of Groningen

11:40 AM	67-116 GHz receiver development for ALMA Band 2+3 9 Pavel Yagoubov	European Southern Observatory (ESO)
12:00 PM	Achieving Ultra-High Sideband Separation in Millimeter and Sub-Millimeter Receivers 10 Patricio Mena	Universidad de Chile
12:20 PM	Back-ends for THz systems: Fast Fourier Transform Spectrometer 11 Bernd Klein	Max-Planck-Institute for Radio Astronomy
12:40 PM	**Lunch Break, 100 min** walk to University "Mensa" dining hall (for those with lunch tickets)	

Invited talk 1 — Chair: Imran Mehdi

2:00 PM	SOFIA, the First Three Years of Full Operation Alfred Krabbe	Deutsches SOFIA Institut, University of Stuttgart

Session M3 — **Space & calibration** — Chair: Andrey Baryshev

2:30 PM	The Far Infrared Spectroscopic Explorer: probing the lifecycle of the ISM in the Universe 12 Dimitra Rigopoulou	University of Oxford
2:50 PM	Spaceborne superconducting sounder (SMILES-2) for the upper-atmosphere observation 13 Satoshi Ochiai	NICT
3:10 PM	RF and thermal aspects of the ground calibration system for the Microwave Sounder Instrument 17 Manju Henry	RAL Space
3:30 PM	Development of Calibration Targets for MetOp-SG Microwave Instruments 18 Arne Schröder	University of Bern
3:50 PM	165-229 GHz Front End Receivers for the Microwave Sounder and Microwave Imager Instruments onboard MetOp-SG satellites - Preliminary Design and Results 20 Simon Rea	STFC RAL Space
4:10 PM	**Afternoon Coffee Break, 30 min, and start of** seminar rooms in front of lecture hall	

Poster Session seminar rooms in front of lecture hall

note: posters can stay up until noon Wednesday

P01 withdrawn

P02 A 350 GHz waveguide coupled MKID design 21
Florian Blauth University of Cologne

P03 Design of simply structured metamaterial filters at sub-THz frequencies 24
Johanna Böhm University of Cologne

P04 1.9 THz balanced superconducting Hot Electron Bolometer mixer fully integrated on chip 27
Sina Fathi University of Cologne

P05 Schottky components for ESA MetOp SG space mission 30
Oleg Cojocari ACST GmbH

P06 Design and optimization of wideband micro-patterned quasi-optical matching structures 31
Jake Connors Harvard University

P07 Micro-Machined Integrated Waveguide Transformers in THz Pickett-Potter Feedhorn Blocks 33
Kristina Davis Arizona State University

P08 The SAFARI grating spectrometer for SPICA 35
Gerhard de Lange SRON

P09 Design of Planar Antenna Arrays for Heterodyne Receivers 36
Duccio Delfini Paris Observatory, PSL

P10 Spectral Domain Simulation of SIS Frequency Multiplication 40
John Garrett University of Oxford

P11 Wideband waveguide power combiner for ALMA Band 7+8 (275-500 GHz) Local Oscillator 44
Alvaro Gonzalez National Astronomical Observatory of Japan

P12 withdrawn

P13 Modelling proximity effects in x-ray Transition Edge Sensors (TESs) for space-based applications 46
Rebecca Harwin University of Cambridge

presenter: Thijs de Graauw

P41 withdrawn

P42 Cryogenic IF Balanced LNAs Based on Superconducting Hybrids for Wideband 2SB THz receivers 105
Erik Sundin — Chalmers University of Technology

P43 An 8-Pixel Compact Focal Plane Array with Integrated LO Distribution Network 108
Boon Kok Tan — University of Oxford

P44 An All Solid-State Receiver at 2 THz for Atmospheric Sounding 111
Jeanne Treutel, — Jet Propulsion Laboratory
presenter: Imran Mehdi

P45 Theoretical consideration of SIS up-converters for frequency division multiplexing 112
Yoshinori Uzawa — NICT

P46 Pre-prototype ALMA Band 2+3 Down-Converter & Local Oscillator System 113
Hui Wang — STFC-Rutherford Appleton Laboratory

P47 Noise Temperature of a Wideband Superconducting HEB mixer 114
Kangmin Zhou — Purple Mountain Observatory

P48 GREAT's Internal Beam Scanner 116
Urs Graf — University of Cologne

P49 Resonant Modes in Parallel Josephson Junction Arrays for Submm Oscillator Applications 117
Faouzi Boussaha — Observatoire de Paris

Tour of institute (part 1) — I. Physikalisches Institut, Zülpicher Str. 77

6:00 PM Walk to institute
7:30 PM End

Tuesday, March 14

Invited talk 2 — Chair: Gregory Goltsman

9:00 AM — Engineering the physics of superconducting hot-electron bolometer mixers
Teun Klapwijk — Kavli Institute of Nanoscience, Delft University of Technology

Session T1 — Superconductor devices — Chair: Sergey Cherednichenko

9:30 AM — THz Heterodyne Sensors Using Superconducting MgB2 119
Boris Karasik — Jet Propulsion Laboratory

9:50 AM — Shot Noise in NbN/AlN/NbN Superconducting Tunneling Junctions 120
Dong Liu — Purple Mountain Observatory

10:10 AM — Effect of local non-uniformities of the tunnel barrier on the specific capacitance of Nb/Al-AlOx/Nb SIS junctions with extremely low RnA product 121
Parisa Yadranjee Aghdam — Chalmers University of Technology

10:30 AM — Titanium nitride for kinetic-inductance detectors: a problematic material or an engineering opportunity? 124
Eduard Driessen — IRAM

10:50 AM — **Morning Coffee Break, 30 min**
seminar rooms in front of lecture hall

Session T2 — HEB device development — Chair: Boris Karasik

11:20 AM — MgB2 THz HEB mixer with an 11GHz bandwidth 125
Sergey Cherednichenko — Chalmers University of Technology

11:40 AM — IF bandwidth of NbN HEB mixers on GaN buffer layer at 2 THz local oscillator frequency 126
Sergey Antipov, presenter: Gregory Goltsman — Moscow State Pedagogical University

12:00 PM — Design of a wideband balanced waveguide HEB mixer employing a GaN buffer-layer for the 1-1.5 THz band 128
Sascha Krause — Earth and Space Science, Chalmers

12:20 PM — MgB2 THz HEB mixer operation from 5K till 20K 132
Evgenii Novoselov — Chalmers University of Technology

12:40 PM **Lunch Break, 100 min**
walk to University "Mensa" dining hall (for those with lunch tickets)

Session T3 **Receivers 2** Chair: Jacob Kooi

2:00 PM The wSMA receivers - a new wideband receiver system for the Submillimeter Array 133
Paul Grimes Harvard-Smithsonian Center for Astrophysics

2:20 PM NOEMA Receivers: Upgrade for simultaneous dual band observations 134
Anne Laure Fontana IRAM

2:40 PM A Cartridge-type Multi-pixel Receiver for the 1.5 THz Frequency Band of GLT 135
Yen-Ru Huang ASIAA

3:00 PM 4GREAT: A multiband extension of GREAT from 490 GHz to 2.7 THz 138
Carlos Duran Max-Planck-Institute for Radioastronomy

3:20 PM Results from the Kilopixel Array Pathfinder Project (KAPPa): a 6mm × 6mm 650 GHz Heterodyne Mixer Pixel with Integrated SiGe LNA and Permanent Magnet 143
Christopher Groppi Arizona State University

3:40 PM **Afternoon Coffee Break, 30 min**
seminar rooms in front of lecture hall

Session T4 **Supra-THz mixers** Chair: Sheng-Cai Shi

4:10 PM Development of a 16-pixel monolithic 1.9 THz superconducting waveguide HEB mixer 144
Jonathan Kawamura Jet Propulsion Laboratory

4:30 PM 4.7 THz flight mixers for upGREAT 145
Patrick Pütz University of Cologne

4:50 PM 4-Pixel Heterodyne Receiver at 1.9 THz using a CMOS Spectrometer 146
Jenna Kloosterman Jet Propulsion Laboratory

5:10 PM Performance of NbN and NbTiN HEB waveguide mixers for GREAT and upGREAT 152
Denis Büchel University of Cologne

5:30 PM Silicon Micromachined Integrated 4-Pixel Heterodyne Receiver at 1.9 THz 153
Goutam Chattopadhyay NASA-JPL/Caltech

Conference dinner Wolkenburg, Mauritiussteinweg 59

7:00 PM	Reception	walk from institute
8:00 PM	Dinner	
Midnight	End	

Wednesday, March 15

Invited talk 3 Chair: Netty Honingh

9:00 AM	Application of Terahertz Technologies in Laboratory Astrophysics Stephan Schlemmer	 I. Physikalisches Institut, University of Cologne

Session W1 **MKID** Chair: Karl Jacobs

9:30 AM	Performance and surface wave reduction in large monolithic kinetic inductance detector arrays 154 Andrey Baryshev	 University of Groningen
9:50 AM	Deep Neural Networks for Tuning MKID Digital Readouts 155 Rupert Dodkins	 University of Oxford
10:10 AM	The effects of changes in bath temperature on Kinetic Inductance Detectors 159 Tejas Guruswamy	 University of Cambridge
10:30 AM	Photon-Counting with KID Resonators for THz/Submillimeter Space Spectroscopy 160 Omid Noroozian	 NRAO / NASA GSFC
10:50 AM	**Morning Coffee Break, 30 min** seminar rooms in front of lecture hall, **note: please remove posters**	

Session W2 **Optics & Waveguide** Chair: Ghassan Yassin

11:20 AM	Compact diffractive optics for THz imaging 161 Linas Minkevicius	 Center for Physical Sciences and Tech.
11:40 AM	Complex Beam Mapping of Large MKID Focal Plane Arrays 163 Kristina Davis	 Arizona State University

12:00 PM A spline-profile diagonal horn with low cross-polarization and reduced side lobes, suitable for integration into waveguide split-block THz devices. 166
Hugh Gibson — Gibson Microwave Design EURL

12:20 PM Measurement and design of a waveguide probe based WR3.4 optically controlled modulator 168
Jake Connors — Harvard University

12:40 PM **Lunch Break, 100 min**
walk to University "Mensa" dining hall (for those with lunch tickets)

Session W3 **Sources** Chair: Jiang-Rong Gao

2:00 PM Ultra-Compact THz Multi-Pixel Local Oscillator Systems for Balloon-borne, Airborne and Space Instruments 174
Jose V. Siles — NASA Jet Propulsion Laboratory

2:20 PM A continuous wave terahertz molecular laser pumped by a quantum cascade laser 175
Jean-Francois Lampin — IEMN/CNRS

2:40 PM Local Oscillator for a 4.7-THz Multi-Pixel Heterodyne Receiver Based on a Quantum-Cascade Laser 176
Heiko Richter — German Aerospace Center (DLR)
presenter:
Heinz-Wilhelm Hübers

3:00 PM A single-mode BCB-embedded antenna-integrated continuous wave quantum cascade laser for heterodyne measurement at 4.745 THz 177
Lorenzo Bosco — ETH Zürich

3:20 PM Solid State Terahertz Sources 178
Thomas Crowe — Virginia Diodes, Inc.

3:40 PM **Afternoon Coffee Break, 30 min**
seminar rooms in front of lecture hall

Session W4 **Semiconductor devices and receivers** Chair: Jeffrey Hessler

---

4:10 PM Local oscillator requirements and noise performance of a cryogenic 360 GHz Schottky diode subharmonic mixer 179
Diego Pardo — RAL Space

4:30 PM Spectroscopy around 245 GHz based on a SiGe Transmitter and

	Heterodyne Receiver 183	
	Nick Rothbart	German Aerospace Center
4:50 PM	The diode heterostructures for THz devices 184	
	Dmitry Pavelyev	Lobachevsky State University N. Novgorod
5:10 PM	Qualification of Direct Detection Technology for ESA MetOp-SG Space Mission 186	
	Matthias Hoefle	ACST GmbH
5:30 PM	**Wrap-up & Farewell**	

Tour of institute (part 2) — I. Physikalisches Institut, Zülpicher Str. 77

6:00 PM Walk to institute
7:30 PM End

Thursday, March 16

Excursion to Effelsberg Radio Telescope

12:15 PM departure behind Physics building, see map

6:00 PM approx. return

The upGREAT THz heterodyne arrays for SOFIA: 1.9 THz and 4.7 THz first results

C. Risacher[1], R. Güsten[1], J. Stutzki[2], H.-W. Hübers[3], D. Büchel[2], U. Graf[2], M. Greiner-Baer[3], S. Heyminck[1], C.E. Honingh[2], K. Jacobs[2], B. Klein[1,4], P. Pütz[2], H. Richter[3], N. Rothbart[3], O. Ricken[1]

1 Max Planck Institut für Radioastronomie, Auf dem Hügel 69, 53121 Bonn, Germany
2 KOSMA, I. Physikalisches Institut, Universität zu Köln, Zülpicher Strasse 77, 50937 Köln, Germany
3 German Aerospace Center (DLR), Institute of Planetary Research, Rutherfordstr. 2, Berlin, Germany

Abstract— We present the status and results of the upGREAT heterodyne array receivers for astronomy, used with the SOFIA airborne observatory, a 2.5-m telescope flying on a NASA/DLR Boeing 747. The low frequency array (LFA) will cover ultimately the 1.9-2.5 THz band with 14 pixels, and the high frequency array (HFA) covers the 4.745 THz line of atomic oxygen [O I] with 7 pixels. The frontend operates superconducting Hot Electron Bolometers (HEB) waveguide mixers. The local oscillators are based on commercial synthesizer driven solid-state multiplier chains for the LFA and a QCL for the HFA. Both receivers are cooled using closed cycle pulse tube refrigerators, reaching temperatures below 4 K.
The upGREAT LFA receiver, with its 14 channels, was successfully commissioned in 2015 covering parts of the frequency range 1.83-2.07 THz and has flown since then dozens of science flights. The HFA receiver was successfully commissioned in November 2016, with the 7 pixels at 4.745 THz reaching 800-1000 K DSB uncorrected receiver noise temperature at 0.5 GHz IF with an IF noise bandwidth of ~3.5 GHz. The two upGREAT receivers were used separately until now and from May 2017, they will be used simultaneously. This new system is already offered to the scientific community from observing Cycle 5 (2017).

Demonstration and Stabilization of a 2x4 HEB Array Receiver at 1.4 THz Based on a Fourier Phase Grating Local Oscillator

J. R. Silva[1,*], X. liu[1,2], B. Mirzaei[3], Y. C. Luo[1,3], D. J. Hayton[1], J. R. Gao[1,3], C. Groppi[4]
[1] *SRON Netherlands Institute for Space Research, Groningen/Utrecht, the Netherlands*
[2] *Microelectronics Department, Delft University of Technology, Delft, the Netherlands*
[3] *Kavli Institute of NanoScience, Delft University of Technology, Delft, the Netherlands*
[4] *School of Earth and Space Exploration, Arizona State University, AZ, USA*
[*]Contact: j.r.g.d.silva@sron.nl

Abstract— This paper is a report on the demonstration and stabilization of a 2x4 NbN hot electron bolometer (HEB) array receiver for GUSTO based on a multiple beam local oscillator (LO). Through the combination of a FIR Gas laser and a Fourier phase grating we were able to generate a 8 beam pattern at 1.4 THz. For the array demonstration, it used a 2x2 HEB mixer array which was translated vertically to match the full 2x4 LO beam pattern. Here two lens with a 60 mm were applied to reduce the beam pattern divergence, allowing for a 12 mm spacing between the beams to be achieved at 130 mm from the grating. To stabilize the array receiver, we introduced a new stabilization method, requiring only a voice coil PID controlled loop between the beam originated from the gas laser and a given pumped pixel. The total power Allan time measurements indicate a five times improvement on a two-pixel system stability.

INTRODUCTION

NbN hot electron bolometers (HEB) are currently the most suitable mixers for heterodyne terahertz astronomy (> 1 THz). The reasons for this are its operating frequencies ranges, low noise temperature and low local oscillator (LO) requirements for operation, although somewhat limited in terms of the intermediate frequency (IF) bandwidth. Thus far, this devices have been used in different types of astronomic telescopes in order to map different lines of terahertz radiation.[1]–[3] Since the mapping process is inefficient for single pixel receivers, multi-pixel arrays are now demanded for airborne (SOFIA) and balloon borne THz observatories (GUSTO), and future space telescopes (FIRSPEC, OST, TST). The use of multi-pixel arrays allows for an increase of the mapping speed while maintaining device performance.

GUSTO is a NASA balloon borne observatory, which will map three astronomic lines: [NI] at 1.4 THz, [CII] at 1.9 THz, and [OI] at 4.7 THz. All three lines will be measured simultaneously across the galactic plane of our Milky Way. This will be achieved using three independent 2x4 HEB array receivers. Until now some array receivers have been reported using different LO solutions: defocused LO[4], [5] or multiplexed LO[6]. Also, another possible solution would be the use of multiple LO sources as is being thought for the lower frequencies bands of GUSTO. In the case of the 4.7 THz array, it requires the use of quantum cascade lasers (QCL) as the LO source. This type of lasers are currently very limited regarding their power output and therefore using it as a divergent LO would require power levels not yet available. On the other hand, having multiple QCL's is also not an option because of the cooling footprint required to run them and the complex setup to frequency locking these devices. Due to these limitations, the best candidate solution is the multiplexing of the LO beam into sub beams, which can be achieved using a Fourier phase grating.

The Fourier phase grating,[7] is a reflective grating whose surface profile is based on the Fourier series expansion and can be tailored to transform a single incident LO beam into any desired beam pattern. In this work, two prototype gratings were designed to obtain a 2x2 and a 2x4 beam pattern respectively at an optimum angle of 25 degrees, for a central operation frequency of 1.4 THz. The simulations and experimental characterization have a good agreement, having the gratings measured efficiency of 66-73%. More details on these gratings can be found in other research.[8]

When dealing with heterodyne receivers, the IF time stability is a very important figure that determines the optimal observation strategy for an instrument.[9] To determine this figure it is employed the Allan variance method from which the Allan time is extracted.[10] The Allan time is defined as the optimum integration time of a given system, after which no improvement is achieved on the quality of the signal being measured. Some detailed studies using this method to characterize HEB heterodyne receivers can be found.[11]–[13] It has been found that HEB's suffer from poor stability performance mainly because of their sensitivity as direct detectors. Therefore, any fluctuations in the LO will be sensed by the device and is one of the main source of instability of the mixers. In this work we study the implementation of a HEB stabilization scheme introduced by Darren et al [13], with the goal of applying it to the stabilization of multi-pixel systems.

Measurement Schemes

Our experiments can be divided into two parts. One is the 2x4 array demonstration, where it used a Fourier phase grating with a 2x4 beam pattern. The second is the stabilization scheme implementation where we used another grating with a 2x2 beam pattern, allowing for more power in each beam, and making it easier to study the stabilization. In terms of measurement setup both are very similar. The setup to measure the Allan time is represented in Fig. 1. A more detailed description of the optical path, IF circuit and PID control loop can be found elsewhere[13], [14]. Using this setup, we make use of the four-beam pattern grating to match the array receiver, while allowing to measure simultaneously the IF power of two HEB. Here a PID feedback loop between one of the pixel and the voice coil can be used. Using this setup, the total power Allan time is measured.

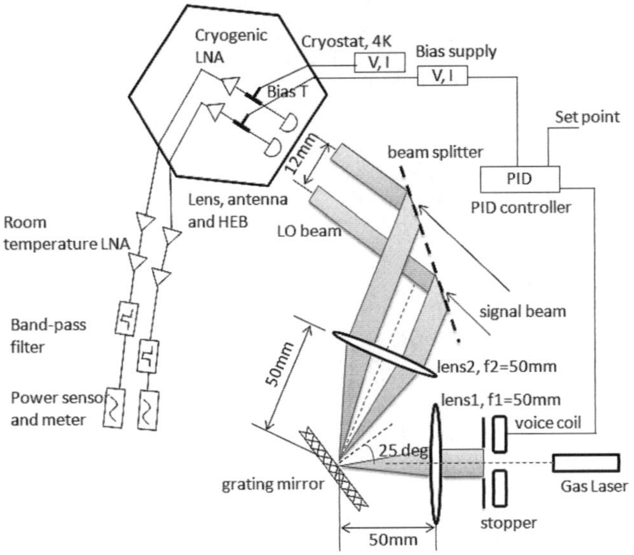

Fig. 1 Allan time measurement scheme. The optical path of the LO is similar for the 2x4 pixel array demonstration, with the devices being pumped directly from the grating instead of using a beam splitter.

A similar setup is used for the array demonstration. Since the beam pattern now is bigger (8 beams instead of 4), it requires an adaptation using lens with bigger diameter. In order to keep an acceptable F#, it was determined the best lenses required would have 60 mm focal length and 55 mm diameter. With this change, the distance from the gratings to the lenses changed to 60 mm. Besides this, the array is being pumped directly without the use of any beam splitter, and only de DC component is measured to obtain the IV curves.

Results

A. 2x4 Array demonstration

Since an eight-pixel array was unavailable, it was used a four-pixel array in a 2x2 configuration instead. This array receiver used for both experiments uses quasi-optical coupled HEBs with an optimum LO power requirement of 89±7 nW across the four pixels. The lenses are 10 mm diameter with a pitch size of 12 mm.

The beam pattern obtained at 130 mm from the grating can be seen in Fig. 2. It shows the two positions where the array was placed, by means of vertical translation of the cryostat. In the same figure, it also shows the IV curves obtained for such positions where it's possible to see an over pump of the devices, in both situations, effectively demonstrating the array. Although an over pumping is achieved, the mismatch of the array to the beam pattern at this position should also be noticed. This happens due to the design of angular offset of the out coming beams of the grating that was not optimized for the array block available, inducing some limits on the control of the beam pattern using optics, causing this mismatch to occur.

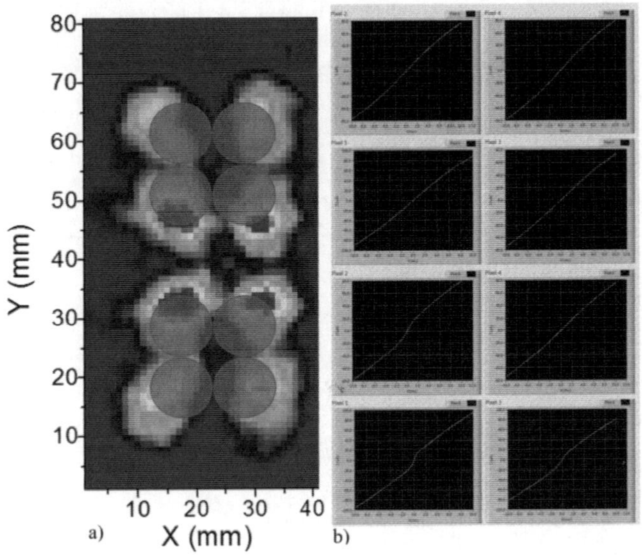

Fig. 2 Array demonstration results. a) Beam pattern at 13 cm from the grating with the placement position of the array to match the bottom and top four beams. The Y axis in the plot represents the Z dimension. b) Pumped IV curves obtained at the given arrays.

B. Array stabilization

When looking at the current of two pixels, both being pumped with a different beam originated at the grating, a similar drift structure together with other smaller effects were noticed. Cognizant of these effects, it was studied the current in frequency domain, to confirm the existence of a correlation. In Fig. 3 it can be seen the FFT for both currents. For higher frequencies, it can be seen some differences in the peaks intensity, but when looking at the first 20 Hz its clearly the existence of a correlation between both pixels.

Since the LO beams generated at the grating are all duplicated from the original, and the two current are found to be highly correlated, the possibility to stabilize both currents using a single PID controller was studied. In order to compare the differences between having both pixels running free (without stabilization) and the case where one is running free and the other being stabilized (using the PID feedback loop), the total power Allan variance was measured. In Fig. 4 the results obtained can be seen.

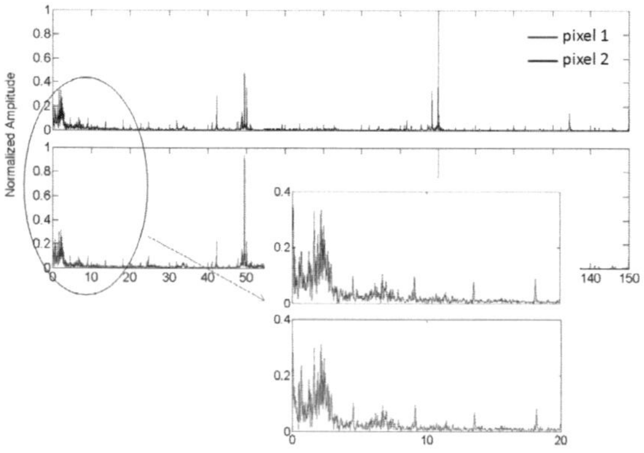

Fig. 3 Measured currents of two free running pixels in frequency domain.

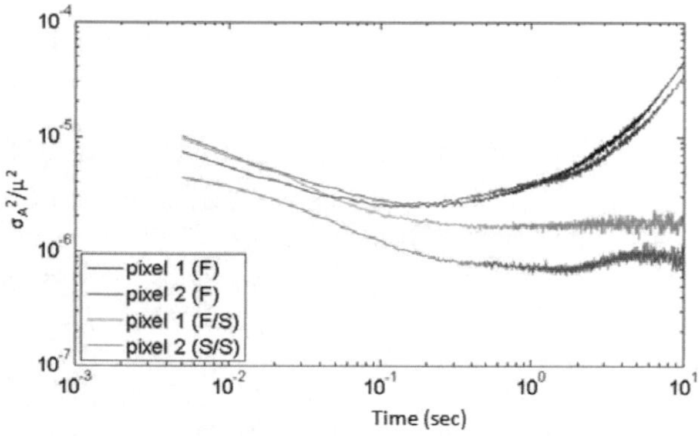

Fig. 4 Allan variance time for two IF chains when the HEB mixers are in optimal pumped stated. For the free running group, red and blue curve, the Allan time is 0.15s, while for the stabilization group, green and pink curves, the Allan time is 0.8s and 1.5 s respectively.

When letting both pixels to run free, without LO stabilization, the Allan time is very short, around 0.15 s. This result was expected since we are using a FIR gas laser, which amplitude is known to be highly unstable, which combined with the direct detection of HEB leads to low stability time. When applying the stabilization to pixel 2 however, there was an improvement of this pixel Allan time to 1.5 s. Although in this case pixel 1 is still running free, it also sees an Allan time improvement of around five times to 0.8 s. Based on this result, it seems the drift noise is highly reduced, but not the 1/f, resulting in a prolonged plateau on the Allan variance plot.

CONCLUSIONS

We successfully demonstrated an eight beam LO based on a Fourier phase grating to pump an 2x4 HEB array receiver at 1.4 THz. We demonstrate that this approach can be an efficient way of multiplexing a single LO beam although the angular offset of the out coming beams must be carefully designed, in order to optimize the match between the beam pattern and the array dimensions while avoiding undesired constrains on the optical path design.

Moreover, we achieved a five times improvement on the Allan time of a two-pixel system through the implementation of a PID control feedback loop between one pixel and the original LO beam, while leaving the second pixel running free. This result is a good indicator of the possibilities of this technique to stabilize an array receiver based on multiple beam LO. The next steps on this work will be the stabilization of the entire four-pixel array in use, and a thorough analysis of factors that critically affect the multi pixel stabilization.

ACKNOWLEDGMENT

One of the authors, J.R. Silva, would like to thank Wouter Laauwen for fruitful discussion on the array demonstration.

REFERENCES

[1] S. Heyminck, U. U. Graf, R. Güsten, J. Stutzki, H. W. Hübers, and P. Hartogh, "GREAT: the SOFIA high-frequency heterodyne instrument," Astron. Astrophys., vol. 542, no. L1, pp. 1–7, 2012.
[2] S. Cherednichenko, V. Drakinskiy, T. Berg, P. Khosropanah, and E. Kollberg, "Hot-electron bolometer terahertz mixers for the Herschel Space Observatory.," Rev. Sci. Instrum., vol. 79, no. 34501, 2008.
[3] D. J. Hayton et al., "A 4.7 THz Heterodyne Receiver for a Balloon Borne Telescope," in Proc. SPIE , Millimeter, Submillimeter, and Far-Infrared Detectors and Instrumentation of Astronomy VII, 2014, paper 91531R.
[4] F. Rodriguez-morales et al., "A Terahertz Focal Plane Array Using HEB Superconducting Mixers and MMIC IF Amplifiers," IEEE Microw. Wirel. Components Lett., vol. 15, no. 4, pp. 199–201, 2005.
[5] L. Liu, H. Xu, A. W. Lichtenberger, and R. M. Weikle, "Integrated 585-GHz hot-electron mixer focal-plane arrays based on annular slot antennas for imaging applications," IEEE Trans. Microw. Theory Tech., vol. 58, no. 7, pp. 1943–1951, 2010.
[6] C. Risacher et al., "First Supra-THz Heterodyne Array Receivers for Astronomy with the SOFIA Observatory," IEEE Trans. Terahertz Sci. Technol., vol. 6, no. 2, pp. 199–211, 2016.
[7] U. U. Graf and S. Heyminck, "Fourier gratings as submillimeter beam splitters," IEEE Trans. Antennas Propag., vol. 49, no. 4, pp. 542–546, 2001.
[8] B. Mirzaei et al., "Efficiency of multi-beam Fourier phase gratings at 1 . 4 THz," Opt. Express, vol. 25, no. 6, pp. 6581–6588, 2017.
[9] V. Ossenkopf, "Optimization of mapping modes for heterodyne instruments," Astron. Astrophys., vol. 495, no. 2, pp. 677–690, 2009.
[10] D. W. Allan, "Statistics of Atomic Frequency Standards," in Proc. IEEE, 1966, vol. 54, no. 2.
[11] J. W. Kooi et al., "Stability of heterodyne terahertz receivers," J. Appl. Phys., vol. 100, no. 64904, 2006.
[12] J. Chen et al., "Stability of superconducting hot electron bolometer receivers," IEEE Trans. Appl. Supercond., vol. 21, no. 3, pp. 667–670, 2011.
[13] D. J. Hayton, J. R. Gao, J. W. Kooi, Y. Ren, W. Zhang, and G. De Lange, "Stabilized hot electron bolometer heterodyne receiver at 2.5 THz," Appl. Phys. Lett., vol. 100, no. 81102, 2012.
[14] X. X. Liu, D. J. Hayton, Y. C. Luo, L. Wei, N. V. J. R. Gao, and C. Groppi, "A 2 x 2 Array Receiver at 1.4 THz based on HEB mixers and a Fourier Phase Grating Local Oscillator," in Proc. 26th International Symposium on Space Terahertz Technology, 2015, paper M2-4.

The Ice Cloud Imager Front End Receivers onboard MetOp-SG satellite – Preliminary Design and Results

B. Thomas*[1], M. Brandt[1], J. Goliasch[1], G. Sonnabend[1], T. Stangier[1], E. Orlandi[1], H. Gibson[1], W. Bauer[1], K. Lichius[1], A. Busch[1], R. Henneberger[1], A. Walber[1], C. Schmöller[2], S. Rea[3], K. Parow-Souchon[3], H. Wang[3], M. Henry[3], B. Moyna[3], B. Ellison[3], A. Andres[4], M. Bergada[4], N. Alcaraz[4], R. Gonzalez[4], M. Gotsmann[5], and U. Klein[6]

[1]*Radiometer Physics GmbH, 53340, Meckenheim, Germany*
[2]*Rohde & Schwarz, 81671, Munich, Germany*
[3]*STFC – Rutherford Appleton Laboratory, OX11 0QX, Didcot, United Kingdom*
[4]*Airbus Defence & Space SA.U., 28022, Madrid, Spain*
[5]*Airbus DS GmbH, 88039, Friedrichshafen, Germany*
[6]*ESA/ESTEC, 2201 AZ, Noordwijk, The Netherlands*
*Contact: Bertrand.Thomas@radiometer-physics.de

Abstract—MetOp-SG is a joint ESA/EUMETSAT program that will provide high quality weather monitoring data from a series of polar-orbiting satellites (A&B) over the 2020-2040 timeframe. One of the instruments onboard satellite-B will be dedicated to the remote sensing and atmospheric retrieval of high altitude ice clouds: the Ice Cloud Imager (ICI) is a conical scanning instrument featuring seven uncooled heterodyne receivers operating at five frequencies between 183 GHz and 664 GHz, two of them being dual-polarized.

RPG is responsible for the overall development of the ICI Front-End (FE) which will be mounted on top of the cylindrical rotating structure. The 183 GHz, 243 GHz and 325 GHz FE receivers development is under the responsibility of RAL, whereas RPG is undertaking the development of 448 GHz and 664 GHz channels, integration of receivers and test of all channels into the Front-End including horn antennas, and verification of the Front-End to Back-End operation. The ICI receivers rely on European technology for the high frequency Low Noise Amplifiers (IAF-Freiburg transistor MMICs), mixers and multipliers (Teratech and ACST Schottky diodes) to be qualified by RPG and RAL, and on US space grade MMIC & hybrid technology for Medium Power Amplifiers (AD/Hittite), IF Low Noise Amplifiers and DROs (L3/Narda/Miteq).

The preliminary design of the ICI FE receivers has been successfully reviewed by Airbus and ESA, and will be presented at this conference. Special focus will be put on the development of high frequency RF modules which were designed in a minimal envelope to meet the 8mm wide footprint imposed by the feed-cluster arrangement. For instance, semi-rigid coaxial waveguide with custom miniature flanges have been developed specifically for this purpose. First light RF results for all channels showing state-of-the-art performance in terms of Noise Figure will be presented. Short term gain stability (i.e. 1/f noise) has also been extensively investigated in order to ensure that this requirement can be met over the mission temperature and lifetime. Demonstration of compliance to these requirements are key to ensure the successful scientific & operational return of this mission.

Stratospheric Terahertz Observatory 2016, Sub-orbital flight from McMurdo, Antarctica

A. Young*[1], C. Walker[1], C Kulesa[1], P Bernasconi[2], R Dominguez[1], J. Siles[3], D. Hayton[3,4], J. R. Gao[4], W. Peters[1], P. Goldsmith[3]

[1]*University of Arizona Dept of Astronomy, 85721, Tucson, AZ, USA*
[2]*Johns Hokpins Applied Physics Laboratory, 20723, Laurel, MD, USA*
[3]*NASA Jet Propulsion Laboratory, 91109, Pasadena, CA, USA*
[4]*SRON Netherlands Institute for Space Research, 3584 CA, Utrecht, Netherlands*
*Contact: young@physics.arizona.edu

Abstract—The Interstellar Medium within our galaxy is constantly mixed by stellar winds and replenished by supernovae. It consists of ionized and neutral atoms, molecules, and dust. Warm diffuse regions form first, followed by denser, cooler regions containing neutral atoms and molecules, such as CO. From these molecular regions, nascent stellar systems form. Typical molecular clouds contain enough matter to form thousands of new star systems. A number of astrophysically important transitions of atoms and molecules occur at terahertz frequencies, beyond the reach of most, if not all ground-based observatories. It is this frequency region that is the focus of the Stratospheric Terahertz Observatory (STO).

We describe here the major subcomponents of STO, including the *cryogenic system*, *detector system* consisting of focal plane and local oscillator, *backend electronics*, and the *telescope and gondola* pointing system. The cryogenic system consists of a Ball Aerospace Lightweight Low-Cost cryostat with a liquid Helium volume of 94 liters and a Sunpower CT cryocooler. The cryogenic hold time in flight was more than 21 days. The detector focal plane was provided by SRON and contained 2x 1.46 THz and 2x 1.9 THz, and 1x 4.7 THz HEB mixers. The lower frequency bands were coupled to the telescope by twin-slot antennas and hemispherical silicon lenses, while the 4.7 THz channel is a similar HEB coupled via a spiral antenna. The local oscillator was delivered by the NASA – JPL Submillimeter-Wave Technology group. The 1.46 and 1.9 THz LO channels are active multiplier chains multiplying a synthesizer frequency by 108 times. At 4.7 THz, the LO was a Quantum Cascade Laser designed and constructed by MIT and packaged with a second Sunpower CT cryocooler at 50 K by SRON.

In this paper we describe the performance of the flight instrument. The telescope optics and gondola pointing system developed by Johns Hopkins Applied Physics Laboratory will also be reviewed. While data reduction is ongoing, preliminary data and first-light results from the commissioning phase are shown.

A High-Performance 650 GHz Sideband-Separating Mixer — Design and Results

Ronald Hesper[1]*, Andrey Khudchenko[1], Andrey M. Baryshev[1,2], Jan Barkhof[1], and F. Patricio Mena[3]
[1]*Kapteyn Astronomical Institute, University of Groningen, 9747 AD Groningen, The Netherlands*
[2]*SRON Netherlands Institute for Space Reasearch, 9747 AD Groningen, The Netherlands*
[3]*Electrical Engineering Department, Universidad de Chile, Santiago 8320000, Chile*
*Contact: R.Hesper@astro.rug.nl

***Abstract*—We designed, built and tested a new sideband-separating mixer assembly for the 600–720 GHz band (ALMA Band 9). By concentrating on the input matching and isolation of the quadrature hybrid and associated waveguide components, rather than on the phase and amplitude balance, we minimized standing waves and especially asymmetric reflection paths, which are highly detrimental to the image rejection ratio (IRR). IRRs in excess of 15 dB are obtained repeatably with different blocks and mixer pairs. At the same time, the SSB noise temperature is increased by not more than 20–30 K with respect to the bare mixer devices, corresponding to a loss of about 0.5 dB in the waveguide structure. A considerable contribution to the IRR are reflections in the IF system. If these are eliminated, *i.e.*, by using highly matched IF amplifiers, we expect worst-case IRRs of 18 dB or better can be reached, even in array configurations. In less demanding cases, the ample margin in IRR on the RF side can be used to build a system with reasonably matched amplifiers that still meets a typical 10 dB IRR specification. These 2SB mixers are intended for future sideband-separating receivers on the APEX (Chile) and LLAMA (Argentina) observatories and for deployment in any other observatory that would benefit from sideband separation in the 600–720 GHz band.**

I. INTRODUCTION

A new waveguide structure for a 600–720 GHz sideband-separating (2SB) mixer was recently presented [1], based on all our findings with the previous generation modular Band 9 2SB mixer [2], [3]. The waveguide structure follows the classical quadrature hybrid architecture, micromachined together with LO couplers and an in-phase LO splitter into a modular waveguide split-block [4]. The mixer devices are the same superconductor-insulator-superconductor (SIS) devices used in the current ALMA Band 9 receivers. The waveguide losses observed in the preceding design were minimized by choosing the waveguide dimensions as large as possible compatible with single-moded operation (400×200 μm), and machining them out of low-loss copper-tellurium alloy without gold plating.

Detailed study [5] reveals that the main mechanisms limiting the image rejection ratio (IRR) are asymmetrical reflection paths inherent in the 2SB architecture. Two such paths exist (Fig. 1): 1) reflections from each SIS device that pass back through the hybrid to interfere constructively at the RF-load port, after which any reflection from the load is redistributed equally over the mixers; and 2) reflections from either SIS device to the other by taking a "U-turn" through the hybrid

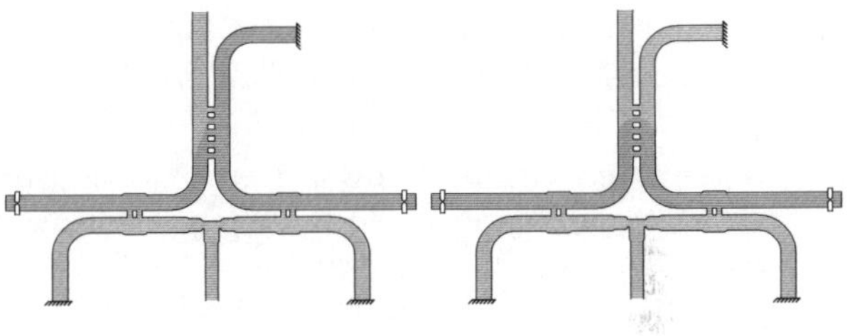

Fig. 1. The two reflection paths that cause most of the deterioration of the image rejection, when the contributions of the amplitude and phase balance are accounted for.

(corresponding to the hybrid's isolation parameter). Although the detailed accounting of ±90° phase shifts is different, the total effect is the same in both cases: whenever due to the overall phase rotation in the system the direct and reflected signals arrive in phase in one mixer, they are in precisely in anti-phase in the other, having a maximum detrimental effect on the the image rejection ratio. To optimize the IRR, both reflection paths were suppressed as much as possible, by improving the hybrid's idle port load, and by making the isolation of the hybrid one of its primary optimization goals.

II. HYBRID DESIGN

Fig. 2 shows a representative set of simulated S-parameters for the hybrid. The crucial isolation factor $|S_{21}|^2$ is below -24 dB within the band, considerably lower than in the previous design. At the same time, the gain and phase errors

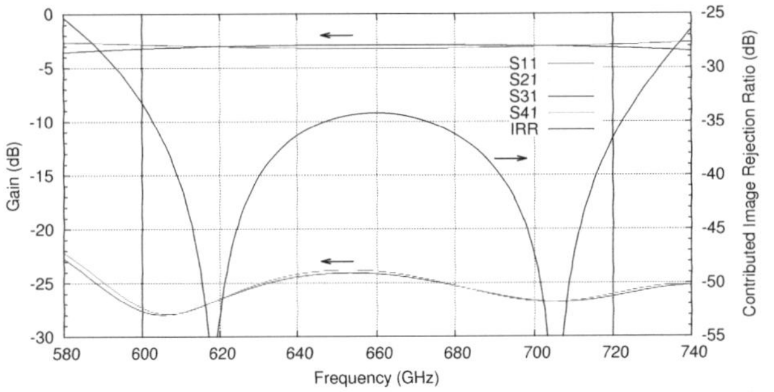

Fig. 2. Simulated S-parameters of the hybrid and the hybrid's contribution to the image rejection ratio ("IRR").

A FULL VERSION OF THIS PROCEEDINGS ARTICLE HAS BEEN SUBMITTED TO THE IEEE TRANSACTIONS OF TERAHERTZ SCIENCE AND TECHNOLOGY (TST) FOR PEER-REVIEW, PLEASE CHECK THE WEBSITE: HTTPS://WWW.MTT.ORG/TERAHERTZ

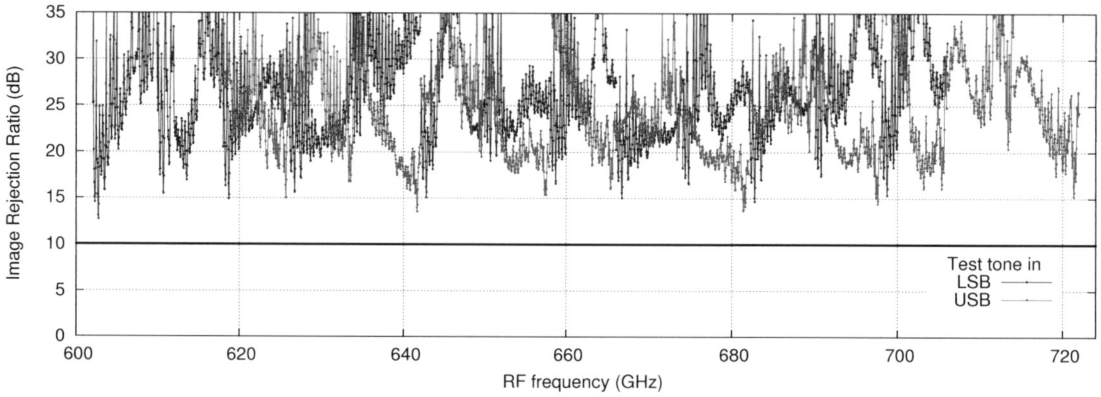

Fig. 3. Image rejection ration (IRR) of the first production hybrid block as function of RF observation frequency (*i.e.*, the frequency of the test tone used to determine the IRR). The typical specification for 2SB ALMA bands (10 dB minimum) is indicated with a horizontal line.

(not shown here) are within ± 0.4 dB and $\pm 0.4°$, respectively. The hybrid's contribution to the IRR, derived from the S-parameters, is plotted as well. The worst-case point in the band is about -33 dB, which gives the upper limit for the overall image rejection possible with this design.

III. Results

Fig. 3 shows the image rejection ratio obtained with one of the four production blocks. The IRR is above 15 dB in almost all points, with ample margin within the typical specification ($\geq$10 dB) of current receivers.

The single-sideband (SSB) noise temperature (Fig. 4), meets ALMA-class specifications with margin. To give an idea of the noise penalty incurred by the waveguide structures, the sum of the DSB noise temperatures of the individual mixer devices is plotted as well, showing an excess of about 20 K, corresponding to about 0.5 dB loss in the waveguide structure.

About a dozen different pairs of SIS mixers were tested. Most were chosen to match gain and normal-state resistance R_N as closely as possible, but some were mismatched on purpose. However, no simple correlation between the mixer gains as determined in DSB measurements (nor normal-state resistance R_N, noise temperature T_n or pumping current) and the obtained IRRs was observed. Geometrical proximity on the production wafer seemed to be the best predictor for high IRR, possibly due to gradients in the SiO_2 dielectric modifying the phase relations of the on-chip filter structures.

Apart from large-scale (order 10 GHz) patterns in the IRR resulting from residual imbalances in the RF circuit, there are persistent small-scale (sub-GHz) ripples, attributable to the IF system. Classically, IF ripples are suppressed by inserting isolators between the mixers and the LNAs. For 4-channel 4–12 GHz ALMA-style receivers, or small multipixel arrays, this becomes unpractical. In a simple experiment, where we removed the isolators from the IF chain and used MMIC LNAs with improved input reflection (≤ -7 dB), the worst-case IRR is reduced by about 5 dB, making it touch the typical 10 dB specification, while the noise temperature was reduced by about 10 K. The observed deterioration of the IRR is not simply due to standing waves, but probably to similar unbalanced interference effects as in the RF circuits.

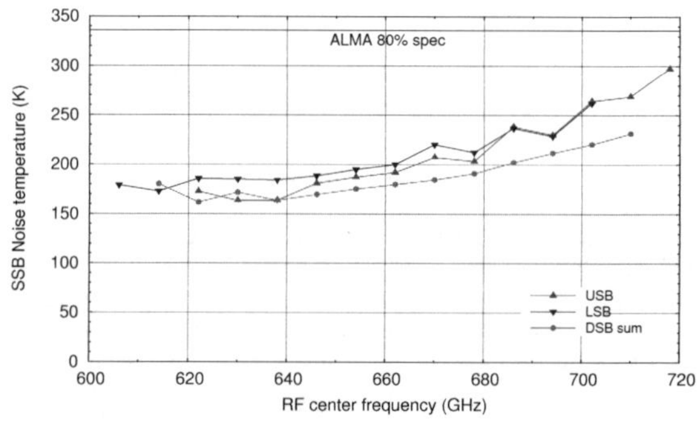

Fig. 4. Single-sideband noise temperature of both upper and lower sidebands as function of the RF frequency, as well as the sum of the DSB noise temperatures of the two individual SIS mixer devices.

References

[1] R. Hesper, A. Khudchenko, A. M. Baryshev, J. Barkhof, and F. P. Mena, "A new high-performance sideband-separating mixer for 650GHz," in *Society of Photo-Optical Instrumentation Engineers (SPIE) Conference Series*, vol. 9914, 2016, pp. 99 140G–99 140G-11. [Online]. Available: http://dx.doi.org/10.1117/12.2233065

[2] A. Khudchenko, R. Hesper, A. Baryshev, F. P. Mena, G. Gerlofma, T. Zijlstra, T. M. Klapwijk, J. W. Kooi, and M. Spaans, "First results of the sideband-separating mixer for alma band 9 upgrade," in *Twenty-Second International Symposium on Space Terahertz Technology, Charlottesville*, 2011, pp. 143–149.

[3] A. Khudchenko, R. Hesper, A. Baryshev, G. Gerlofma, J. Barkhof, J. Adema, P. Mena, T. Klapwijk, and M. Spaans, "Sideband separating mixer for 600-720 ghz for alma band 9 upgrade," in *Society of Photo-Optical Instrumentation Engineers (SPIE) Conference Series*, vol. 8452, sep 2012, p. 14.

[4] R. Hesper, G. Gerlofsma, F. P. Mena, M. C. Spaans, and A. M. Baryshev, "A sideband-separating mixer upgrade for alma band 9," in *Twentieth International Symposium on Space Terahertz Technology, Charlottesville*, apr 2009, pp. 257–260.

[5] A. Khudchenko, R. Hesper, A. M. Baryshev, J. Barkhof, and F. P. Mena, "Modular 2sb sis receiver for 600–720 ghz: Performance and characterization methods," *IEEE Transactions on Terahertz Science and Technology*, vol. PP, no. 99, pp. 1–8, 2016.

67-116 GHz receiver development for ALMA Band 2+3

P. Yagoubov[1], A. Gonzalez[2], V. Tapia[3], N. Reyes[3], F.P. Mena[3], R. Nesti[4], F. Cuttaia[5], S. Ricciardi[5], F. Villa[5], H. Wang[6], B. Ellison[6], G. Fuller[7], D. George[7], D Cuadrado-Calle[7], I. Lapkin[8], M. Fredrixon[8], V. Belitsky[8]

[1]*European Southern Observatory (ESO), Garching, Germany*
[2]*National Astronomical Observatory of Japan (NAOJ), Mitaka, Tokyo, Japan*
[3]*Universidad de Chile (U.Ch), Santiago, Chile*
[4]*Istituto Nazionale di Astrofisica (INAF/OAA), Arcetri, Italy*
[5]*Istituto Nazionale di Astrofisica (INAF/IASF), Bologna, Italy*
[6]*STFC-Rutherford Appleton Laboratory, Oxfordshire, OX11 0QX, UK*
[7]*The University of Manchester, Manchester, M13 9PL, UK*
[8]*GARD, Chalmers University of Technology, Gothenburg, Sweden*
*Contact: pyagoubo@eso.org

Abstract— The ALMA telescope is already a functional instrument delivering great science. However, it is not equipped with all receiver bands, and particularly band 2 has not been approved for construction yet. Recent technological developments in cryogenic MMICs open an opportunity to extend the originally planned bandwidth of this receiver (67-90 GHz), and potentially combine the two ALMA bands 2 and 3 (84-116 GHz) within a single receiver.

In this paper we report the design and first test results of the ALMA system compatible wideband 67-116 GHz receiver. Two types of the optics, feed horns and OMTs, have been designed to couple to the ALMA telescope beam using a modified Fresnel lens. Both types of hardware have been manufactured and tested in a near field beam scanner. The measured beam patterns and optical efficiencies are compliant to ALMA specifications and in good agreement with simulations.

The receiver system uses cryogenic low noise amplifier (LNA) at its input, developed using the state-of-art 35 nm gate length InP HEMP process of NGC. The MMIC design process was performed with individual simulation of the different matching networks using the electromagnetic (EM) simulator Momentum; a tool included with the Keysight ADS package. The best LNAs tested so far show a noise temperature less than 28 K from 70 GHz to 110 GHz at cryogenic ambient temperature of 20 K.

The amplified by the LNA signal is frequency translated (down converted) to the ALMA intermediate frequency (IF) range of 4-12GHz. The down-convertor comprises a Schottky barrier based subharmonic sideband separating mixer, local oscillator (LO) and IF amplifier chain. Design simulations indicate the expected mixer noise and conversion loss performance to be approximately 1000K (SSB) and -8dB respectively for a typical LO power level of +8dBm.

A fully compatible to ALMA electrical and mechanical interfaces, cartridge type cold receiver has been designed and built to accommodate receiver optics and LNAs, while the downconverter is located in the warm cartridge assembly at room temperature. The system has been assembled and its characterization is currently underway, results of the measurements will be reported at the Conference.

Achieving Ultra-High Sideband Separation in Millimeter and Sub-Millimeter Receivers

R. Finger[1], R. Rodríguez[1], A. Alvear[1], F. P. Mena[2*], R. Fuentes[1], F. Curotto[1], A. Baryshev[3,4], A. Kudchenko[3], R. Hesper[4], J. Barkhof[4] and L. Bronfman[1]

[1]*Astronomy Department, Universidad de Chile, Camino El Observatorio 1515, 7591245 Santiago, Chile*
[2]*Electrical Engineering Department, Universidad de Chile, Avenida Tupper 2007, 8370451 Santiago, Chile*
[3]*SRON Netherlands Institute for Space Research, Postbus 800, 9700 AV Groningen, The Netherlands*
[4]*NOVA/Kapteyn Astronomical Institute, University of Groningen, Postbus 800, 9700 AV Groningen, The Netherlands*
[*]Contact: fpmena@u.uchile.cl

Abstract—During the last years we have been working in implementing real-time calibrated sideband-separating (2SB) receivers using FPGA technology under different conditions. Briefly speaking, this technique consists in replacing the functionality of the analogue IF hybrid of a 2SB receiver by a digital tool. In this way it is possible to calibrate out the phase and amplitude imbalances that limit the sideband-rejection ratio (SSR) of all-analogue 2SB receivers. After a first demonstration at very low frequencies (below 4 GHz), we have implemented successfully this technique, under laboratory conditions, at the millimeter (80−110 GHz) and sub-millimeter (600−720 GHz) ranges. In all these cases we have achieved sideband rejection ratios above 35 dB across the whole pertinent RF band. More recently, we have been working towards two different goals. First, the implementation of this technique at operating conditions. For that purpose we have used our own 3-mm survey telescope demonstrating that a high SRR can be obtained. The second goal is to correct amplitude and phase imbalances in 2SB receivers containing analogue IF hybrids. We have implemented this correction in a full-analogue ALMA Band-9 receiver, which demonstrates that this technique can be implemented in any existing receiver with no changes at the front-end side. In this paper we will summarize all these efforts and demonstrate the robustness of the technique.

We would like to acknowledge CONICYT funding through grants Center of Excellence in Astrophysics and Associated Technologies (CATA) PFB06, Gemini-CONICYT 32120004, ALMA-CONICYT 31150012, FONDECYT 11140428 and Quimal 140002.

Back-ends for THz systems: Fast Fourier Transform Spectrometer

Bernd Klein[*,1,2], Stefan Hochgürtel[1], Ingo Krämer[1], Andrea Bell[1], Klaus Meyer[1], and Rolf Güsten[1]

[1]*Max-Planck-Institut für Radioastronomie, Bonn, Germany*
[2]*University of Applied Sciences Bonn-Rhein-Sieg, Sankt Augustin, Germany*
[*]Contact: bklein@mpifr-bonn.mpg.de

Abstract—Since a few years digital Fast Fourier Transform Spectrometers (FFTS), based on high-speed analog-to-digital converter (ADC) and high-performance field-programmable gate array (FPGA) chips, have become a standard for heterodyne receivers, particularly in the mm and submm wavelength range. They offer high instantaneous bandwidths with many thousands spectral channels and have been proven to be extremely reliable and robust with Allan-stability times of several 1000 seconds.

At the Max-Planck-Institut für Radioastronomie, the FFTS technology has been advanced over the last 13 years from 50 MHz to 4 GHz instantaneous bandwidth today. Our current wide-band FFTS4G spectrometer board offer 4 GHz bandwidth with 65536 (64k) spectral channels. For the first time, this novel FFTS allows base-band (0 - 4 GHz) and direct IF-sampling (4 – 8 GHz, in the 2nd Nyquist band) without IF mixing. Currently two FFTS4G systems are installed: a 16 board system aboard SOFIA and an eight board FFTS4G at APEX to serve the new PI 230 GHz receiver.

The dual-FFTS4G is our newest development and offer 2 x 4 GHz instantaneous bandwidth with up to 128k spectral channels on one single 160 x 100 mm euro-sized board. This new spectrometer will serve the upcoming detector arrays for SOFIA (upGREAT and 4GREAT) as well as the new receiver generation for APEX.

The announcement of a new 12-bit ADC with even higher sample rate and wider analog input bandwidth, together with the still increasing processing capability of future FPGA chips, make it very likely that FFT spectrometer can be extended to even broader bandwidth with adequate numbers of spectral channels in the near future. Our next FFTS development will be the design of a spectrometer board that will allow analyzing 13 GHz of bandwidth.

The Far Infrared Spectroscopic Explorer: probing the lifecycle of the ISM in the Universe

Dimitra Rigopoulou[*], Brian Ellison[2], M. Caldwell[2], G. Yassin[1], M. Wiedner[3], V. Ossenkopf-Okada[4] and the FIRSPEX Consortium

[1]*University of Oxford, OX1 3RH, Oxford, UK*
[2]*RAL Space, STFC, OX11 0QX, Didcot, UK*
[3]LERMA, Observatoire de Paris,F74014, France
[4] I. Physikalisches Institut, Universität zu Köln,50937, Köln, Germany
[*]Contact: dimitra.rigopoulou@physics.ox.ac.uk

Abstract— The Far Infrared Spectroscopic Explorer (FIRSPEX) is a novel European-led astronomy mission concept developed to enable large area ultra high spectroscopic resolution surveys in the THz regime. FIRSPEX opens up a relatively unexplored spectral and spatial parameter space that will produce an enormously significant scientific legacy by focusing on the properties of the multi-phase ISM, the assembly of molecular clouds in our Galaxy and the onset of star formation; topics which are fundamental to our understanding of galaxy evolution. The mission uses a heterodyne instrument and a ~1.2 m primary antenna to scan large areas of the sky in a number of discreet spectroscopic channels from L2. The FIRSPEX bands centered at [CI] 809 GHz, [NII]1460 GHz, [CII]1900 GHz and [OI]4700 GHz have been carefully selected to target key atomic and ionic fine structure transitions difficult or impossible to access from the ground but fundamental to the study of the multi-phase ISM in the Universe. The need for state-of-the-art sensitivity dictates the use of superconducting mixers configured either as tunnel junctions or hot electron bolometers. This technology requires cooling to low temperatures, approaching 4K, in order to operate. The receivers will operate in double sideband configuration providing a total of 7 pixels on the sky. FIRSPEX will operate from L2 in both survey and pointed mode enabling velocity resolved spectroscopy of large areas of sky as well as targeted observations. FIRSPEX has been submitted in response to ESA's M5 call for proposals and is currently under review.

Spaceborne superconducting sounder (SMILES-2) for upper-atmosphere observation

Satoshi Ochiai[1,*], Philippe Baron[1], Yoshinori Uzawa[1], Yoshihisa Irimajiri[1], Toshiyuki Nishibori[2], Makoto Suzuki[2], Takeshi Manabe[3], Hiroyuki Maezawa[3], Akira Mizuno[4], Tomoo Nagahama[4], and Masato Shiotani[5]

[1]*National Institute of Information and Communications Technology, Koganei, Tokyo 184-8795, Japan*

[2]*Japan Aerospace Exploration Agency, Tsukuba, Ibaraki 305-8505, Japan*

[3]*Osaka Prefecture University, Sakai, Osaka 599-8531, Japan*

[4]*Institute for Space-Earth Environmental Research, Nagoya University, Nagoya, Aichi 464-8601, Japan*

[5]*Research Institute for Sustainable Humanosphere, Kyoto University, Uji, Kyoto 611-0011, Japan*

*Contact: ochiai@nict.go.jp

***Abstract*—The Superconducting Submillimeter-Wave Limb-Emission Sounder 2 (SMILES-2) is a proposed satellite mission for the comprehensive observation of Earth's atmosphere with superconducting receivers. The SMILES-2 mission will have four SIS/HEB receivers with bands near 487, 527 GHz, 557, 576 GHz, 623, 653 GHz, and 1.8, 2 THz to observe spectral lines of atomic oxygen, OH, O_3, O_2, H_2O, CO, NO, NO_2, N_2O, ClO, HCl, and BrO. The atmospheric limb will be scanned in an altitude range from 20 km to more than 160 km in a minute or shorter time. The low-noise superconducting receivers will provide highly qualified limb spectra of a variety of atmospheric species, from which we can retrieve temperature and wind over the whole altitude range encompassing the stratosphere, mesosphere, and lower thermosphere. The global temperature and wind profiles of the whole atmosphere measured by a single instrument will be the most scientifically useful data for studying the lower and upper atmosphere coupling. The wind precision of SMILES-2 measurement is estimated to be better than 3 m/s with an altitude resolution of 3 km in the upper stratosphere and mesosphere, and 10 m/s with an altitude resolution of 5 km in the lower thermosphere. We are now studying the feasible design of the mission to be boarded on a Japanese small satellite.**

Introduction

The millimeter- and submillimeter-wave limb-sounding technique has been used for more than 2 decades to measure Earth's atmosphere from space. It allows us to measure various chemical compositions in the stratosphere and mesosphere, as well as measuring temperature profiles with high altitude resolution and, as demonstrated recently, horizontal winds in a wide altitude range. The demand for temperature measurement in the stratosphere and mesosphere has recently been increasing, because the tops of atmospheric reanalysis models have been extended to the mesosphere, and the upper atmospheric measurements can be assimilated into models or compared with the model representations [1][2]. The unique feature of millimeter and submillimeter limb sounding is that it can provide temperature profiles in the upper atmosphere where microwave sounders on operational meteorological satellites cannot measure or can only measure with very low altitude resolution. Wind is the key parameter for describing atmospheric phenomena in the mesosphere and lower thermosphere. The measurement of the global distribution of wind will reveal the energy and momentum transports between the lower atmosphere and the thermosphere, which are induced by gravity and planetary waves (e.g., [3]). In the thermosphere, the neutral atmosphere, which can be measured by a submillimeter limb sounder, also interacts with ions and is affected by energetic particles coming from above. The observations of wind, atomic oxygen, and other components such as NO will contribute to the study of such couplings between the upper atmosphere and space.

The Superconducting Submillimeter-Wave Limb-Emission Sounder (SMILES) was operated aboard the Japanese Experiment Module (JEM) of the International Space Station (ISS) during the period from October 2009 to April 2010 and successfully observed submillimeter limb emissions [4]. The SMILES receivers at 625 and 650 GHz use superconductor-insulator-superconductor (SIS) mixers cooled below 4.6 K by a mechanical cryocooler and showed a system noise temperature of 297 K (SSB) in orbit [5]. Because of the ISS non-sun-synchronous orbit, a diurnal variation of atmospheric compositions can be derived from a 1–2-month compilation of the SMILES observation. The SMILES measurement suggests that the diurnal variations of ozone cannot be neglected even in the stratosphere when interpreting and combining measurements taken at different local time [6]. Moreover, SMILES demonstrated the successful measurement of wind between 8 and 0.01 hPa (~35–80 km) [7].

We are proposing an upgraded version of SMILES, named SMILES-2. The main objectives of SMILES-2 are to measure temperature and wind in the stratosphere, mesosphere, and lower thermosphere, and to measure atomic oxygen, ozone, NO, and other molecules. In this paper, we give an overview of our SMILES-2 mission concept, the progress in technological development, and the status of our proposal to the Japan Aerospace Exploration Agency (JAXA).

Observation requirements for SMILES-2

The coverage of a wide measurement altitude range and high-precision measurement are the most important features of SMILES-2 to fulfill the scientific requirements defined by the SMILES-2 working group. The frequency bands, shown in Table I, are selected for optimizing the retrieval of temperature and wind in the entire altitude range from the stratosphere to the lower thermosphere. The retrieval of temperature and wind depends on strong emission lines in the observed bands. The cluster of ozone emission lines around 655 GHz is efficient in retrieving wind in the altitude range where the ozone signal is strong, i.e., in an altitude range of 40–70 km. In this range, wind can be measured with a precision better than 2 m/s and a vertical resolution of 3 km [8]. SMILES-2 will use the atomic oxygen line at 2.06 THz and can measure wind with a precision better than 10 m/s with a vertical resolution of 5 km in an altitude range between 110 and 160 km [8]. The O_2 line at 487.2 GHz and the H_2O line at 556.9 GHz will contribute to the temperature and wind measurements in an altitude range between 70 and 110 km. Figure 1 shows the estimated single-scan random errors of wind and temperature retrievals synthesized from the 4 frequency bands shown in Table I. The receiver noises are assumed to be 120, 135, 150, and 990 K for the SIS1, SIS2, SIS3, and HEB bands, respectively. The integration time and the tangent height step of the limb scanning are assumed to be 0.25 s and approximately 1.2 km, respectively. The estimated error does not include other errors, e.g., antenna pointing, intensity calibration, frequency, and model parameter errors.

The bands in Table I are also chosen so that the emission lines of the important molecules are included in these bands. The molecules to be observed include atomic oxygen, OH, O_3, O_2, H_2O, CO, NO, NO_2, N_2O, ClO, HCl, BrO, CH_3Cl, H_2CO, and isotopes of some of these molecules. A sensitivity study of the SMILES-2 observation of these molecules is described in [10]. Among these molecules to be observed, atomic oxygen is particularly important in knowing the energy budget of the mesosphere and lower thermosphere. Atomic oxygen in the ground electronic state has only two bright emissions at 2.06 and 4.74 THz. The 4.74 THz line has been measured from balloon, airplane, and spaceborne platforms. Because the 4.74 THz is optically thick in the limb-viewing geometry, it is difficult to retrieve the oxygen density below 130 km using this line with a low-sensitivity broadband detector from space [11]. An indirect measurement of atomic oxygen density in an altitude range of 65–105 km was reported with a larger uncertainty [12]. Recently, THz limb sounder projects, which will observe the oxygen line with heterodyne receivers, have been proposed [13][14]. The SMILES-2 mission will have a 2.06 THz receiver with better sensitivity than that of other projects. We estimate the measurement precision of atomic oxygen to be better than 30 % at an altitude above 100 km.

A non-sun-synchronous orbit with an orbit inclination of 66° and an altitude of about 550 km is expected for the SMILES-2 platform. With this orbit, the measurement local time at a given latitude shifts by 24 h during about 90 days. The highest latitude of the measurement will be up to 80° in one hemisphere and 50° in another hemisphere. To observe northern and southern polar regions, yaw maneuvering of the satellite will be needed because the SMILES-2 antenna sees the atmosphere in only one side of the satellite. The orbit will meet the scientific requirements of both diurnal variation measurement and polar region observation.

TABLE I
SMILES-2 BANDS

SIS1	LO=507 GHz	IF=18-22 GHz	O_2, H_2O, O_3, HO_2 BrO, NO_2
SIS2	LO=566.5 GHz	IF=8.5-10.5 GHz	H_2O, O_3 O_3, CO, ClO
SIS3	LO=638.15 GHz	IF=11.15-19.15 GHz	O_3, HCl, N_2O, BrO O_3, ClO, HO_2, NO
HEB	LO=1836.05 GHz and 2058.8 GHz	IF=1-2 GHz	OH atomic-O

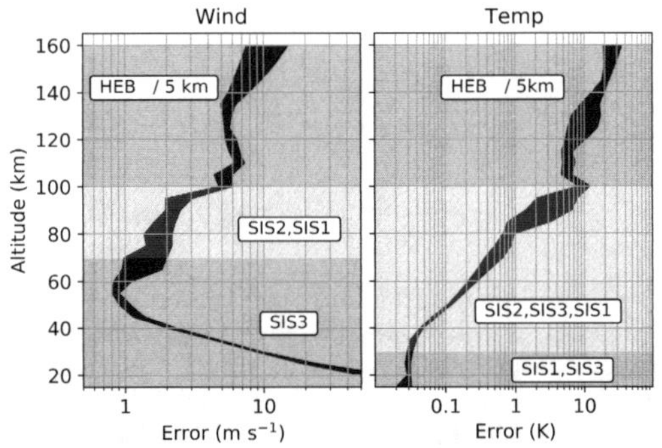

Fig. 1. Random error in wind (left) and temperature (right) retrievals [9]. The vertical retrieval resolution is 3 km below 100 km and 5 km above 100 km. The name of the band which contributes most to the retrieval in a certain altitude range, is indicated. The line thickness shows the range of error variation depending on the local time and latitude assumed in this error estimation.

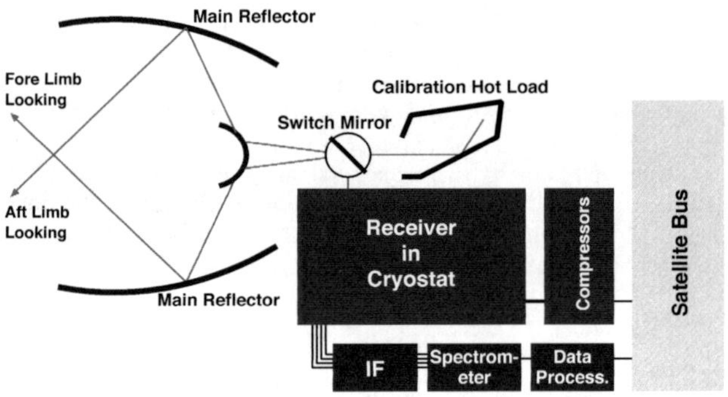

Fig. 2. SMILES-2 schematic block diagram

SMILES-2 PAYLOAD

The block diagram of the SMILES-2 payload is schematically shown in Fig. 2. The SMILES-2 payload is assumed to be aboard a small satellite equipped with a JAXA standardized bus system. The main instruments of the SMILES-2 payload are large twin antennas and a 4 K cooled superconducting receiver.

A. Antenna and limb scanning

Two offset Cassegrain-type antennas are used to see the atmospheric limb in the directions of 45° and 135° from the spacecraft velocity. The antennas are not steerable but can be pointed toward the atmospheric limb and scanned from the bottom of the atmosphere (below 20 km) to the upper atmosphere (above 200 km) by maneuvering the satellite attitude. The scene at the uppermost tangent height is used as a cold reference. During the gap between downward scanning and upward scanning, the receiver beam is switched toward the calibration hot load. The forward looking antenna is used during the upward scan, and backward looking antenna is used on the downward scan. Each scan takes 40–50 s so that one observation sequence is completed in about 100 s. By looking toward the 45° and 135° directions periodically in this way, the same atmospheric vertical column is observed from two perpendicular directions at an interval of 8 min, so that the horizontal wind vector can be derived. The side to observe with respect to the orbital plane, that is, the north side or the south side, depends on the direction of the sun in order to optimize the power generation by a solar array panel and radiation cooling of the instrument.

The main reflector has an aperture diameter of about 1 m. The vertical beam width at 487 GHz corresponds to a vertical resolution of 2 km at the tangent point. To make the large antenna lightweight, a carbon-fiber-reinforced plastic (CFRP) panel will be used for the antenna material. The mirror surface is formed with plasma-sprayed aluminum.

B. Receiver front end

The SMILES-2 mission has three SIS receivers and a hot electron bolometer (HEB) receiver. These are installed in a vacuum cryostat as schematically shown in Fig. 3. The atmospheric signal collected by the antenna is fed into the cryostat via a single window. The signal is split by a frequency-selective surface and a wire grid into the signals that are fed into the corrugated horns of SIS and HEB mixers Two or three SIS mixers share a wide-band corrugated horn. Each mixer is supplied with each local frequency signal via a stainless steel waveguide or a quasi-optical waveguide. Each local signal is generated with a multiplier chain. The local frequencies for SIS mixers are fixed as shown in Table I. The local frequency for the HEB mixer switches to either the frequencies for atomic oxygen of 2.06 THz or that for OH of around 1.8 THz, or occasionally to other frequency. All mixers work in double-sideband mode in order to obtain better precision in temperature and wind measurements by using emission lines in both sidebands for the retrieval.

One of the design images of integrated receiver components in a cryostat is shown in Fig. 4. The design of the cryostat and submillimeter optics is based on that of SMILES. Compared with SMILES, the number of mixers, including the HEB mixer, is doubled in SMILES-2. The diameter of the cryostat cylinder in Fig. 4 is also doubled in comparison with the SMILES cryostat. A much smaller cryostat than that shown in Fig. 4 is being designed by applying the technology of a waveguide multiplexer [15]. The development of the HEB mixer is also ongoing in Japan (e.g., [16]).

C. Cooling system

The SIS and HEB mixers are cooled below 4.8 K with a He closed-cycle mechanical cooler. The 4 K cooler consists of a Joule–Thomson (JT) cycle cooler and a two-stage Stirling cooler as a precooler. The 4 K cooler by Sumitomo Heavy Industries, Ltd. (SHI) has been successfully demonstrated in space by SMILES and a Japanese X-ray astronomy satellite ASTRO-H [17]. The JT cooler of SMILES unfortunately stopped after 9 months of operation in space. It was speculated that the cause of the failure is that the He closed cycle was clogged with solidified CO_2 contamination. After the investigation of the SMILES cooler failure, the JT cooler was improved. The JT cooler used in ASTRO-H has a longer lifetime compressor and a getter to trap impurity molecules in the He closed cycle. A ground test of the improved JT cooler has demonstrated a lifetime of more than 3 years [17].

The SHI JT cooler has a cooling capacity of 40 mW at 4.5 K. When the JT cooler is used with a large heat load, large cooling capacities are required for the precooler at the 100 and 20 K stages. In the case of the cryostat shown in Fig. 4, the heat load at the 4 K stage is estimated to be more than 30 mW, so two two-stage Stirling coolers are required for precooling the JT cooler and for cooling the thermal shields on the 100 and 20 K stages. SMILES was able to operate with one two-stage Stirling cooler because its 4 K heat load was about 10 mW and less cooling capacity was required for cooling the 100 K shield, which had a comparatively small surface area. The input power to the SMILES-2 cooling system including

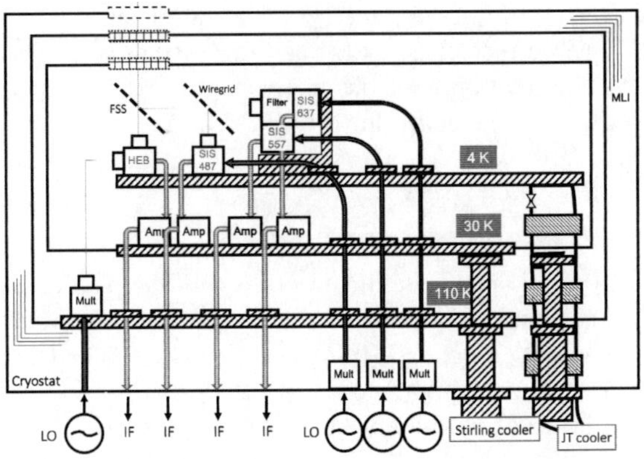

Fig. 3. Block diagram of the SMILES-2 receivers installed in a cryocooler

Fig. 4. CAD image of the SMILES-2 cryoreceiver

the driver electronics for the cooler is estimated to be about 410 W, while that of SMILES cooling system was less than 300 W. The total mass of the receiver front end, cryostat, and cooling system including the driver electronics is estimated to be around 105 kg.

Targeted opportunity to launch

JAXA has a small satellite program with the plan to launch a scientific small satellite via the Epsilon rocket every two years. We will propose SMILES-2 as a mission to be launched in 2023 or 2025. A JAXA small satellite can afford to carry a mission instrument with a weight of 200 kg or more. The total weight of the SMILES-2 mission will meet this requirement. A standard bus of a JAXA small satellite supplies power of only 300 W on average, which may not be sufficient for the SMILES-2 coolers. Studying the possibilities of power system modification on the bus and reduction in the power consumption by the cooling system is ongoing. The proposal of the SMILES-2 mission will be prepared before March 2018 with a solution to the power problem.

Conclusions

The SMILES-2 mission aims to contribute to a broad range of atmospheric studies in the stratosphere, mesosphere, and lower thermosphere by providing observations of the atmospheric parameters over an altitude range between 20 and 160 km and a latitude range between 80S and 80N. SMILES-2 will provide new products, such as a horizontal wind vector in an altitude range between 30 and 160 km and the concentration profile of ground-state atomic oxygen, using SIS/HEB receivers from 487 GHz to 2.06 THz. The SMILES-2 receiver will take over the spaceborne superconducting technology with a 4 K cooling system, which was successfully demonstrated by the SMILES mission. The lifetime of SMILES-2 is expected to be 3 years.

SMILES-2 is under pre-phase A stage at present. The SMILES-2 working group is preparing a mission proposal to be submitted to JAXA. The largest issue in the mission feasibility is whether the power consumption of the SMILES-2 cooling system is consistent with that of the small satellite.

Acknowledgment

The authors would like to acknowledge the SMILES-2 working group. The hardware development and feasibility studies are funded by the National Institute of Information and Communications Technology (NICT) and the Japan Aerospace Exploration Agency (JAXA). The cryostat design and the estimation of the cooling capacity were made by Sumitomo Heavy Industries. Ltd. (SHI).

References

[1] R. Gelaro, *et al.*, "The Modern-Era Retrospective Analysis for Research and Applications," *J. Climate*, **30**, 5419–5454, 2017, doi:10.1175/JCLI-D-16-0758.1.

[2] Y. Harada, *et al.*, "The JRA-55 Reanalysis: Representation of Atmospheric Circulation and Climate Variability," *J. Meteor. Soc. Japan*, **94**, 3, 269–302, 2016, doi:10.2151/jmsj.2016-015.

[3] Y. Miyoshi, *et al.*, "A global view of gravity waves in the thermosphere simulated by a general circulation model," *J. Geophys. Res. Space Physics*, **119**, 5807–5820, 2014, doi:10.1002/2014JA019848.

[4] K. Kikuchi, *et al.*, "Overview and early results of the Superconducting Submillimeter-Wave Limb-Emission Sounder (SMILES)," *J. Geophys. Res.*, **115**, D23306, 2010, doi:10.1029/2010JD014379.

[5] S. Ochiai, *et al.*, "Receiver Performance of the Superconducting Submillimeter-Wave Limb-Emission Sounder (SMILES) on the International Space Station," *IEEE Trans. Geosci. Remote Sensing*, **51**, 7, 2013, doi:10.1109/TGRS.2012.2227758.

[6] T. Sakazaki, *et al.*, "Diurnal ozone variations in the stratosphere revealed in observations from the Superconducting Submillimeter-Wave Limb-Emission Sounder (SMILES) on board the International Space Station (ISS)," *J. Geophys. Res. Atmos.*, **118**, 2991–3006, 2013, doi:10.1002/jgrd.50220.

[7] P. Baron, *et al.*, "Observation of horizontal winds in the middle-atmosphere between 30S and 55N during the northern winter 2009–2010," *Atmospheric Chemistry and Physics*, **13**, 13, 6049, 2013, doi:10.5194/acp-13-6049-2013.

[8] P. Baron, *et al.*, "Measurement of stratospheric and mesospheric winds with a submillimeter wave limb sounder: results from JEM/SMILES and simulation study for SMILES-2," *Proc. SPIE 9639, Sensors, Systems, and Next-Generation Satellites XIX*, 96390N, 2015, doi:10.1117/12.2194741.

[9] S. Ochiai, *et al.*, "SMILES-2 mission for temperature, wind, and composition in the whole atmosphere," submitted for publication.

[10] M. Suzuki, *et al.*, "Sensitivity study of smiles-2 for chemical species," *Proc. SPIE 9639, Sensors, Systems, and Next-Generation Satellites XIX*, 96390M-15, 2015, doi:10.1117/12.2194832.

[11] K. U. Grossmann, *et al.*, "A global measurement of lower thermosphere atomic oxygen densities," *Geophis. Res. Lett.*, **27**, 9, 1387–1390, 2000.

[12] M. G. Mlynczak, *et al.*, "Atomic oxygen in the mesosphere and lower thermosphere derived from SABER: Algorithm theoretical basis and measurement uncertainty," *J. Geophys. Res.: Atmospheres*, **118**, 5724–5735, 2013, doi: 10.1002/jgrd.50401.

[13] S. P. Rea, *et al.*, "The Low-Cost Upper-Atmosphere Sounder (LOCUS)," *Proc. 26th Int. Symp. Space THz Technol.*, M1-3, 2015.

[14] D. L. Wu, *et al.*, "THz limb sounder (TLS) for lower thermospheric wind, oxygen density, and temperature," *J. Geophys. Res. Space Physics*, **121**, 7301–7315, 2016, doi:10.1002/2015JA022314.

[15] T. Kojima, *et al.*, "Design and Development of a Hybrid-Coupled Waveguide Multiplexer for a Multiband Receiver," *IEEE Trans. RHz Sci. Tech.*, **7**, 1, 10–19, 2017, doi:10.1109/TTHZ.2016.2627220.

[16] Y. Irimajiri, *et al.*, "Development of a Superconducting Low-Noise 3.1-THz Hot Electron Bolometer Receiver," *IEEE Trans. THz Sci. Tech.*, **5**, 6, 1154–1159, 2015.

[17] K. Narasaki, *et al.*, "Lifetime Test and Heritage On-Orbit of SHI Coolers for Space Use," *Cryocoolers* **19**, 613–622, 2016.

RF and thermal aspects of the ground calibration system for the Microwave Sounder Instrument

Manju Henry*[1], F. Cahill[1], R. Green[1], P. G. Huggard[1], B. Ellison[1], K. Ostoljic[1], P. Henshall[1], J. Mann[1], T. Brooke[1], D. Peters[1], D. Smith[1], A. Caldwell[1], G. Hutchinson[1], M. Winser[2], G. Tennant[2], P. Campbell[2], and V. Kangas[3]

[1]*RAL Space, STFC Rutherford Appleton Laboratory, Didcot Ox11 0QX, UK*
[2]*Airbus Defence and Space Ltd, Anchorage Road, Portsmouth, Hampshire, PO35PU*
[3]*European Space Agency, ESTEC, Keperlaan, 2201 Noordwijk, The Netherlands.*

*Contact: manju.henry@stfc.ac.uk

Abstract—We present the RF and thermal design of the ground calibration targets for the Microwave Sounder (MWS) instrument series. MWS is a cross track scanning passive radiometer, operating from 23 to 229 GHz, and is one of the three millimetre and sub millimeter wave instruments of the EUMETSAT/ESA MetOp 2nd Generation programme. These instruments will provide critical information on atmospheric temperature and water vapour content, a major input to numerical weather prediction. Accurate calibration of MWS to reduce the returned temperature uncertainty to ± 0.1 K is required by the data users. A consortium, comprising the RAL Space department at the STFC Rutherford Appleton Laboratory and Magna Parva Ltd., are working under contract to Airbus Defence and Space (UK) to provide ground calibration apparatus for MWS. This will permit pre-launch instrument level calibration under thermal vacuum conditions. The calibration system contains a "cold" black body target, operating at close to 80 K, which replaces the view of cold space and a "Earth" target, which is variable in temperature from 80 to 315K. The cold target is moveable along a ±3° arc, and Earth target can be moved from ± 50° with respect to nadir to simulate different angles on incidence on the planet's surface during the scan. The targets' absorbing structures are made from 41 mm high aluminium pyramids coated with an Eccosorb CR series iron loaded epoxy. Ansys HFSS is used to optimise the pyramid dimensions and coating thickness to reduce the normal incidence, polarisation independent, return loss to better than -45 dB. Three blade servers each with 512 GB RAM were used to reduce computational times during the optimisation process. In house VNA reflectivity measurements using a quasi-optical network have confirmed that the required return loss is achieved. The target's diameter of 480 mm accommodates the -35 dB contour of the largest, lowest frequency, MWS beam. It is necessary to place the targets at the bottom of temperature controlled cylindrical baffles to reduce thermal gradients in the Eccosorb arising from incoming thermal infrared radiation [1]. These cylinders are 700 mm in depth, and their infrared absorbing walls are held at the same temperature as the calibration loads by means of a common liquid nitrogen/helium gas gap/ radiation shield [2] arrangement. Varying the pressure of the helium gas controls the thermal conductance, and so reduces the electric heater power required for intermediate temperatures. This novel approach eliminates the thermal instabilities arising when using a combination of liquid nitrogen and electrical heaters to achieve temperatures above 80 K. Our presentation will compare the predicted RF return loss and results, as well as the design approach used to minimise thermal gradients in the absorber and the ≈0.1 K level effects on the equivalent radiometric temperatures of the calibration loads.

[1]: Axel Murk *et al.*, "Temperature/Absorption Cross Integrals and the Validation of Radiometric Temperatures for Space-Based Radiometers", Proc. 10th European Conf. Antennas and Propagation (EuCAP), pp. 1 - 3 (2016)

[2]: N. Melzack *et al.*, "Variable Temperature Blackbodies via Variable Conductance: Thermal Design, Modelling and Testing" Int J Thermophys (2017) DOI 10.1007/s10765-016-2167-5

Development of Calibration Targets for MetOp-SG Microwave Instruments

Arne Schröder[1]*, Axel Murk[1], Richard Wylde[2], Karl Jacob[1], Mike Winser[3], Marc Bergada Pujades[4], and Ville Kangas[5]
[1]*University of Bern, 3012, Bern, Switzerland*
[2]*Thomas Keating Ltd, RH14 9SH, Billingshurst, United Kingdom*
[3]*Airbus Space and Defence, PO3 5PU, Portsmouth, United Kingdom*
[4]*Airbus Space and Defence, 28022, Madrid, Spain*
[5]*European Space Agency, 2201 AZ, Noordwijk, Netherlands*
Contact: arne.schroeder@iap.unibe.ch, axel.murk@iap.unibe.ch

***Abstract*—This work presents the development of microwave calibration units for the second generation of meteorological operational satellites (MetOp-SG), developed by ESA and EUMETSAT. In particular, we focus on the design and manufacturing of on-board calibration targets (OBCT) for the microwave sounder (MWS) and the ice cloud imager (ICI), respectively. MWS operates in seven bands between 23 GHz and 230 GHz, while ICI covers five bands between 180 GHz and 670 GHz. As the OBCTs act as temperature reference for microwave radiometers, they are required to exhibit a low electromagnetic reflectivity and a uniform temperature distribution. This paper summarizes the electromagnetic and thermal design and analyses of the MWS and ICI OBCTs.**

I. Introduction

MetOp-SG is the second generation of European meteorological operational polar orbit satellites which are developed by ESA and EUMETSAT [1]. MetOp-SG will consist of two series of satellites, planned to be operational between 2021 and 2048. The Earth's observation data provided by various on-board remote sensing instruments will enhance future numerical weather prediction and monitoring. The microwave sounder (MWS) and the ice cloud imager (ICI) are two of the instruments on board the MetOp-SG satellites. MWS is a cross-track scanning radiometer observing temperature and humidity profiles as well as cloud water in 24 channels between 23 GHz and 230 GHz [2]. ICI is a conical scanning radiometer intended to measure the cloud ice water path, the cloud ice effective radius, and the cloud ice mean altitude 13 channels between 180 GHz and 670 GHz [3].

Both instruments, ICI and MWS, will be continuously calibrated during operation using a two-point hot-cold calibration. The cold reference point will be provided by cold space 2.7 K measurements. The hot reference point will be provided by an on-board calibration target (OBCT). Two common classes of calibration targets are usually applied in radiometers: Cavity absorbers and pyramid absorbers. Cavity absorbers, such as conical [4], [5] or wedge geometries [6] can exhibit excellent electromagnetic and thermal performance. But, the large dimensions of those cavities often restrains the application in compact instruments. Pyramid absorbers provide a good electromagnetic performance, but their thermal performance is inferior. Nevertheless, due to the compact dimensions, spaceborne radiometers are commonly equipped with pyramid calibration targets. Pyramid absorbers are therefore foreseen for the OBCTs of the MWS and ICI instrument.

The development of spaceborne calibration units raises a wide variety of engineering challenges: A high electromagnetic emissivity and a high temperature uniformity at the same time are required to guarantee an accurate calibration. Moreover, a low coherent backscattering is needed. In particular, the diffuse reflectivity for MWS shall be below -35 dB while the coherent backscatter shall be below -45 dB. For ICI, a diffuse reflectivity of at most -45 dB and a coherent backscatter of at most -50 dB is required. Beside these requirements, a mechanically robust design is needed to ensure mechanical shock and vibration resistivity. The target has to operate in a wide range of temperatures. Moreover, temperature sensors deployed in the target need to measure the physical OBCT temperature within an accuracy of a few mK.

During the design of OBCTs, we concentrated on design aspects which haven't been considered in previous works. In particular, we analyzed the impact of the metal backing geometry on the brightness temperature and electromagnetic properties of a target. Based on these studies, a novel pyramid layout was developed. We analyzed and applied multi-layer absorber coatings for broad band applications. As previous works neglected potential manufacturing artifacts, we studied the impact of most common artifacts on the reflectivity and remedies to the most critical manufacturing imperfections have been proposed. Prototypes of the numerically designed pyramid targets have been manufactured and examined experimentally. This abstract summarizes the OBCT development for MWS and ICI. Details on the development can be found in [7].

II. Numerical Design and Experimental Validation

A pyramid geometry has been chosen as absorbing structure for the MWS and ICI OBCTs. Both OBCTs consist of a metallic pyramid array coated with an absorbing medium. We use Stycast doped with carbonyl iron powder (CIP) as

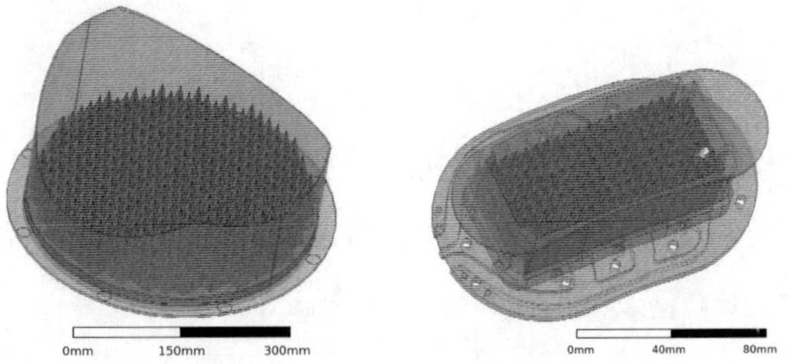

Fig. 1. Illustration of MWS and ICI pyramidal calibration targets. CAD model of MWS prototype (*left*) and ICI prototype (*right*).

absorber. Although other available absorbing materials exhibit superior electromagnetic properties, Stycast turned out to be more robust in terms of adhesion and cracking. The electromagnetic properties of different Stycast-CIP compositions were determined by measurements which have been carried out in a broad frequency range between 18 GHz and 620 GHz [8], [9]. The design of the OBCTs was performed by numerical simulations. We have employed the finite element method (FEM) and modeled the pyramid targets as infinite periodic array to reduce the computational effort. To study the reflectivity in the high frequency regime where the target is electrically very large, we have implemented a geometrical optics (GO) approach.

Using GO simulations, we chose a pyramid geometry with a width to high ration of 1:3.5 as starting point for the design process. A main challenge is to find a pyramid geometry which exhibits a good electromagnetic performance and a good thermal performance. In particular, temperature gradients in the absorber should be as small as possible in order to provide a defined brightness temperature for radiometric calibrations. We have developed a curved pyramid kernel geometry which, compared to conventionally used linear kernels, exhibits additional absorber volume at the bottom section of the pyramids. With that the electromagnetic performance is significantly improved while the temperature distribution is only slightly degraded. We worked on a numerical procedure to analyze the brightness temperature of pyramid targets [10]. Employing this approach, we have demonstrated that reducing temperature gradients results in an improvement of the brightness temperature and hence in a more accurate calibration.

For the ICI OBCT, a single absorber layer turned out to be sufficient. For the MWS OBCT, a two-layer absorber structure has been considered. In order to provide low reflectivity in the low frequency bands of MWS, we coat the metal backing structure with a highly absorbing Stycast-CIP composition. A less absorbing second layer is deposited on top. Numerical analyses predicted that this absorber composition exhibits a very good broad band performance and is hence superior to a single layer design. Figure 1 illustrates the designed MWS and ICI OBCTs.

After developing the pyramid absorbers, we have investigated the impact of potential manufacturing artifacts on the electromagnetic performance of the absorber. In particular, imperfect absorber and metal tips were studied. Moreover, radii at the pyramid notches and material property variations have been considered. While small variations in the material properties are negligible, imperfect pyramid tips decrease the absorptivity of the target. However, the studies revealed that round pyramid notches is the most severe artifact as radii downgrade the reflectivity significantly. We have developed a novel notch design which turned out to be insusceptible to manufacturing precision related issues.

Prototypes of the calibration targets were manufactured and we have conducted measurements in the selected frequency bands between 20 GHz and 670 GHz. It has been shown that the performance of both calibration targets fulfill the MetOp-SG specific radio frequency requirements.

III. Conclusion and Outlook

We have developed OBCT prototypes for the MWS and ICI instruments. The manufactured prototypes were analyzed experimentally and we have proven that the prototypes fulfill the electromagnetic specifications. The designed prototype geometries serve as starting point for the development of engineering qualification models (EQM) in the next stage of the project.

References

[1] EUMETSAT, "EUMETSAT Polar System - Second Generation," 2016, [Online; accessed 18-August-2016]. [Online]. Available: http://www.eumetsat.int/website/home/Satellites/-FutureSatellites/EUMETSATPolarSystemSecondGeneration/index.html

[2] V. Kangas, S. D'Addio, M. Loiselet, G. Mason, M. Buckley, G. Tennant, and P. Campbell, "Microwave sounder instrument for MetOp second generation," in *2014 13th Specialist Meeting on Microwave Radiometry and Remote Sensing of the Environment (MicroRad)*, March 2014, pp. 232–235.

[3] V. Kangas, S. D'Addio, U. Klein, M. Loiselet, G. Mason, J. C. Orlhac, R. Gonzalez, M. Bergada, M. Brandt, and B. Thomas, "Ice cloud imager instrument for metop second generation," in *2014 13th Specialist Meeting on Microwave Radiometry and Remote Sensing of the Environment (MicroRad)*, March 2014, pp. 228–231.

[4] A. Murk, P. Yagoubov, R. Wylde, G. Bell, and I. Zivkovic, "Development of conical calibration targets for ALMA," in *21st International Symposium on Space Terahertz Technology*, mar 2010.

[5] A. Schröder and A. Murk, "Numerical design and analysis of conical blackbody targets with advanced shape," *IEEE Transactions on Antennas and Propagation*, vol. 64, no. 5, pp. 1850–1858, May 2016.

[6] S. Fernandez, A. Murk, and N. Kämpfer, "Design and characterization of a Peltier-cold calibration target for a 110-GHz radiometer," *Geoscience and Remote Sensing, IEEE Transactions on*, vol. 53, no. 1, pp. 344–351, Jan 2015.

[7] A. Schröder, A. Murk, R. Wylde, K. Jacob, M. Winser, M. B. Pujades, and V. Kangas, "Electromagnetic Design of Calibration Targets for MetOp-SG Microwave Instruments," *IEEE Transactions on Terahertz Science and Technology, in review.*

[8] D. A. Houtz et al., "Electromagnetic Design and Verification of the NIST Conical Microwave Blackbody Calibration Target," *submitted for publication.*

[9] K. Jacob, A. Schröder, and A. Murk, "Design, Manufacturing, and Characterization of Conical Calibration Targets with Optimized Profile," *submitted for publication.*

[10] A. Schröder, A. Murk, R. Wylde, D. Schobert, and M. Winser, "Brightness temperature computation of microwave calibration targets," *IEEE Transactions on Geoscience and Remote Sensing*, 2017, in review.

165-229 GHz Front End Receivers for the Microwave Sounder and Microwave Imager Instruments onboard MetOp-SG satellites – Preliminary Design and Results

S. Rea[1*], B. Moyna[1], K. Parow-Souchon[1], M. Henry[1], H. Wang[1], M. Oldfield[1], N. Brewster[1], P. Hunyor[1], M. Merritt[1], R. Green[1], D. Smith[1], B. Davis[1], A. Obeed[1], D. Klugmann[1], M. Phillips[1], M. Beardsley[1], J. Hampton[1], B. Ellison[1], C. Howe[1], G. Burton[1], A. Marshall[1], B. Thomas[2], M. Brandt[2], M. Philipp[2], G. Sonnabend[2], T. Stangier[2], R. Hammett[3], T. Decoopman[4], V. Kangas[5], S. D'Addio[5], P. Piironen[5]

[1]*RAL Space, STFC-Rutherford Appleton Laboratory, OX11 0QX, Didcot, United Kingdom*
[2]*Radiometer Physics GmbH, 53340, Meckenheim, Germany*
[3]*Airbus Defence & Space, PO3 5PU, Portsmouth, United Kingdom*
[4]*Airbus Defence & Space, 31030, Toulouse, France*
[5]*ESA/ESTEC, 2201 AZ, Noordwijk, The Netherlands*
[*]Contact: Simon.Rea@stfc.ac.uk

Abstract—The European contribution to operational meteorological observations from polar orbit has been provided by the first generation of MetOp satellites since 2006. The MetOp Second Generation (MetOp-SG) series of satellites will provide continuity and enhancement of these observations in the 2020-2040 timeframe. The MetOp-SG space segment will consist of two satellites, Sat-A and Sat-B, comprising a suite of active and passive Earth observing instrumentation covering a broad spectral range from microwave to visible wavelengths. RAL Space is contracted to provide space-qualified front-end heterodyne receivers operating between 165-325 GHz for three instruments on the mission, namely the Microwave Sounder (MWS), Microwave Imager (MWI) and Ice Cloud Imager (ICI); and is supporting Radiometer Physics GmbH (RPG) on the development of front-end receivers at 448 and 664 GHz for ICI. This paper will focus on the preliminary design and results of front-end receivers operating between 165-229 GHz for the MWS and MWI instruments. A companion paper (B. Thomas *et. al.*) will focus on the ICI front-end receivers.

The preliminary design of front-end receivers at 165 GHz, 183 GHz and 229 GHz for the MWS, and at 165 GHz and 183 GHz for the MWI, is complete. Although there is an overlap in the spectral range of the instruments the accommodation constraints of the receivers are significantly different. The receiver front-ends use core technology developed within Europe: mm-wave low-noise amplifier MMICs from IAF-Freiburg, and Schottky diodes from Teratech Components Ltd and ACST GmbH. The packaging and qualification of these devices into LNA, mixer, multiplier and integrated receiver modules is the responsibility of RAL and RPG. In addition, the front-end receivers are supported by US technology: power amplifier MMICs from ADI/Hittite, and IF low-noise amplifiers and dielectric resonator oscillators from L3/Narda/Miteq.

The preliminary design and breadboard test results of the front-end receivers for the MWS and MWI will be presented at the conference. Critical aspects related to the design and performance tuning of the front-end receivers will be addressed.

Design of Waveguide Coupled MKID Detectors

F. Blauth*, K. Jacobs, C. E. Honingh
I. Physikalisches Institut, Universität zu Köln
contact: blauth@ph1.uni-koeln.de

***Abstract*—We investigate the sensitivity and the noise of Microwave Kinetic Inductance Detector (MKID) devices that are coupled to a waveguide utilizing a proven waveguide probe antenna. The RF signal is split into two coplanar waveguide (CPW) absorber lines with a 40 nm thick aluminum centerline as the absorbing part and Niobium ground planes. The CPW centerline forms the inductive section of a 2.5 GHz superconducting resonator, which is capacitively coupled to the readout line. Since the deposition of the centerline is a separate lithography step, the design also enables us to exchange the aluminum and use the circuit as a testbed for kinetic inductance and loss measurements of other absorber materials. This design will be compared to an quarter wave transmission line resonator with identical resonance frequency fabricated on the same wafer. The simulated signal input reflection of both designs is below -15 dB over a frequency band of 100 GHz centered at 350 GHz.**

I. Introduction

Microwave Kinetic Inductance Detectors (MKID, [1]) are direct detectors which utilize the change of the kinetic inductance of superconducting material caused by the breaking of Cooper pairs by incoming radiation for detection. The material is usually part of a resonator with a resonance frequency in the GHz range and changes in the impedance cause a change of the resonance frequency. MKID can be read out using frequency multiplexing by attaching many resonators with different resonance frequencies to a single readout line. This facilitates the construction of cameras [2] or filter bank spectrometers with many channels [3]. MKID can operate over a wide frequency band ranging from the pair breaking energy of the superconductor, approximately given by $2\Delta = 73$ GHz $\cdot T_c/1$K[4] for many superconductors, up to the X-ray regime for applications in nuclear physics [5], [6], [7].

We plan to study the sensitivity and noise performance of a single MKID resonator with a well known signal coupling consisting of a horn antenna and a waveguide probe which have proven to give quantum limited sensitivity in 350 GHz SIS mixer measurements [8]. This paper will present the design of a 350 GHz MKID detector coupled to the waveguide.

II. Design

From free space, the 350 GHz radiation is coupled into a rectangular waveguide using a spline profile horn antenna. The waveguide dimensions are 630 μm by 320 μm . A waveguide to microstrip transition [9] extracts the fundamental mode and guides it to a coplanar waveguide (CPW) transmission line on a 9 μm silicon membrane substrate. Superconducting bridges using 400 nm SiO_2 as support dielectric are used to ensure a proper CPW mode. The radiation is then coupled to the aluminum absorber section of the MKID for which two different design approaches were taken. The ground and readout connections are made using 3 μm thick electroplated gold beamleads.

A. Transmission line resonator KID

A transmission line resonator KID is a proven design, as similar resonators were already fabricated and tested [10]. The design is shown in Fig. 1. The impedance of the line behind the antenna is set to 50Ω which results in a CPW line which are reliable to fabricate. To couple a 2.5 GHz transmission line resonator to the readout line capacitively, the open end of the resonator has to be at the coupling capacity and the shorted end at the antenna. Since waveguide probe is an open circuit at the readout frequency, two 1/8 wave tuning stubs are used to provide the short circuit to the readout signal. These stubs transform their short at the end to an open at 350 GHz and thus do not obstruct the propagation of the 350 GHz signal on the CPW transmission line. Due to the large frequency difference between 350 GHz and the 2.5 GHz readout, the stubs are just short circuits at the readout frequency. Two stubs are used to maintain the symmetry of the circuit. The stubs are made of superconducting niobium, since no absorption is intended here.

The resonator consists of the absorber line section which is 740 μm long that consists of a CPW with an aluminum center conductor and niobium ground planes. The width of the center conductor is 2 μm and the gaps are 1 μm , while the aluminum thickness is 40 nm. Since the 350 GHz input signal is above the pair breaking energy, the signal is absorbed and cooper pairs are broken. The readout signal is far below the gap energy, therefore absorption is much smaller but can still happen due to heating or multi photon events at high readout signal powers [11]. The absorber line ensures an estimated absorption of about 30 dB of the signal before the second section of the resonator starts which is made of niobium only. Since the total resonator length is about 1.8 cm, only the front part is intended for absorption of radiation and thus made of thin and narrow aluminum. Most of the length has uncritical dimensions of the center conductor and the gaps. For coupling the resonator to the readout line, a line is branched off the main readout line that goes parallel to the open end of the resonator as shown in Fig. 1 c). The coupling strength can be set by the length of the section and the gap width and is designed for coupling quality factors of 10000 or 100000.

The RF side of the circuit is simulated using CST^{TM}microwave studio. The input port is located at the waveguide entrance, so the horn antenna is not included

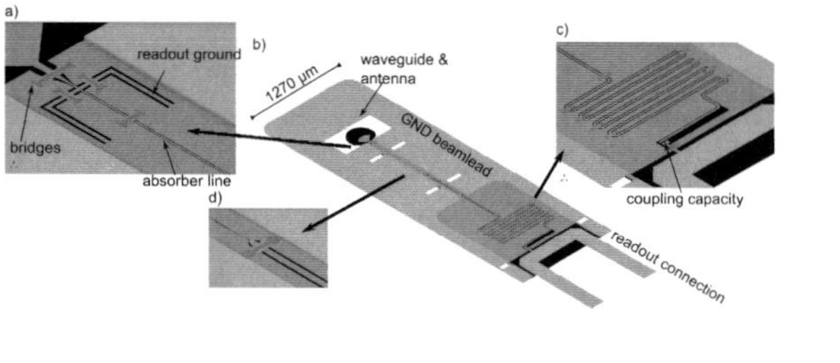

Fig. 1. CAD rendering of the transmission line resonator design. a) Waveguide antenna and choke. b) Overview of the chip. Dimensions are 1.2 by 3.6 mm. c) Meandered resonator line and coupling to the readout. d) Transision from aluminum absorber to the niobium resonator line.

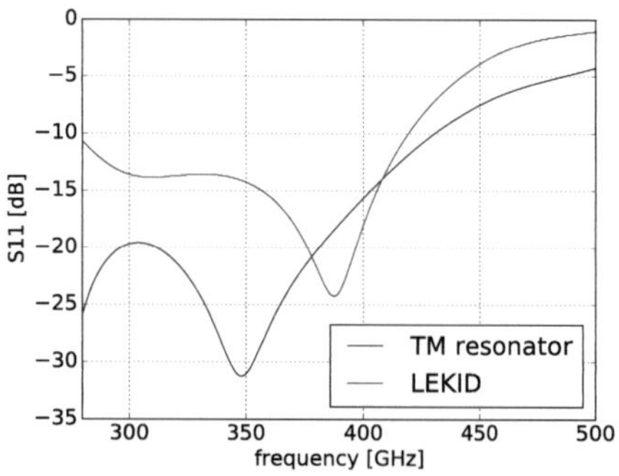

Fig. 2. CST^{TM}microwave studio simulation results of the input reflection (S11) seen from the waveguide for the high frequency signal input for the transmission line resonator design (TM) and the lumped element design.

in the simulation. The frequency dependent impedance calculated from the operating temperature and the measured resistivity of 40 nm aluminum films for absorber line were included into the simulation to estimate the absorption. For measured Lift-Off films the resisitivity at 4.2 K is 1.35 $\mu\Omega$ cm. The simulated performance of the antenna and choke section is shown in Fig. 2 and shows a good input coupling up to at least 400 GHz.

B. Lumped Element resonator KID

The alternative design for the KID part of the detector follows the lumped-element kinetic inductor design [12]. In this design, the resonator is made of an inductor and capacitor as circuit elements instead of a traveling wave on a transmission line. The advantage of this design is that the L/C ratio can be specifically designed and that the current distribution on the inductor (the absorber) is uniform. This design is shown in Fig. 3. The waveguide probe is the same as in the design mentioned above except that the CPW line impedance of the attached line is about 35 Ohm. Behind the probe the signal is split onto two absorber lines which each have a line impedance of 70 Ohm to maintain a good match. Those impedances are chosen to be still managable in fabrication for the 35 Ω line and still keep the CPW fundamental mode for the 70 Ω line. The width of the absorber lines is 1.6 μm and the gaps are 2 μm . Bridges across the line are necessary again to keep a proper ground potential on the part between the absorber lines. Splitting the CPW line is necessary to form a loop out of the

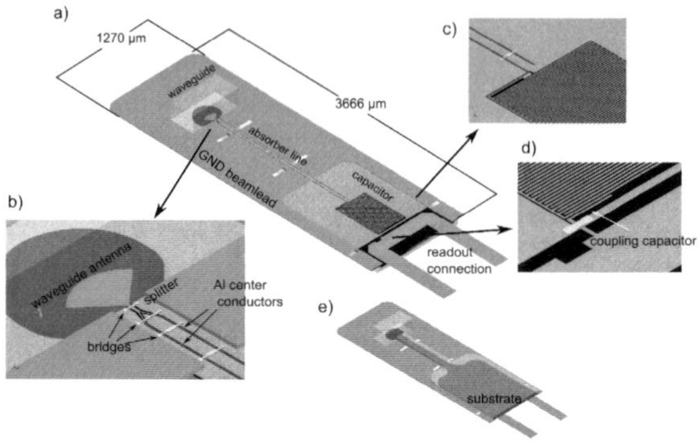

Fig. 3. CAD rendering of the lumped element resonator design. a) Overview b) Waveguide probe section with splitter. c) Transition from absorber line to capacitor. d) Couplig to the readout line.

absorber line which acts as the inductor of an LC resonator. At the end of the inductor which has a length of about 1700 μm , an interdigital capacitor is attached. The capacitor is again capacitively coupled to the readout line and is designed to coupling Q factors of 10000 or 100000.

The signal part of this circuit was also simulated using CST^{TM}microwave studio as shown in Fig. 2. The simulation is done the same way as for the TM resonator design mentioned above. The input bandwidth if this design goes up to 400 GHz.

III. Summary

We designed two single pixel waveguide coupled MKID detectors for 350 GHz with about 100 GHz of bandwidth. Those detectors can be used to examine the absorption efficiency and act as a testbed for investigation of different active materials since the deposition of the absorber lines is designed as a seperate lithography step. Since the absorber lines will be written using electron beam lithography, they can be adapted to an absorber material with different resistivity.

The devices will be measured in an ADR cryostat at a base temperature below 100 mK with an blackbody load as radiation source. Measurements of the detector response and noise at different radiation levels will allow determination of an optical noise equivalent power (NEP) in dependence of the optical loading level for a relevent characterization of the performance.

Acknowledgment

MKID detector development is carried out within the Collaborative Research Centre 956, funded by the Deutsche Forschungsgemeinschaft (DFG).

References

[1] P. K. Day, H. G. LeDuc, B. A. Mazin, A. Vayonakis, and J. Zmuidzinas, "A broadband superconducting detector suitable for use in large arrays," *Nature*, vol. 425, no. 6960, pp. 817–821, 2003, 10.1038/nature02037.

[2] Monfardini, A., Swenson, L. J., Bideaud, A., Désert, F. X., Yates, S. J. C., Benoit, A., Baryshev, A. M., Baselmans, J. J. A., Doyle, S., Klein, B., Roesch, M., Tucker, C., Ade, P., Calvo, M., Camus, P., Giordano, C., Guesten, R., Hoffmann, C., Leclercq, S., Mauskopf, P., and Schuster, K. F., "NIKA: A millimeter-wave kinetic inductance camera," *A&A*, vol. 521, p. A29, 2010. [Online]. Available: http://dx.doi.org/10.1051/0004-6361/201014727

[3] A. Endo, P. van der Werf, R. Janssen, P. de Visser, T. Klapwijk, J. Baselmans, L. Ferrari, A. Baryshev, and S. Yates, "Design of an integrated filterbank for DESHIMA: On-chip submillimeter imaging spectrograph based on superconducting resonators," *Journal of Low Temperature Physics*, vol. 167, pp. 341–346, 2012, 10.1007/s10909-012-0502-1. [Online]. Available: http://dx.doi.org/10.1007/s10909-012-0502-1

[4] M. Tinkham, *Introduction to superconductivity*, 2nd ed. Dover, 2004.

[5] B. Mazin, M. Eckart, B. Bumble, S. Golwala, P. Day, J. Gao, and J. Zmuidzinas, "Optical/UV and X-Ray Microwave Kinetic Inductance Strip Detectors," *Journal of Low Temperature Physics*, vol. 151, pp. 537–543, 2008, 10.1007/s10909-007-9686-1. [Online]. Available: http://dx.doi.org/10.1007/s10909-007-9686-1

[6] J. Zmuidzinas, "Superconducting microresonators: Physics and applications," *Annual Review of Condensed Matter Physics*, vol. 3, no. 1, pp. 169–214, 2012. [Online]. Available: http://www.annualreviews.org/doi/abs/10.1146/annurev-conmatphys-020911-125022

[7] J. Baselmans, "Kinetic inductance detectors," *Journal of Low Temperature Physics*, vol. 167, no. 3, pp. 292–304, May 2012. [Online]. Available: http://dx.doi.org/10.1007/s10909-011-0448-8

[8] S. Andree-Labsch, K. Jacobs, J. Stutzki, M. Schultz, and C. E. Honingh, "Near quantum limited Nb-Al-AlOx-Nb mixers on 9 μm thick silicon substrates around 350 GHz," *Journal of Infrared, Millimeter, and Terahertz Waves*, vol. 35, no. 3, pp. 300–317, 2014. [Online]. Available: http://dx.doi.org/10.1007/s10762-014-0052-5

[9] J. W. Kooi, G. Chattopadhyay, S. Withington, F. Rice, J. Zmuidzinas, C. Walker, and G. Yassin, "A full-height waveguide to thin-film microstrip transition with exceptional rf bandwidth and coupling efficiency," *International Journal of Infrared and Millimeter Waves*, vol. 24, no. 3, pp. 261–284, 2003. [Online]. Available: http://dx.doi.org/10.1023/A:1021903132609

[10] R. M. J. Janssen, J. J. A. Baselmans, A. Endo, L. Ferrari, S. J. C. Yates, A. M. Baryshev, and T. M. Klapwijk, "High optical efficiency and photon noise limited sensitivity of microwave kinetic inductance detectors using phase readout," *Applied Physics Letters*, vol. 103, no. 20, p. 203503, 2013. [Online]. Available: http://dx.doi.org/10.1063/1.4829657

[11] P. J. de Visser, S. Withington, and D. J. Goldie, "Readout-power heating and hysteretic switching between thermal quasiparticle states in kinetic inductance detectors," *Journal of Applied Physics*, vol. 108, no. 11, p. 114504, 2010. [Online]. Available: http://dx.doi.org/10.1063/1.3517152

[12] S. Doyle, P. Mauskopf, J. Naylon, A. Porch, and C. Duncombe, "Lumped element kinetic inductance detectors," *Journal of Low Temperature Physics*, vol. 151, pp. 530–536, 2008, 10.1007/s10909-007-9685-2. [Online]. Available: http://dx.doi.org/10.1007/s10909-007-9685-2

Design of simply structured metamaterial filters at sub-THz frequencies

Johanna Böhm[1,*], Karl Jacobs[1] and C.E. Honingh[1]
[1] *I. Physikalisches Institut, Universität zu Köln, 50937 Cologne, Germany*
*Contact: jboehm@ph1.uni-koeln.de

***Abstract*—We present simulation results of polarization sensitive filters for the sub-THz regime. The filters are designed following the principles of metamaterials.**

I. Introduction:

The increasing interest in Metamaterials (MM) combined with the development of highly efficient numerical simulation programs has caused an abundance of designs for various applications in recent years like perfect absorbers, negative refractive index material and invisible cloaks [1]–[4]. As the dimensions of single MM elements scale inversely with frequency the fabrication of designed MM devices are especially challenging for applications at higher frequencies. MMs are composed of sub-wavelength unit cells. Typically, they follows a periodically arrangement and consists of metallic resonators on dielectric layers. Considering the unit cells as "meta-atoms", their electromagnetic response is summed up in terms of homogenized parameters, e.g. the electric permittivity and magnetic permeability. This electromagnetic behavior strongly depends on the design of the sub-wavelength structures. Offering electromagnetic properties which are to a certain limit definable, MMs are especially interesting for applications at frequencies where the lack of suitable materials leads to challanging issues in the development of receiver and transmitter technologies [3]. Among others this includes the THz gap between 0.1 THz and 10 THz. The usage of MM based filters in this regime enables to significantly improve characteristics such as near 100 % in-band transmission which is important for the receiver applications, our filter designs are intended for. Our filter designs offer filter bands at about 300 GHz to 800 GHz. To enable a preferably straightforward fabrication process with reproducible results we use simple MM structures and restricts the number of available substrate layers to one, instead of multilayer stacked designs.

In the following section, we introduce the concept of our filter designs and present simulation results of one bandpass and one bandstop filter.

II. Design

Our filter designs are based on L-shaped resonators (LR). The excitation of the LR resonances strongly depends on the spatial alignment of the resonator arms to the E-field of the radiation source [5]. The two resonance modes, we concentrate on, can be easily distinguished by their oscillating surface currents as depicted in Fig.1. Mirror symmetric surface currents in both resonator arms indicate the symmetric mode. An incident **E**-field parallel to the axis of symmetry of the LR results in an exclusive excitation of this mode. Accordingly, the asymmetric mode is excited by an incident **E**-field aligned perpendicularly to the axis. An arbitrary aligned radiation causes a superposition of asymmetric and symmetric current flows in the resonator. Such an interplay of oscillations is able to yield polarization conversion [6]. The field amplitude **E** of the refracted radiation is cross-polarized (CR) if it is perpendicular to that of the incoming radiation. Co-polarized (CO) denotes the radiation with polarization direction parallel to the incident radiation.

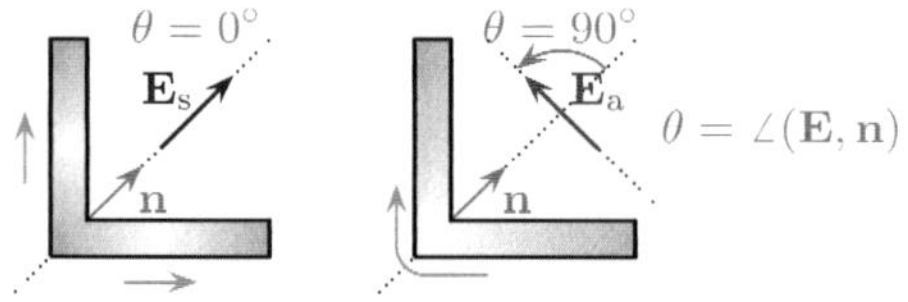

Figure 1. Sketch of two different resonance modes of the LR. The yellow arrows indicate the flows of current for the respective resonance (left: symmetric mode, right: asymmetric mode) The **E**-field vector indicates the relative orientation of an incoming radiation to the LR that is needed to only excite the specific mode. Θ is the alignment angle between the axis of symmetry of the LR and the **E**-field vector.

Table I
Dimensions of simulated filters

Filter	l_1 (µm)	p (µm)	h (µm)
Fig. 2(a)	82.7	100	9
Fig. 2(b)	79.3	92.8	14

$a = 5$ µm and $d = 2.5$ µm for all filter designs

III. Simulation Results

All calculations of the designed MM structures have been performed with the commercial simulation program CST Studio Suite [7] using the frequency domain solver of Microwave Studio. The results are given in terms of scattering parameters (S-parameters).

The filter design based on LRs provides at least two different filter bands. Changing the relative alignment of the filter with respect to the polarization of the incoming

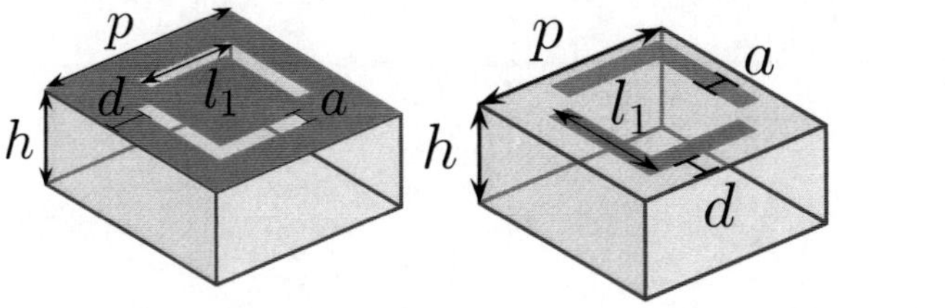

(a) (b)

Figure 2. Dimensions of parameters for the bandpass filter design (a) and bandstop design(d). The dark-gray colored areas in the sketches indicate metal, the substrate is colored green.

radiation one can switch between the filter bands as depicted by the polarization alignment angle Θ in Fig. 1. We have designed one bandstop and one bandpass filter. To maximize the transmission behavior we use silicon membranes as substrate layers. They can be considered to be loss free within the sub-THz regime. The substrate is coated with LR structures on the front side (see Fig. 2). LRs placed on a loss free substrate, totally reflect an incident, linear polarized plane wave if only a single mode is excited. This provides the basis for the bandstop filter design. Following Babinet's principle [5], the reflective behavior can be transposed into a transmittive behavior when coating the substrate with the complementary image of the metal structures. Following this principle we realize the bandpass filter using LR slots instead of filled LRs. As a consequence the corresponding modes are excited when for the LR slots the **H**-field vector of the incoming radiation is positioned as the **E**-field vector for the LRs. A summary of the designs is given in Fig. 2 and Table I.

We have simulated the filtering performance using aluminum (conductivity 2.5×10^7 S/m) for all metal parts of the filters. For optimum alignment of the filter structures with the polarization of the incoming radiation, the simulation results predict a close to 100 % transmission of co-polarized radiation at the center frequency of the passbands as shown in Fig. 3 (a)-(b). The results for the bandstop filter given in Fig. 3 (c)-(d) show a suppression of the transmission of the co-polarized radiation of about −70 dB at the center frequency by reflecting all incoming radiation. The cross-polarized contributions of the radiation are not included in the graphics as they are less than −50 dB and can ne neglected in good approximation. Within an angle tolerance of ±10° there is no significant change in the filter performance.

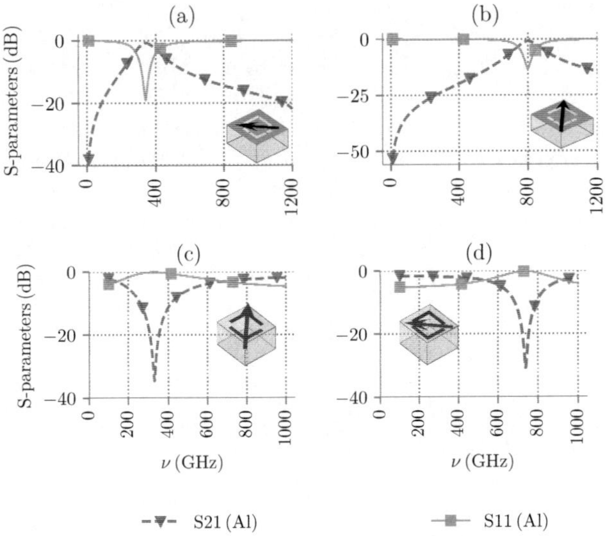

Figure 3. Simulation results of one bandpass (Fig. 2 (a)-(b)) and one bandstop filter design (Fig. 2 (c)-(d)). The left-hand column depicts the normalized amplitude magnitude of transmission S21 (dashed line with triangles) and reflection S11 (solid line with squares) of the lower filter bands and the column on the right those of the upper filter bands. The unit cells used for the filter designs are directly given in each diagram. An arrow shows the alignment of the filter geometries to the **E**-field vector (red) in case of the bandpass or the **H**-field vector (black) in case of the bandstop filter of the incoming linear polarized radiation.

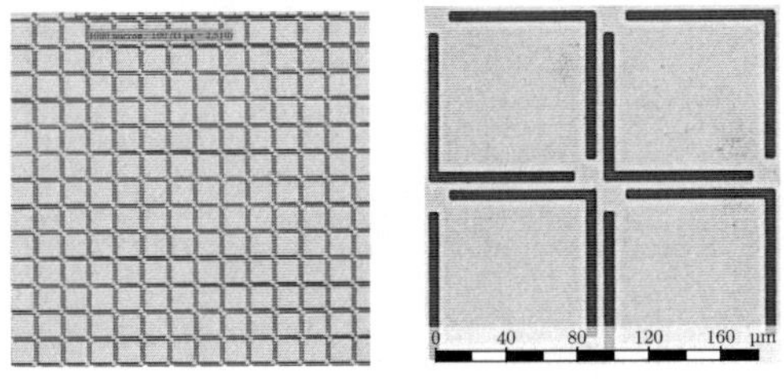

(a) (b)

Figure 4. Photos of the bandpass filter for different zoom levels [9]. Dark brown colored: silicon layer. Bright yellow colored: aluminum.

A. Preliminary fabrication test

To study the stability of the filter and the design's suitability for fabrication we have first fabricated the filter structure of the bandpass filter on an available 15 µm thick substrate before processing on thinner wafers. The fabrication is performed on SOI wafers with a high resistivity silicon device layer (5 kΩ cm) on a 300 µm thick handle wafer. The filter structures are defined by standard photolithographic lift-off processes and deposition of 280 nm aluminum on the front side. The handle wafer is etched away from the back side using Deep Reactive Ion etching in an SF_6 plasma, similar to what has been developed for our HEB mixers [8]. The final bandpass filter is circular with a diameter of 22.5 mm. It consists of more than 65 000 resonators (see Fig. 4) with only non-truncated LRs on its edges to prevent fringe effects.

IV. Conclusion

Based on simple L-shaped resonator geometries we have successfully designed a MM bandpass and bandstop filters for the sub-THz regime. Simulations of our designs predict very high transmission for the pass bands and low transparency for the stop filter.

Comparing commercially available filters for THz radiation [10] with the simulation predictions, the MM filters offer a higher performance with both an above-average transmission and larger bandwidths. In comparison to MM filters that have been investigated for similar frequency ranges [11], [12] we find at least comparable, if not higher, filtering performance while using a significantly simpler design concept. An analysis of the alignment tolerances demonstrates a good practicability of the filters.The filters are currently being fabricated using SOI substrate technique and standard photolithography. Preliminary fabrication tests confirm processability even for huge numbers of L-resonators in the order of 10^4.

References

[1] H.-T. Chen, J. F. O'Hara, A. K. Azad, and A. J. Taylor, "Manipulation of terahertz radiation using metamaterials," *Laser & Photonics Reviews*, vol. 5, no. 4, pp. 513–533, 2011.

[2] N. I. Zheludev and Y. S. Kivshar, "From metamaterials to metadevices," *Nature materials*, vol. 11, no. 11, p. 917, 2012.

[3] H. Tao, W. J. Padilla, X. Zhang, and R. D. Averitt, "Recent progress in electromagnetic metamaterial devices for terahertz applications," *IEEE Journal of Selected Topics in Quantum Electronics*, vol. 17, no. 1, pp. 92–101, 2011.

[4] V. Veselago, L. Braginsky, V. Shklover, and C. Hafner, "Negative refractive index materials," *Journal of Computational and Theoretical Nanoscience*, vol. 3, no. 2, pp. 189–218, 04 2006.

[5] D. Lee and D.-S. Kim, "Light scattering of rectangular slot antennas: parallel magnetic vector vs perpendicular electric vector," *Scientific reports*, vol. 6, 2016.

[6] J. Ding, B. Arigong, H. Ren, J. Shao, M. Zhou, Y. Lin, and H. Zhang, "Mid-infrared tunable dual-frequency cross polarization converters using graphene-based l-shaped nanoslot array," *Plasmonics*, vol. 10, no. 2, pp. 351–356, 2014.

[7] CST - Computer Simulation Technology AG. CST - Computer Simulation Technology. [Online]. Available: https://www.cst.com/

[8] P. Pütz, K. Jacobs, M. Justen, F. Schomaker, M. Schultz, S. Wulff, and C. Honingh, "Nbtin hot electron bolometer waveguide mixers on Si3N4 membranes at thz frequencies," *IEEE Transactions on Applied Superconductivity*, vol. 21, no. 3, pp. 636–639, 2011.

[9] J. Böhm, "A metamaterial absorber for the sub-thz regime at cryogenic temperatures," *Master Thesis*, 2016.

[10] T. K. Ltd and Q. I. Ltd. Qmc instruments. [Online]. Available: http://www.terahertz.co.uk/qmc-instruments-ltd/thz-optical-components/multi-mesh-filters

[11] Q. Li, X. Zhang, W. Cao, A. Lakhtakia, J. F. O'Hara, J. Han, and W. Zhang, "An approach for mechanically tunable, dynamic terahertz bandstop filters," *Applied Physics A*, vol. 107, no. 2, pp. 285–291, 2012.

[12] M. Lu, W. Li, and E. R. Brown, "Second-order bandpass terahertz filter achieved by multilayer complementary metamaterial structures," *Optics letters*, vol. 36, no. 7, pp. 1071–1073, 2011.

1.9 THz balanced superconducting Hot Electron Bolometer mixers fully integrated on chip

S. Fathi*, K. Jacobs, M. Schultz, P. Pütz, D. Büchel, S. Selig, C. E. Honingh
I. Physikalisches Institut, Universität zu Köln
contact: fathi@ph1.uni-koeln.de

***Abstract*—We present the development of a balanced HEB mixer operating at 1.9 THz. The mixer is a waveguide-based and its complete RF circuit is integrated on a chip employing a planar 180° RF hybrid-ring coupler. Preliminary measurement results with a first fabricated batch of devices at 4.2 K shows well balanced pumping of the two separated micro-bridges of a mixer device, using a 1.9 THz VDI multiplier chain or a QCL LO.**

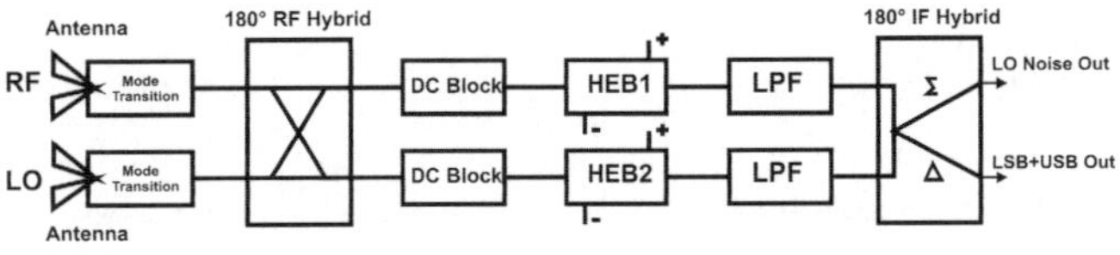

Fig. 1. HEB Balanced mixer schematic

I. Introduction

HOT -electron-Bolometer (HEB) mixers are currently the most sensitive heterodyne detectors above the 1THz. They are applied for example in the GREAT instrument on SOFIA flight observatory to observe the [CII] line at 1.9 THz in the interstellar medium (ISM)[1]. In order to achieve high resolution and time efficient observations focal plane arrays (FPAs) are used, where a balanced mixer configuration can be an appealing approach, see Fig.1. A balanced mixer removes the need for the existing diplexer or beam-splitter for combining the LO and sky signal, thereby allowing an efficient use of local oscillator power. Putting the complete RF circuit on chip avoids the development of waveguide circuitry with dimensions considerably smaller than the waveguide itself. While this is still possible with precision CNC mechanical fabrication at lower frequencies, THz waveguide circuitry is commonly defined and fabricated using some sort of photolithographic technique. There have been two groups that reported about the 1.3 [2] and 2.7 THz balanced HEB mixers [3] based on 90 ° waveguide hybrid couplers. First work uses the technique of putting metal over a photo-lithographically patterned thick SU-8 [4]. The latter case is based on waveguide structures etched in Silicon. Our group has already developed a balanced on chip mixer around 460 GHz [5], showing that the on chip approach can be successful. In this work, we present the development of 1.9 THz balanced HEB mixers where we are using a 180° ring RF-hybrid coupler. Using a standard through waveguide for the LO and RF input, the designed isolation between the LO and sky signal is at least -30 dB in the final mixer circuit. All the RF elements are defined directly by the E-beam lithography on a SOI substrate. This brings advantages of reducing the complexity in the fabrication and improving re-producibility which is especially important in the development of THz-FPAs.

II. Design

We used CST^{TM}studio suite to design and simulate the THz mixer circuit established in Au coplanar (CPW) and slot lines on a 3μm Silicon substrate. The complete design of a 1.9 THz balanced HEB and its RF performance are shown in the Fig.2a and Fig.2b, respectively. The circuit consists of a co-planar

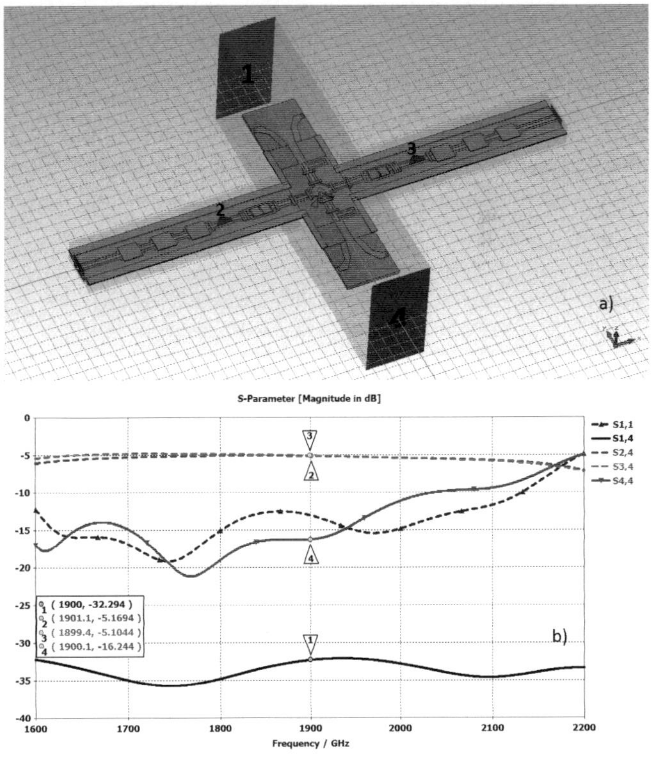

Fig. 2. a) Complete design of a 1.9 THz balanced HEB mixer on a 3 μm silicon b) S-parameter results of the simulated circuit in 'a' showing isolation (S1,4) between port 4 and port 1(opposite to the port 4), reflections at the port 4 (S4,4) and 1 (S1,1), and transmission from port 4 to the two HEBs (discrete ports 2 and 3)

ring hybrid coupler to equally distribute the LO and sky signals with 180° and zero phase differences to the two HEB bridges. Two slotline E-plane planar antennas are implemented to couple the signal and LO from the waveguide channels (50×100 μm^2) to the chip. Two NbN HEB microbridges which are integrated oppositely in gaps of the CPW transmission lines show the impedance of about 120 Ohm at 20 K for a dimension of 3 μm width, 200 nm length and about 4.5 nm thickness. Two blocking capacitors prevent on chip coupling between the two HEBs at the IF and at DC. S-parameter results show at 1.9 THz the amplitude imbalance between two signals is almost zero and reflection at waveguide ports is less than -10 dB over a Band from 1.6 to slightly above 2 THz with an isolation better than -35 dB the whole band.The phase imbalance is about ±2 degrees.

III. Fabrication and DC-Characterization

All mixer circuit elements on one chip are defined and written by the E-beam lithography. A detailed description of the Cologne HEB fabrication is given in [6]. The measured critical temperature (Tc) of about 4.5 nm sputtered NbN film on the SOI wafer is about 10.5 K. An important characteristic in the balance mixer circuit is that the two HEBs on one chip show very similar DC - responses.

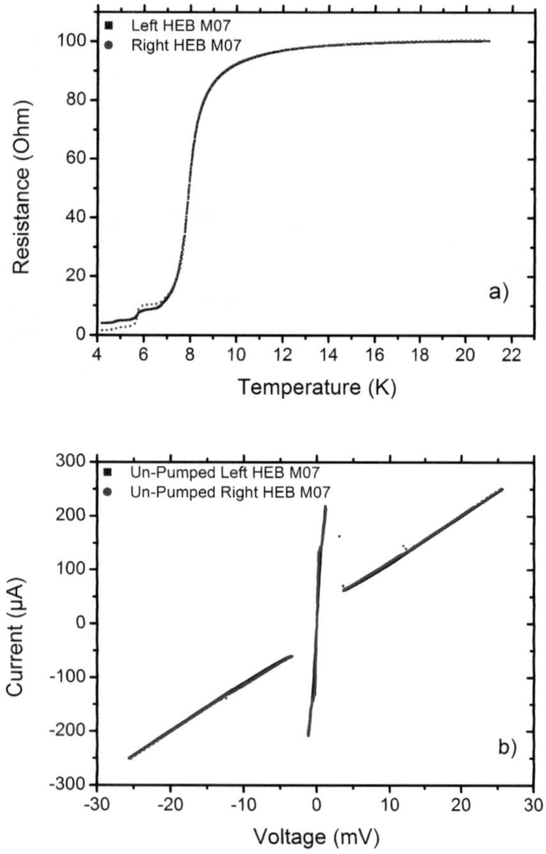

Fig. 3. a)and b) Measured R-T and I-V of the device M07 in liquid Helium before separation and SOI back side etching

The R versus T measurement in liquid Helium of the final micro-fabricated device before SOI back side etching and device separation, shows a Tc of about 8 K Fig.3a. The Ic of the same device is around 220 μA Fig.3b. This device is selected from the sector M and row number seven of the lithography fabrication mask. RN (HEB impedance at 20 K) is about 100 Ohm which is lower than 120 Ohm design value. However, the tolerance analysis in CST shows that we can accept such an impedance difference of the HEB bridges.

IV. RF Measurement

The mixer device is mounted in a machined CuTe alloy E-plane split waveguide block.The signal and LO are coupled by two waveguide horns. Two intermediate microstrip IF boards and two SMA connectors are used to extract the generated IF signals, and supply the DC bias at the output ports. In order to pump the mixer, 1.9 THz multiplier LO chain from VDI or an in-house developed QCL LO [7] is applied. In Fig.4a pumped IVs of two HEBs for the device M04 are shown. The relative power imbalance between the two HEBs is estimated by the Isothermal approach to be about 10% at a maximum, based on the single mixer results of the LFA development [8]. This should be acceptable for the balanced mixer performance. Unfortunately, due to technical difficulties, we could not apply

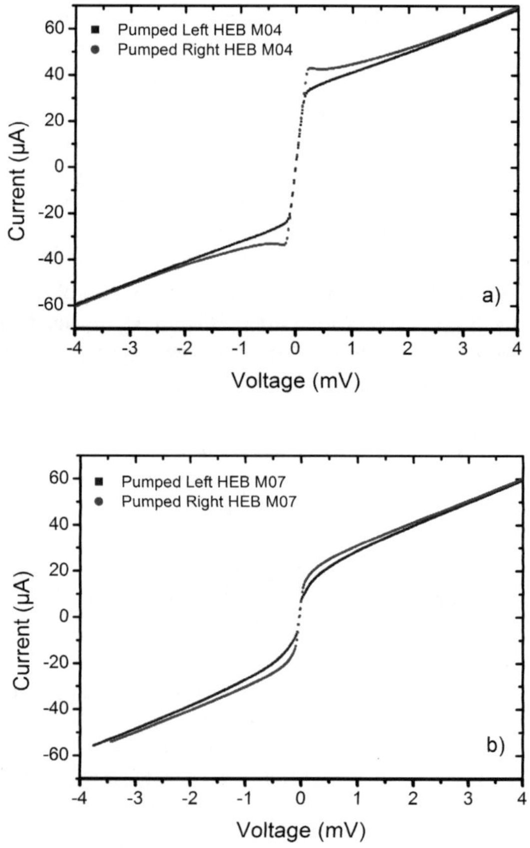

Fig. 4. a) Pumped IV curve of the device M04 in dewar with the VDI multiplier chain,b) Pumped IV curve of the device M07 in dewar with the QCL LO

RF signal from both ports.This prohibited any real balanced

mixer measurement. To check if the circuit itself works also from the side that we were not able to apply RF, a new device (M07), with a flipped direction was built in. In this way, the other side of the chip is pointing towards the horn that is able to supply RF-signal. The measured pumped IV-curves for this configuration are shown in Fig.4b. To get a preliminary idea about the sensitivity, we performed a noise temperature measurement from the one accessible port, using a 24 μm thick Mylar beam-splitter in vertical polarization. The measured noise temperature, over a 1-2 GHz IF bandwidth is approximately 7000 K. This is the noise temperature of one of the mixers with two HEBs. Each HEB mixer receives only half of the input power unavoidably due to the coupling via the ring coupler. Therefore, a proper estimate for the noise temperature of a single mixer would be around 3500 K.

V. Summary

We have designed, fabricated and characterized the 1.9 THz balanced HEB mixer on one chip using a 180° planar ring hybrid. The initial RF measurement shows possible pumping of two HEB bridges with about 10% power imbalance. Further on, balance heterodyne measurement to obtain IF noise temperature is not executed due to technical difficulties. Single mixer heterodyne measurement shows a noise temperature of about 3500 K.

Acknowledgment

Development of 1.9 THz balanced HEBs is carried out within the Collaborative Research Centre 956, sub-project D3, funded by the Deutsche Forschungsgemeinschaft (DFG).

References

[1] C. Risacher, R. Güsten, J. Stutzki, H. W. Hübers, D. Büchel, U. U. Graf, S. Heyminck, C. E. Honingh, K. Jacobs, B. Klein, T. Klein, C. Leinz, P. Pütz, N. Reyes, O. Ricken, H. J. Wunsch, P. Fusco, and S. Rosner. "First Supra-THz heterodyne array receivers for astronomy with the SOFIA observatory,". *IEEE Transactions on Terahertz Science and Technology.*, Vol.6, March.2016.

[2] D. Meledin, V. Desmaris, M. Fredrixon S.-E. Ferm, D. Henke, I. Lapkin, O. Nystrm, M. Pantaleev, A. Pavolotsky, M. Strandberg, E. Sundin, and V. Belitsky. "APEX band T2: A 1.25 1.39 THz waveguide balanced HEB receiver,". *IEEE Transactions on Microwave Theory and Techniques.*, Vol.57, Jan.2009.

[3] F. Bousaha, J. Kawamure, J. Stern, and C. Jung-Kubiak. "2.7 THz balanced waveguide HEB Mixer,". *IEEE Transactions on Terahertz Science and Technology.*, Vol.4, Sept.2007.

[4] V. Desmaris, D. Meledin, A. Pavolotsky, R. Monje, and V. Belitsky. "All-metall micromachining for the fabrication of sub-milimeter and THz waveguide components and circuits ,". *Journal of Micromechanics and microengineering.*, (Num.9), July.2008.

[5] M. P. Westig, M. Justen, K. Jacobs, J. Stutzki, M. Schultz, F. Schomacker, and C.E. Honingh. "A 490 GHz planar circuit balanced Nb-Al2O3-Nb quasiparticle mixer for radio astronomy: Application to quantitative local oscillator noise determination,". *Journal of Applied Physics.*, 2012.

[6] P. Pütz, D. Büchel, K. Jacobs, M. schultz, and C. E. Honingh. "1.9 THz waveguide HEB mixers for the upGREAT low frequency arrayr,". *26th International Symposium on Space Terahertz Technology.*, March.2015.

[7] M. Justen, K. Otani, D. Tur¨cinkov¨a, F. Castellano, M. Beck, U.U. Graf, D. Büchel, M. Schultz, and J. Faist. "Waveguide embedding of a double metal 1.9 THz Quantum Cascade Laser: design, manufacturing and results,". *IEEE Transactions on Terahertz Science and Technology.*, Oct.2016.

[8] D. Büchel. *"Hot Electron Bolometer mixers for THz arrays."*. PhD thesis, Univ. zu Köln, 2017.

Schottky components for ESA MetOp SG space mission

Oleg Cojocari[1*], Ion Oprea[1], Martin Rickes[1], Matthias Hoefle[1]

[1]*ACST GmbH, 63457, Hanau, Germany*
[*]Contact: oleg.cojocari@acst.de

Abstract—ACST GmbH is a leading European commercial supplier of Schottky diodes for millimeter wave and sub millimeter wave applications. With core business in the Schottky diode fabrication technology, ACST has extended its activity towards assembly and environmental testing of THz components and modules for Space missions. This allows gathering experience and know-how in different critical processes in house and provides opportunity for linkage between different stages within the development process of Space THz-electronics. Being involved in several space-related activities, ACST has become a strong partner in the area of development and qualification of THz electronics for space applications.

Several types of ACST Schottky diodes are currently considered for high-frequency receivers of the MWS, MWI, and ICI instruments as part of the second European meteorological operational satellite program MetOp-SG. This contribution will focus on fabrication technology of Schottky components for space applications and will update the status of procurement activity in the framework of ESA MetOp-SG space mission.

Design and Optimization of Micro-patterned Quasi-optical Impedance Transformers

Jake A. Connors[1*], C.-Y. E. Tong[2], Paul K. Grimes[2] and Scott N. Paine[2]
[1]*Physics Department, Harvard University, Cambridge, MA 02138*
[2]*Smithsonian Astrophysical Observatory, Cambridge, MA 02138*
*Contact: jconnors@physics.harvard.edu

Abstract— Presented is method to model and optimize multi-layer micro-patterned impedance matching structures. For a number of geometric patterns in silicon, effective dielectric constants are calculated in the long wavelength limit using an effective capacitance method. The S-parameters of multi-layer impedance transformers are then computed using the transfer matrix method. A stochastic method known as simulated annealing is used to optimize impedance transformers and incorporate fabrication constraints. Lastly, the effective capacitance method is used to model a square pyramidal impedance taper.

INTRODUCTION

Artificial dielectric materials, such as expanded Teflon sheets, have long been used as impedance matching layers for millimeter and sub-millimeter vacuum windows, filters and lenses[1]. More recently, there has been concentrated effort to design and produce micro-patterned anti-reflection layers in silicon through deep-reactive ion etching[2]. Micro-patterned layers, also known as metamaterials, are composed of a repeated structure on sub-wavelength scales and can be characterized by an effective dielectric constant. The effective dielectric constants for grooves[3], both parallel and perpendicular to the electric field, and square posts/holes[4] have been analytically studied. We expand upon the effective capacitance method of Biber[4] to study several simple patterns in silicon. Using a stochastic optimization routine, we design a simple micro-patterned impedance transformer. Lastly, we use our method of calculating effective dielectric constants to study a square pyramidal impedance taper.

EFFECTIVE DIELECTRIC CONSTANTS

Micro-patterned materials, consisting of a repetitive pattern on scales much smaller than the wavelength of interest, can be characterized by an effective dielectric constant. This dielectric constant determines both the propagation constant of an EM wave in a layer of the material as well as the reflection and transmission at an interface between two layers of differing pattern. The effective dielectric constant of a pattern is in general polarization and frequency dependent. In the long-wavelength limit, the effective dielectric constant can be calculated using the capacitance of a single unit cell of the pattern. Fig. 1 plots the results of calculations of the effective dielectric constant

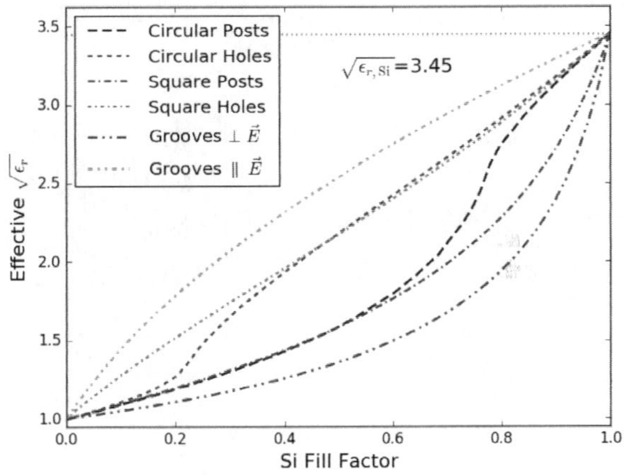

Fig. 1 Effective refractive index versus fill factor for several different patterns in silicon as calculated in the long-wavelength limit. Circular patterns experience a transition around the fill factor at which they begin to overlap with nearby holes/posts.

(plotted as the effective refractive index) for several different patterns in silicon and air as a function of silicon fill factor. As can be seen in the plot, patterns which have a continuous path of silicon along the polarization direction, such as grooves parallel to $\vec{E}$ or any variety of hole pattern, have a distinctly larger effective dielectric constant than patterns which do not have continuous silicon paths at the same fill factor.

OPTIMIZATION OF IMPEDANCE TRANSFORMERS

The S-parameters of a multi-layer micro-patterned impedance transformer can be calculated as a function of each layer's pattern geometric parameters and thickness. From the pattern parameters, polarization-dependent effective dielectric constants for each layer are calculated as described previously. Assuming a normal incidence plane wave as the impinging EM mode, we calculate the S-matrix of the multi-layer structure using the transfer matrix method[5] over the polarizations and frequency band of interest. Using the calculated S-parameters, a cost function can defined for the transformer optimization, such as a band-averaged return loss or band-worst transmission. This cost function is then minimized using a stochastic algorithm known as simulated annealing where a Metropolis-Hastings

A FULL VERSION OF THIS PROCEEDINGS ARTICLE HAS BEEN SUBMITTED TO THE IEEE TRANSACTIONS OF TERAHERTZ SCIENCE AND TECHNOLOGY (TST) FOR PEER-REVIEW, PLEASE CHECK THE WEBSITE: HTTPS://WWW.MTT.ORG/TERAHERTZ

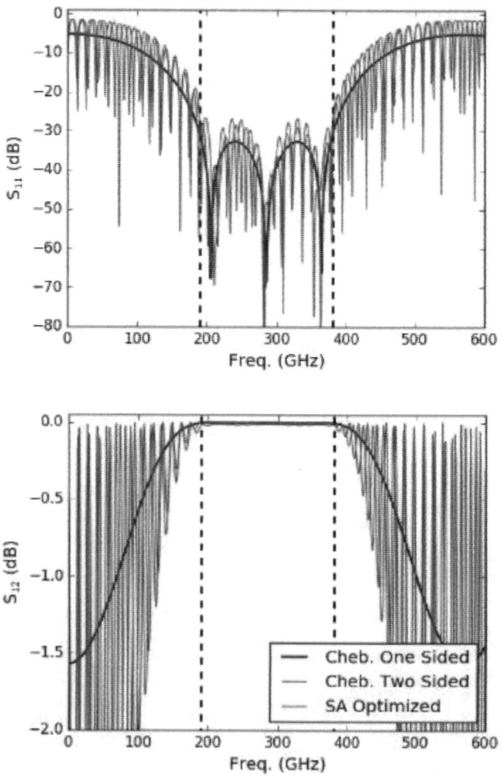

Fig. 2 Example of the results of a simulated annealing optimization of a 3-layer impedance transformer designed on a 3mm thick silicon wafer for 190-380GHz. The return loss (top) and insertion loss (bottom) are very similar to a Chebyshev transformer in this simple case, with minor improvement coming from tuning the thickness of the wafer.

algorithm is used to update layer parameters and thicknesses while the effective system temperature is slowly decreased. In this way, the layer parameters and thickness which minimize the cost and produce an 'optimal' transformer can be found.

Results of such an optimization are shown in Fig. 2 for a 3-layer two sided anti-reflection coating covering the 190-380GHz band on a ~3mm thick silicon wafer intended as a vacuum window. As can be seen, the result is very similar to a Chebyshev transformer in this simple case. In more complicated scenarios where the fabrication of a Chebyshev transformer is not feasible, the stochastic approach still allows for the design of an 'optimum' impedance transformer. Such a situation can simple arise from fabrication constraints, such as a maximum achievable aspect ratio during pattern etching.

Modeling Impedance Tapers using the Effective Capacitance Method

Tapered impedance transformers provide the ultimate in wideband performance at the cost of increased transformer length and manufacturing difficulty. Tapered square pyramidal patterns, studied in [6], can be machined in

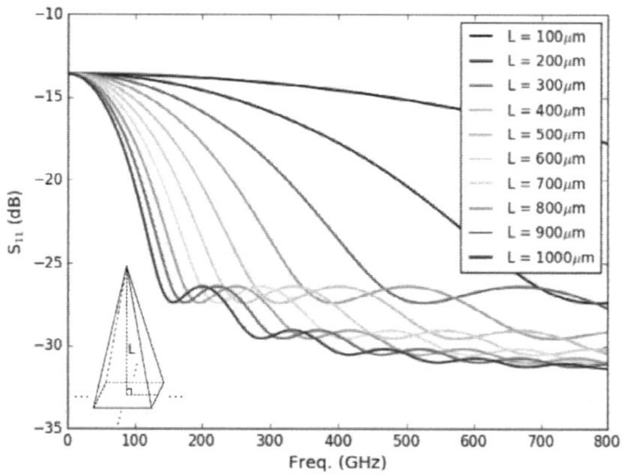

Fig 3. Simulated return loss of a 100μm pitch square pyramidal taper in HDPE as cut by a custom slitting saw. As the length of the taper increases, the cut-off frequency systematically decreases. The in-band return loss of the taper is limited by the size of the saw tip, in this simulation set to 6μm.

plastics using a custom-ground slitting saw. In order to model such a taper, the effective dielectric constants for an array of sizes of square posts was first calculated. The S-matrix of the taper was then calculated using the transfer matrix method, treating the taper as a large number of infinitesimally thin steps of square posts. In Fig. 3, we plot the simulated return loss for a square pyramidal taper of varying length. The cut-off frequency is controlled by the length of the taper, while the in-band return loss is controlled by the size of the tip of the slitting saw.

Conclusions

We have presented a simple method for the modelling and design of micro-patterned impedance transformers. In layered transformers, an effective dielectric constant is calculated for each layer. Using a stochastic approach known as simulated annealing, multi-layer transformers can be optimized over the space of layer geometric properties and thicknesses. Lastly, this method was used to study square pyramidal impedance tapers.

References

[1] D. Koller, A.R. Kerr, G.A. Ediss, D. Boyd, "Design and Fabrication of Quartz Vacuum Windows with Matching Layers for Millimeter-Wave Recievers," *12th International Symposium on Space Terahertz Technology Proceedings,* San Diego, CA, Feb. 2001.

[2] R. Datta et. al., "Large-aperture wide-bandwidth antireflection-coated silicon lenses for millimeter wavelengths," Appl. Opt. 52, 8747-8758 (2013)

[3] S.M. Rytov, "Electromagnetic Properties of a Finely Stratified Medium", *JETP*, Vol. 2, No. 3, p. 466 (1956)

[4] S. Biber et. al., "Design of artificial dielectrics for anti-reflection-coatings," *33rd European Microwave Conference Proceedings*, Vol. 3, pp. 1115-1118, 2003.

[5] D. Pozar, "Microwave Engineering 4th Edition," John Wiley & Sons, Inc, 2012.

[6] W.H. Southwell, "Pyramid-array surface-relief structures producing antireflection index matching on optical surfaces," *J. Opt. Soc. Am. A,* Vol. 8, No. 3, March 1991

Micro-Machined Integrated Waveguide Transformers in THz Pickett-Potter Feedhorn Blocks

Kristina K. Davis[1,*], Jennifer Kloosterman[2], Chrisopher Groppi[1], Jonathan Kawamura[2], and Matthew Underhill[1]
[1] *School of Earth and Space Exploration, Arizona State University, Tempe, AZ, USA*
[2] *Submillimeter Wave Advanced Technology, Jet Propulsion Laboratory, Pasadena, CA, USA*
*Contact: Kristina.Davis.1@asu.edu

Abstract— We present laboratory measurements of a circular-to-rectangular waveguide transformer integrated into a 1.9 THz Pickett-Potter feedhorn detector block. This design is applicable for instruments where circularly symmetric feedhorns are required to mate with rectangular waveguide fed receiver devices. Compared to previous transformer segments machined into separate blocks or machined into split-block segments, we ensure axial alignment along the waveguide segments at the cost of rounding the edges of the rectangular waveguide. This architecture was fabricated by direct metal micro-machining, which offers significant advantages over competing techniques in complexity, timescale, and cost of manufacturing. All machining passes during manufacture can be made from the front of the block including the final **waveguide** segment. We compared **simulations of the waveguide circuit performance using multiple electromagnetic software packages to finalize the dimensions of the optimized transformer module. A single pixel feedhorn-transformer module was manufactured on a 3-axis CNC milling machine. We tested integrated feedhorn-transformer modules using waveguide-fed hot electron bolometer mixers designed and fabricated at the Jet Propulsion Laboratory using a liquid helium-cooled cryostat. Beam patterns of the Pickett-Potter modules were measured using a high-power 1.9 THz multiplication chain as the source. We find good agreement between the simulated and laboratory beam pattern.**

INTRODUCTION

Recent work in terahertz (THz) heterodyne instruments for astronomical research have utilized rectangular waveguide-fed receiver devices [1], [2], [3]. For single-polarization receivers, it is generally desirable to use a feedhorn type that maximizes coupling to the detector and minimizes cross-polarization. Several feedhorn profiles offer ease in fabrication and have low cross-polarization [4], [5] [6], but have circular exit waveguides requiring a waveguide transformer to integrate them with existing rectangular waveguide-fed devices.

Circular-to-rectangular (CTR) waveguide transformers (WGT) have been demonstrated in a laboratory setting using direct-metal micromachining techniques [5] [6]. In both cases, the transformer segment was machined into separate pieces of the receiver modules. These transformer segments become increasingly difficult to manufacture at short wavelengths (high frequencies). With multiple receiver segments, the mated system is more susceptible to misalignment between the transformer, the detector housing, and feedhorn blocks. One solution is to integrate the transformer into the feedhorn block using a split-horn fabrication technique, though this decreases the alignment advantage of using circularly symmetric feedhorns because there can be misalignment between the two halves of the split-block upon mating.

We offer the solution of integrating the CTR WGT directly into the feedhorn block. Considering future design and fabrication of large focal plane arrays, decreasing the number of individual components saves valuable machining time and has the additional benefit that misalignment errors do not tend to propagate between pixels in the array.

TRANSFORMER DESIGN

The motivation for this work is based on instrument development to survey large regions of the galaxy in the astrophysically important [CII] cooling line of the interstellar medium at 158 μm (1.9 THz). Pickett-Potter horns have highly symmetric E and H planes, <-28 dB side lobe levels, and cross-polarization coupling less than -25dB relative to the main beam [4]. The fractional bandwidth is 10-20%, which is suitable for emission line surveys. Pickett-Potter horns have a flat-topped conic profile, allowing them to be machined with custom-ground tools without requiring electroforming, adhesives, or etching methods.

The design of the CTR WGT is based on [7], with the dimensions of the cross section scaled in frequency for operation at 158 μm. The feedhorn choice defined the input CWG dimensions at the input of the CTR WGT. Fig. 1 shows a diagram of the transformer module profile, and the dimensions for this particular module at 1.9 THz are givin in Table I.

MANUFACTURING TECHNIQUES

It is possible to machine all waveguide circuitry from the aperture of the Pickett-Potter horn with just three tools; one custom tapered tool with a flat end for the feedhorn, and two additional modified endmills. To achieve the proper dimensions at 1.9 THz, the horn tool was ground at a 6.5-degree half angle and the tip was flattened to match the diameter of the step between the horn and the CWG. The first endmill is slightly smaller in diameter than the CWG (here 150 μm) at the exit of the feedhorn, and the second endmill has a diameter less than that of the short dimension of the output RWG (here 40 μm). The diameter of these endmills is standard but both were recessed an the tool neck in order to fit in the horn aperture. The recessing was done at the Jet Propulsion Laboratory by grinding the tool at the appropriate clearance angle.

A FULL VERSION OF THIS PROCEEDINGS ARTICLE HAS BEEN SUBMITTED TO THE IEEE TRANSACTIONS OF TERAHERTZ SCIENCE AND TECHNOLOGY (TST) FOR PEER-REVIEW, PLEASE CHECK THE WEBSITE: HTTPS://WWW.MTT.ORG/TERAHERTZ

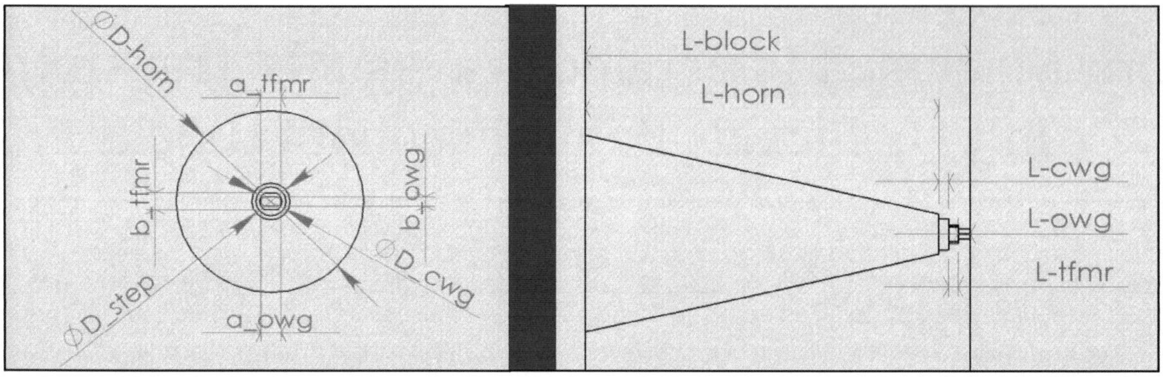

Fig. 1. Critical dimensions of the design model of the waveguide circuitry of the horn-transformer module as seen looking down the aperture (a) and a cut across the profile (b). Dimensions of these features are listed in Table I. The horn is located in the middle of a 20 mm square block of 145 copper (ultra-high purity) copper alloy. A set of four guide pins spaced on the flange outside the frame of this model image ensured precise alignment to the mixer backend.

TABLE I
MODULE DIMENSIONS

Dimension	Design (mm)	Machined (mm)
D_{horn}	1.011	1.015
D_{step}	0.205	0.208
D_{CWG}	0.161	0.159
a_{tfmr}	0.117	0.114
w_{OWG}	0.100	0.108
b_{tfmr}	0.093	0.098
h_{OWG}	0.050	0.053
ℓ_{tfmr}	0.046	0.035

The designed and as-machined dimensions of the feedhorn-transformer module. The as-machined dimensions were verified using a microscope with sub-micrometer precision. The tested unit was measured without cutting it open, so there is $^{+}$/.0.008 mm uncertainty in the measurements due to diffraction of optical light within the horn.

TABLE II
BEAMWAIST

z (mm)	ω_o measured (mm)
480	0.287
600	0.296
800	0.310
Average	0.298
Theory	0.281

The calculated beamwasit for all three measurement scans is slightly larger than the theoretical value calculated in [4]. This is an expected phenomenon since we do not correct for the beam size of the source probe.

RESULTS

We measured the radiation pattern of Picket-Potter feedhorn-transformer module to ensure there were no distortion effects produced by the integration of the WGT into the feedhorn module, and the results are presented in Table II. The beam angle of the receiver is determined by the feedhorn design, but deviation between the expected and measured radiation patterns can be used to verify the transformer performance and diagnose alignment errors. All three measurements show close agreement to the theoretical beamwaist of 0.281 mm.

CONCLUSIONS

We have modeled, fabricated, and tested a CTR WGT, which can be easily machined directly into a feedhorn block with a circularly symmetric feedhorn profile. Pickett-Potter feedhorns are more desirable than the traditional diagonal feedhorns for their lower cross-polarization properties, so these modules provide an attractive alternative in both performance and machining simplicity. The advantages of an integrated architecture are in the more precise alignment of waveguide circuitry features, reduction in the number of independently machined segments, and minimization of the mismatch risk. The performance of the integrated transformer is well matched to physical optics simulations, confirming both simulation and fabrication techniques. These new technological advances can serve as a pathway toward the implementation of large monolithic focal plane array units.

ACKNOWLEDGMENT

The authors would like to acknowledge Maria Alonso for her advice on beam pattern measurement implementation and analysis.

REFERENCES

[1] C. K. Walker, "SuperCam: A 64 pixel SIS receiver array for submillimeter-wave astronomy," *2015 IEEE MTT-S International Microwave Symposium*, Phoenix, AZ, 2015, pp. 1-4. doi: 10.1109/MWSYM.2015.7166873

[2] C. K. Walker, C. A. Kulesa, C. E. Groppi, *et al.* , "The Stratospheric Terahertz Observatory (STO)" Proceedings of the 19th ISSTT, Groningen", pp. 28-32, April, 2008.

[3] Risacher *et al.*, "First Supra-THz Heterodyne Array Receivers for Astronomy With the SOFIA Observatory," in *IEEE Transac. on THz Science and Tech.*, vol. 6, no. 2, pp. 199-211, March 2016. doi: 10.1109/TTHZ.2015.2508005

[4] H. Pickett, J. Hardy, J. Farmhoomand, "Characterization of a Dual-Mode Horn for Submillimeter Wavelengths," *IEEE Transac. On Microwave Theory and Tech.*, vol. 32, no. 8, pp. 936-937. August, 1984.

[5] N. Chahat, T. Reck, C. Jung-Kubiak, T. Nguyen, R. Sauleau, G. Chattopadhyay, "1.9-THz Multiflare Angle Horn Optimization for Space Instruments," *IEEE Transac. On THz Sci. and Tech.*, vol. 5, no. 6, pp 914-921, November, 2015.

[6] J. M. Munier, A. Maestrini, M. Salez, M. Guillon, B. Marchand. "Novel, Low-cost Technique to Manufacture Submillimeter Corrugated Feedhorns," *S Proc. of SPIE*, vol. 4015, no. 53, pp. 500-514, Munich, March 27-31, 2000.

[7] S. Stanislaw, A. Kraszewski, "Wide-Band Rectangular to Circular Waveguide Mode and Impedance Transformer," IEEE Transac. on Microwave Theory and Tech.

The SAFARI grating spectrometer for SPICA

Gert de Lange[1], Peter Roelfsema[1,2], Hiroshi Shibai[3,4]

on behalf of the SPICA-SAFARI consortium

[1]SRON Netherlands Institute for Space Research, Groningen, The Netherlands; [2]Kapteyn Astronomical Institute, Groningen, The Netherlands; [3]Osaka University, Osaka, Japan; [4]ISAS, Tokyo, Japan;

*Contact: g.de.lange@sron.nl

Abstract**—** The far-infrared spectrograph SAFARI, on the joint European-Japanese space telescope SPICA (proposed in the ESA M5 Call), will provide the most sensitive view ever of the cool, obscured universe. By cooling the SPICA telescope to below 8 K its thermal emission is decreased to such low levels that the latest generation of ultra-sensitive Transition Edge Sensor (TES) detectors can be utilized to their full potential. With this combination of low background and extreme detector sensitivity SPICA will be able to look far deeper into space than was possible with any of its predecessors. SPICA/SAFARI is the only facility that will fill the gap in the wavelength domain between the other great observatories, and is as sensitive as both the James Webb Space telescope and the ALMA radio observatory – only with SPICA/SAFARI we will complete the view on the star-formation history of our universe.

The current baseline SAFARI design uses a beam steering mirror (BSM) that forwards the incoming signal to the dispersing and detection optics. The BSM is used to select sky or calibration signals and forward that to a nominal R~300 (low) resolution optics chain or to a R~11000 (high) resolution optics chain. The low resolution is obtained by dispersion through a diffraction grating illuminating a line of TES detectors. For the high resolution mode the signal is first pre-dispersed using a Martin-Puplett interferometer before entering the grating. The full 35-230 μm wavelength range is split in to several different bands, each with its own grating and TES detectors. The baseline design has for each of the bands three separate spatial pixels, to provide background reference measurements, but also to provide some imaging capability.

With SPICA's cold, 2.5 meter telescope and the baseline TES NEP of $2x10^{-19}$ W/ √ Hz, for the new grating based SAFARI the sensitivity of the R~300 mode will be about 5 x 10^{-20} W/m^2 (5 σ , 1hr). With this high sensitivity astronomers will e.g. be able to detect the [OIV] line in relatively average galaxies out to a redshift z~3. Thus the evolution of galaxies can be followed through their most active periods in cosmic time from about 10 billion years ago to what they look like today. Additionally we will be able to observe dust features from even earlier epochs, out to redshifts of z~7-8, thus providing insight into dust formation in the very early phases of the universe.

Investigation into Possible Planar Heterodyne Receiver Arrays with large (n≥100) number of pixels.

Duccio Delfini[1*], Martina Wiedner[1], Massimiliano Casaletti[2], Julien Sarrazin[2], Yan Delorme[1] and Hugh Gibson[3]
[1]LERMA, Observatoire de Paris, 75014 Paris, France
[2]L2E, UPMC – Paris 6, Paris, 75005, France
[3]Gibson Microwave Design, Antony 92160, France
**Contact: duccio.delfini@obspm.fr*

Abstract—Heterodyne receivers are widely used in astronomy for observations of high spectral resolution. Most heterodyne receivers have a single pixel but thousands of frequency channels. To obtain spectra over a larger area of the sky, the first small (few to tens of pixels) heterodyne array receives have been built, consisting of a group of individual receivers. For large arrays (hundreds of pixels) we propose to radically rethink the design of heterodyne arrays and to develop planar heterodyne arrays that can be easily fabricated and need minimal assembly. In a first step, we want to replace the horn or the lens that are difficult to fabricate and assemble for large arrays by one planar structure. As a first attempt we have used FEKO to design an array of patch antennas to create a well-collimated beam using 256 patches. This array would replace one horn/ lens. Arrays of such patch arrays could be fabricated in a single process for array receivers. We describe the design approach, as well as the results for a test array at 600 GHz. The simulations show that the concept works, that a narrow, near Gaussian beam can be obtained and that losses can be kept low. In this paper we will also discuss the difficulties encountered, in particular, that a ground plane at 13 microns from the patches. The current design also has very low RF bandwidth. Nevertheless, we think that the concept might be interesting for future large receiver arrays.

INTRODUCTION

Heterodynes receivers are widely used in astronomy and the first arrays of heterodyne receivers have been developed, e.g. CHAMP[1], HERA[2], SMART[3], CHAMP+[4], HARP[5], Supercam[6], STO[7], upGREAT[8]. Some effort has been made to simplify heterodyne receivers for arrays (e.g. split block for several mixers[6]), but currently most array receivers consist more or less of an assembly of single pixel heterodyne receivers. We propose to radically rethink heterodyne array receivers and to try to build planar array receivers with integrated components, where a whole array can be produced on one monolithic chip akin to CCD cameras.

In a first step, we wanted to eliminate the horn or lens of the receiver. Instead, we aim to use a planar antenna structure. A simple single antenna will have a too divergent beam, and will require refocusing on a pixel basis. However, a narrow beam can be synthesized by an array of antennas, similar to flat radar arrays.

To our best knowledge, this paper is the first to investigate large phased arrays, with no active components, at high frequencies (> 500 GHz).

THE ANTENNA ARRAY

The array consists of many antennas connected together in phase. Due to the antennas working in phase, the beam becomes narrow. Antenna arrays are a very well-known principle. The classical array theory to steer the beam of the antenna arrays can be found in the most important books of antenna theory[9]. Antenna arrays are used in radar systems with active components to modulate and steer the beam, but such arrays are difficult to reproduce at high frequencies. However, no one has designed large arrays above 500 GHz with no active components, and as far as we know we are the first to attempt a design.

We have used FEKO to design an array of patch antennas to demonstrate the possibility of creating a well-collimated beam at 600 GHz. In our design, we used 256 patches that are connected with microstrip lines. The 16X16 microstrip patch array will be described below. A Hot Electron Bolometer (HEB) is located near the centre of the array and is connected by superconducting microstrips to the 256 patch antennas. The dimension of the patch has been optimized for the 600 GHz radio frequency (RF). The microstrip feedlines also provide impedance matching. The microstrip feedlines connect the patches such that the electrical path length from the mixer to any patch is the same. An intermediate frequency (IF) output line with filters blocking the RF is added. Two very small gaps on the transmission line near the centre of the array block the IF signal to avoid it going towards the patches. The design is a monolithic block, that does not require micron-level machining or optical alignment (e.g. lenses).

A. The Feeding Network

The microstrip lines connect the hot electron bolometer to the 256 patches of the array. All the lines have the same electrical length so that each antenna is in phase for our first design (fig 1 on the left). The microstrip lines are optimized for the RF frequency. It is therefore necessary to block the IF signal generated in the HEB. The blockage is realized by overlaying two microstrip lines separated by silicon (fig. 1 on the right) so that only a high frequency signal is allowed to pass.

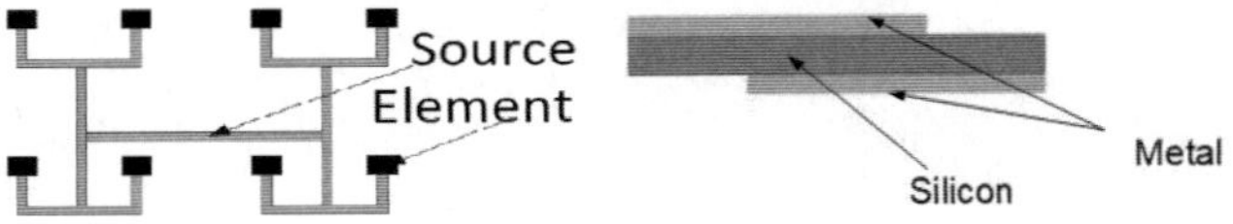

Fig. 1 The microstrip lines. Single block of a 4X4 array (left) and IF blockage (right)

B. The Losses

The simulations show that the dielectric losses for silicon at 600 GHz are negligible, hence we focus only on metallic losses. The losses in a microstrip line are directly related to the resistivity. Since the resistivity increases with the frequency, the losses become very important at THz frequencies. A loss of 2dB/mm is computed for a gold microstrip line of 1μm width and 5μm dielectric thickness at 600GHz and T=4K. The efficiency of the antenna results degraded down to 20%. NbN superconductors cooled down to 4K are therefore necessary to avoid such losses.

C. Array Size and resulting Beam

By connecting the patches in phase with the microstrip lines the beam becomes narrow. Actually the far field of an array is given by the product of the far field of the single element times the array factor *(AF)*, which is in general very narrow. The width of the array factor decreases with the number of patches and the distance between them. However the distance cannot be greater than the wavelength because in such case the side lobes increase. We aim to space two neighboring pixels at 2 FWHM beams apart. Which corresponds to a close packaging of Gaussian beams. Due to this goal and by assuming that no break out mirrors are employed we derived a maximum pixel size in the focal plane of about 5mm using typical focal ratios.

It can be numerically shown that the product of the *AF* times the far field of a single element is almost Gaussian, with 90% Gaussivity. However the microstrip lines radiate and generate sidelobes, which decrease the Gaussivity. The main lobe is however almost Gaussian.

D. The Filter

The mixer is positioned in the center of the array of patch antennas and connected to them by microstrips. In order to extract the IF signal we employ a strip line with a low pass filter to block the RF signal. A stepped impedance filter is chosen for this purpose, as is commonly used in microstrip designs (fig. 2).

Fig. 2 Stepped impedance filter. The narrow part of the microstrip line corresponds to the high impedance pad, the large part of the microstrip line corresponds to the low impedance pad. Such parts in series make a low pass

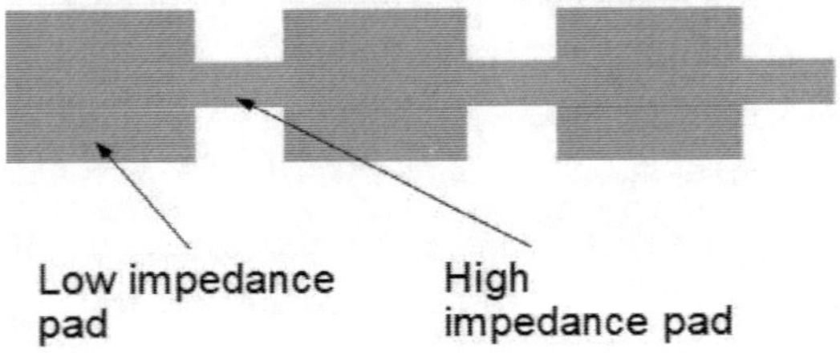

filter

E. Description of the array

The figure 3 shows the whole array. The sky signal is

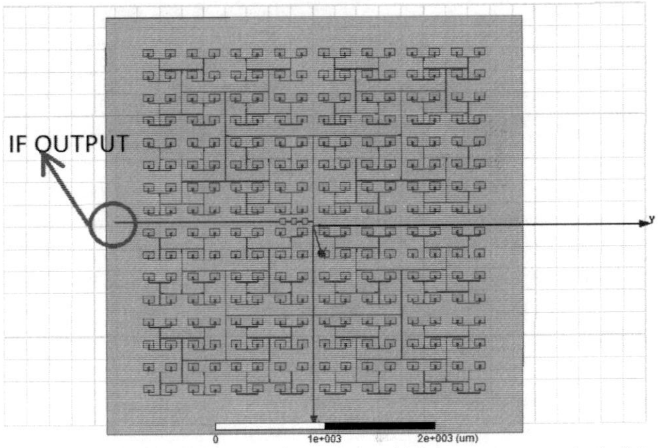

Fig 3 The whole array of 256 patch antennas which represents one pixel of the heterodyne

received by the 256 patches and feeds the HEB near the center of the array. The IF signal passes through the line on the left of the array to be amplified and detected. Two very small gaps on the transmission line near the center of the array block the IF signal to avoid it to go towards the patches (fig 1 on the right). In the same way the stepped impedance filter on the IF output on the left of the array blocks the RF signal from going to the amplifiers

RESULTS

A silicon substrate with ε=11,7 is used and has a thickness of 13μm. Superconductors are used. The dimension of the 18X18 patch array is approximately 16 mm^2. The results are very sensitive to the design parameters. Very small variations lead to great variations of the results, and this make the design very difficult. A genetic algorithm is used to adjust the design in order to obtain impedance matching in the RF band. In the simulation, the bolometer is set as the source of the RF signal. The S11 parameter and the far field is calculated accordingly. Fig 4 shows the S11 parameter.

The band is very narrow because of the resonating nature of the patches. The bandwidth would be sufficient for dedicated receivers to observe a specific line, but the bandwidth is too narrow for general purpose receivers.

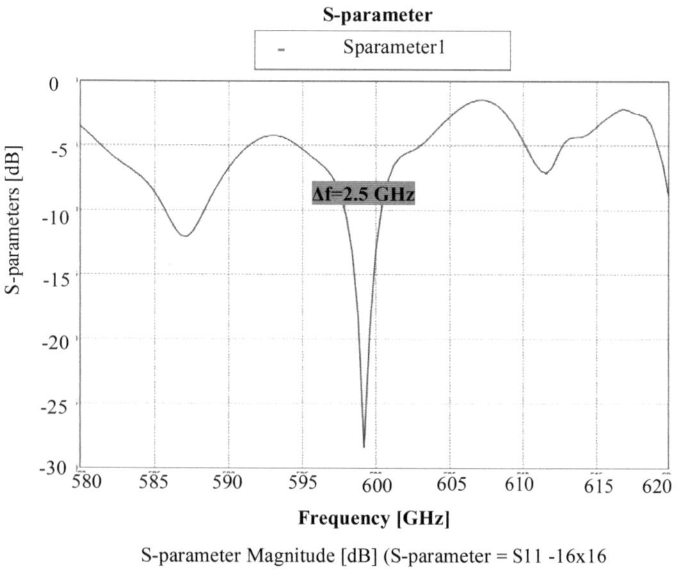

Fig 4 S11 plot, -10 dB bandwidth of 0.4%.

The far field is shown in fig. 5 and 6 in the XZ and YZ plane respectively (coordinate system of fig.3 where the z axis is normal to the array and the y axis is parallel to the IF output line with opposite direction). The largest side lobes are 12 and 14 dB below the main beam, which is acceptable in most cases. The half power full beam width is 7.6 and 6.1 degrees respectively. This collimation of the individual beams is good enough to allow reimaging of the entire array with common optical elements.

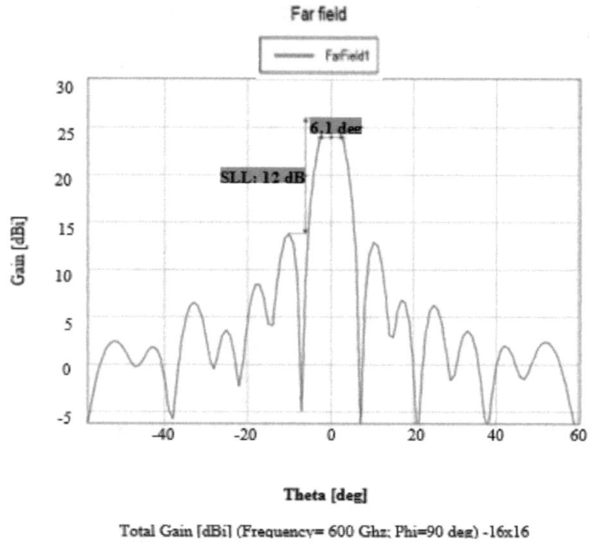

Fig. 5 Far field in the XZ plane. The gap between the main lobe and the side lobe is 12db. The HPBW is 6.1°

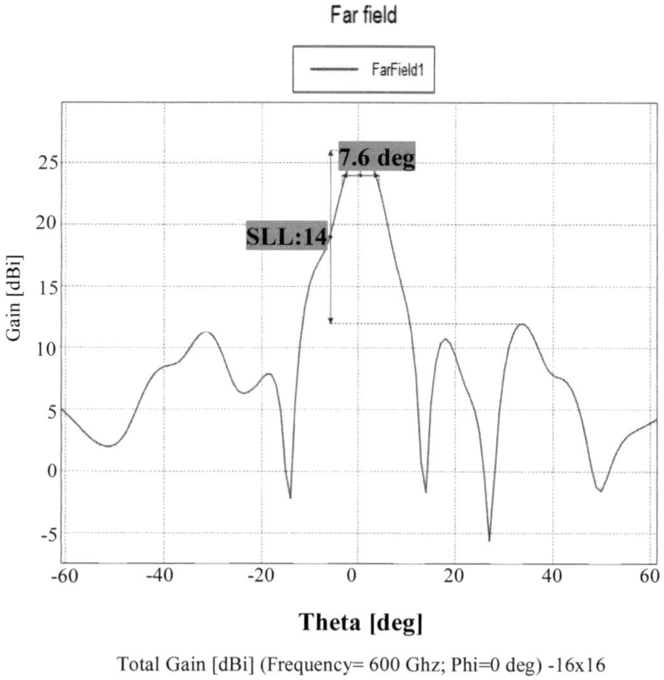

Fig. 6 Far field in the YZ plane. The gap between the main lobe and the side lobe is 14db. The HPBW is 7.6°

To understand the functioning of the array at the IF frequency, the bolometer is set as a 2 GHz source and the S12 parameter is calculated between the bolometer and a 50 Ohm resistance at the end of the transmission line which connects the filter to the IF chain (the IF output of fig. 4). The Measured S12 is acceptable, of the order of -0.15dB, while S11 =-15dB.

The simulation has shown that the phased array produces the narrow beam as expected, that the beams are nearly Gaussian and the side lobes are adequate. When a superconductor is used for the microstrip, line losses are also low. However, the array we present has also its shortcomings: the narrow RF bandwidth is acceptable only for receivers designed for specific observations; the microstrip line only works as a superconductor up to its gap frequency; it is difficult to fabricate the ground plane for a so thin substrate (too thin for etching and too thick for chemical deposition). The results are also very sensitive to the design parameters, which makes the fabrication difficult. We realize that the phased array could be further optimized, for example side lobes can be reduced by changing the illumination of the array from a top hat to a more Gaussian amplitude distribution and by adjusting the phase of the individual patches [10].

CONCLUSIONS

A 600 GHz phased array has been designed in our laboratory. The simulations show a beam with a narrow opening angle, and low loss. However, currently our design using patch antennas is too difficult to fabricate due to the necessity of a ground plane at a distance of 13μm. More work is required to find an entirely satisfactory solution, but we hope to have motivated the attractiveness of planar heterodyne arrays and have shown the first steps towards them as well as indicated the difficulties.

REFERENCES

[1] R. Güsten, G. Ediss, F. Gueth, K. Gundlach, H. Hauschildt, C. Kasemann, T. Klein, J. Kooi, A. Korn, I. Krämer, R. LeDuc, H. Mattes, K. Meyer, E. Perchtold, M. Pilz, R. Sachert, M. Scherschel, P. Schilke, G. Schneider, J. Schraml, D. Skaley, R. Stark, W. Wetzker, H. Wiedenhöver, S. Wongsowijoto, F. Wyrowski, in *Advanced Technology MMW, Radio, and Terahertz Telescopes*, ed. by T.G. Phillips (Proceedings of SPIE Vol. 3357, 1998)

[2] K.F. Schuster, C. Boucher, W. Brunswig, M. Carter, J.Y. Chenu, B. Foullieux, A. Greve, D. John, B. Lazareff, S. Navarro, A. Perrigouard, J.L. Pollet, A. Sievers, C. Thum, H. Wiesemeyer, Astronomy and Astrophysics **423**, 1171 (2004)

[3] U.U. Graf, S. Heyminck, E.A. Michael, S. Stanko, C.E. Honingh, K. Jacobs, R.T. Schieder, J. Stutzki, B. Vowinkel, in *Millimeter and Submillimeter Detectors for Astronomy, Proceedings of the SPIE*, vol. 4855, ed. by T.G. Phillips, J. Zmuidzinas (2003), *Proceedings of the SPIE*, vol. 4855, pp. 322– 329

[4] C. Kasemann, R. Güsten, S. Heyminck, B. Klein, T. Klein, S.D. Philipp, A. Korn, G. Schneider, A. Henseler, A. Baryshev, T.M. Klapwijk, in *Millimeter and Submillimeter Detectors and Instrumentation for Astronomy III, Society of Photo-Optical Instrumentation Engineers (SPIE) Conference Series*, vol. 6275, ed. by Z. J., W.S. Holland, S. Withington, W.D. Duncan (2006), *Society of Photo-Optical Instrumentation Engineers (SPIE) Conference Series*, vol. 6275.

[5] J.V. Buckle, R.E. Hills, H. Smith, W.R.F. Dent, G. Bell, E.I. Curtis, R. Dace, H. Gibson, S.F. Graves, J. Leech, J.S. Richer, R. Williamson, S. Withington, G. Yassin, R. Bennett, P. Hastings, I. Laidlaw, J.F. Lightfoot, T. Burgess, P.E. Dewdney, G. Hovey, A.G. Willis, R. Redman, B. Wooff, D.S. Berry, B. Cavanagh, G.R. Davis, J. Dempsey, P. Friberg, T. Jenness, R. Kackley, N.P. Rees, R. Tilanus, C. Walther, W. Zwart, T.M. Klapwijk, M. Kroug, T. Zijlstra, Monthly Notices of the Royal Astronomical Society **399**(2), 1026 (2009)

[6] C. Groppi, C. Walker, C. Kulesa, D. Golish, J. Kloosterman, S. Weinreb, G. Jones, J. Barden, H. Mani, T. Kuiper, J. Kooi, A. Lichtenberger, T. Cecil, G. Narayanan, P. Pütz, A. Hedden, in *Proceedings of the*20th *International Symposium on Space Terahertz Technology* (Charlottesville, 2009), pp. 90–96

[7] C. Walker, C. Kulesa, P. Bernasconi, H. Eaton, N. Rolander, C. Groppi, J. Kloosterman, T. Cottam, D. Lesser, C. Martin, A. Stark, D. Neufeld, C. Lisse, D. Hollenbach, J. Kawamura, P. Goldsmith, W. Langer, H. Yorke, J. Sterne, A. Skalare, I. Mehdi, S. Weinreb, J. Kooi, J. Stutzki, U. Graf, M. Brasse, C. Honingh, R. Simon, M. Akyilmaz, P. Pütz, M. Wolfire, in *Society of Photo-Optical Instrumentation Engineers (SPIE) Conference Series*, vol. 7733 (2010), vol. 7733

[8] Risacher, C.; Güsten, R.; Stutzki, J.; Hübers, H.-W.; Bell, A.; Buchbender, C.; Büchel, D.; Csengeri, T.; Graf, U. U.; Heyminck, S.; Higgins, R. D.; Honingh, C. E.; Jacobs, K.; Klein, B.; Okada, Y.; Parikka, A.; Pütz, P.; Reyes, N.; Ricken, O.; Riquelme, D.; Simon, R.; Wiesemeyer, H. 2016, A&A, 595, 34

[9] Constantine A. Balanis, Antenna Theory Analysis and Design, 3nd ed., John Wiley & Sons, Inc, Ed. Hoboken New Jersey, 2005.

[10] K. Bahadori; Y. Rahmat-Samii EEE Transactions on Antennas and Propagation (Volume: 53, Issue: 11, Nov. 2005)

Spectral Domain Simulation of SIS Frequency Multiplication

J.D. Garrett[1*], H. Rashid[2], V. Desmaris[2], A. Pavolotsky[2], V. Belitsky[2], and G. Yassin[1]

[1]*University of Oxford, Oxford, OX1 3BD, United Kingdom*

[2]*Chalmers University of Technology, Gothenburg, 412 58, Sweden*

*Contact: john.garrett@astro.ox.ac.uk

Abstract— In this paper, we compare simulations to experimental results for a new SIS frequency multiplier. To simulate these devices, we developed software based on spectral-domain analysis, which is ideal for simulating higher-order harmonics such as those present in a multiplier. In addition, we included the embedding circuit and interpolated the experimental I-V curve to allow the simulation to capture the experimental system as closely as possible. For the experimental data, results were taken from a new SIS frequency multiplier that has recently been developed at the Chalmers University of Technology. Previously, these experimental results were compared to simulations based on Tucker theory. Here, we compare these results to spectral-domain simulations. Qualitatively, the inclusion of embedding impedances and the use of spectral-domain analysis improves the agreement between simulation and experiment. The software can now be used to design multipliers with high output power and high conversion efficiency.

INTRODUCTION

The nonlinear behavior of quasiparticle tunneling in superconductor-insulator-superconductor (SIS) junctions is ideal for creating mixers and frequency multipliers. Frequency multipliers created from SIS junctions have many potential benefits including improved conversion efficiency and creating on-chip multipliers. These benefits will become even more important as superconducting receivers are pushed to higher frequencies.

Recently, the Chalmers University of Technology has developed new distributed SIS frequency multipliers (DSMs, [1-3]). The name comes from their elongated shape (1 um x 20 um), which helps to reduce the effect of the junction's intrinsic capacitance. Compared to frequency multipliers that use arrays of SIS junctions in series, DSMs do not have the same problems with local heating and harmonic intermixing.

In [1] and [2], experimental results from a DSM were compared to Tucker theory [4] with good agreement. In this work, we hope to expand upon these advances by comparing experimental results to spectral-domain simulations that include the complex embedding impedance for each harmonic.

SIS MULTIPLIER SIMULATION

Tucker theory [4] provides a quantum mechanical model of SIS mixer operation. Typically, when this theory is applied however, the local-oscillator signal is assumed to be purely sinusoidal with all higher-order modes short-circuited by the junction's intrinsic capacitance. This is only accurate when the ωRC product is less than ~4 [5]. Since the goal of this work is to simulate the higher-order currents, we must use a different technique to simulate the tunneling current.

With this in mind, we developed a software package based on spectral-domain analysis [5] at the University of Oxford. The spectral-domain technique calculates higher-order harmonic currents without assuming anything about the junction's properties. In addition, this technique can include complex embedding impedances and solve the resultant nonlinear circuit using harmonic balance. We also added several other features in order to recreate experimental data as closely as possible. This includes the ability to import and interpolate experimental I-V curves to fully capture the junction's properties (e.g., subgap current, transition linearity, proximity effect, etc.). We wrote the software package in Python with heavy use of the Numpy library, making it fast, flexible and portable to other systems.

SIMPLE MULTIPLIER SIMULATION

As a basic demonstration of the software, we simulated the AC quasiparticle tunneling current for the first two harmonics (Fig. 1). The frequency of the local-oscillator (LO) signal was set to one quarter of the gap frequency, i.e., $\tilde{V}_{ph} = 0.25$ where $\tilde{V}_{ph}$ is the normalized photon voltage. The junction drive level was set to $\alpha = \tilde{V}_J/\tilde{V}_{ph} = 1$ where $\tilde{V}_J$ is the normalized voltage applied across the junction. For the sake of keeping the simulation simple, no embedding circuit was included and therefore no harmonic balance was needed. All currents and voltages in Fig. 1 are normalized to the gap voltage v_{gap} and normal resistance R_N of the junction.

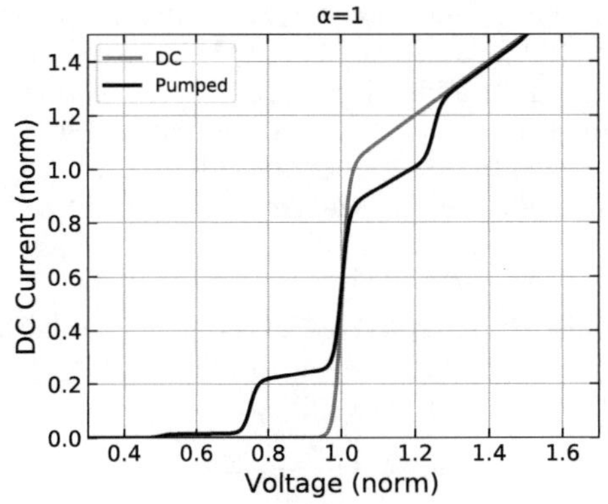

(a) DC tunneling current

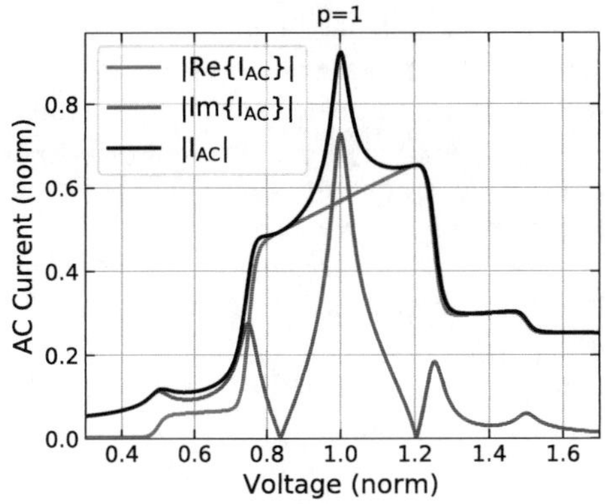

(b) AC tunneling current at the first harmonic

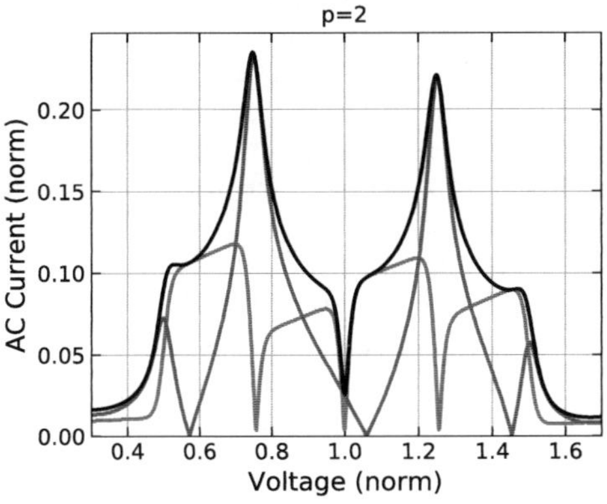

(c) AC tunneling current at the second harmonic

Fig. 1 Simulated tunneling current in a simple SIS frequency multiplier.

The current of the second harmonic is seen in Fig. 1c. This is the multiplied signal as its frequency is twice that of the original LO signal. Since power is proportional to $|I_{AC}|^2$, these simulations predict peak output power at a bias voltage equal to $1 - \tilde{V}_{ph}$ although this could change depending on the pump level and embedding circuit.

Comparison to Experimental Results

The experimental results in this section were previously reported in [1, 2]. They come from a DSM that has recently been developed at the Chalmers University of Technology. The experimental results include the DC I-V curve (both with and without the LO present) and the output power level at the second harmonic (i.e., the multiplied signal). These were measured for two different pump levels (α~0.85 and α~3.0) with the LO frequency set to 93 GHz.

From the experimental DC I-V curve, the gap voltage and normal resistance were found to be v_{gap} = 2.93 mV and R_N = 0.81 Ω, respectively. The embedding impedances were recovered by fitting simulations to the experimental pumped I-V curves (Fig. 2). For the first and second harmonic, the normalized impedances were found to be $Z_1 = 0.7 - j0.3$ and $Z_2 = 0.015$, respectively. To improve the matching, the characteristic I-V curves were rounded slightly. This was done by convolving the I-V curve with a Gaussian function.

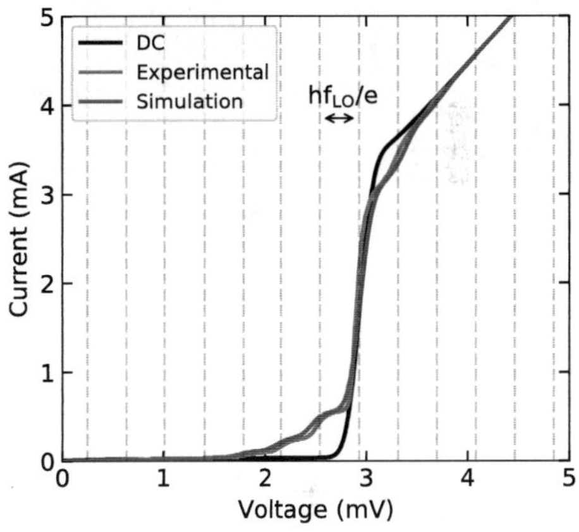

(a) Pumped I-V curve for α~0.85

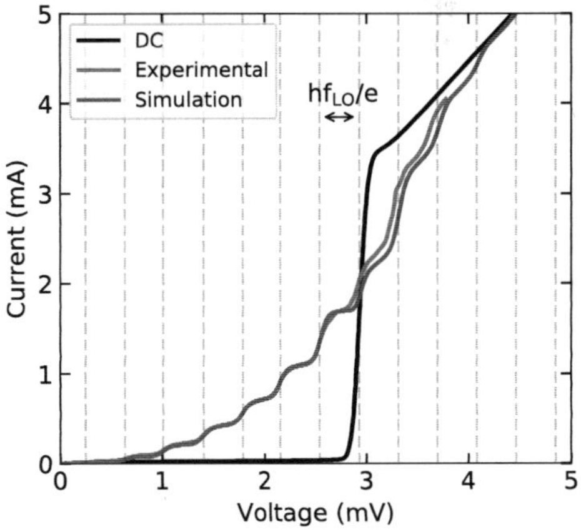

(b) Pumped I-V curve for α~3.0

Fig. 2 Simulated DC tunneling current compared to experimental results.

With the embedding circuit and characteristic I-V curve, we were then able to simulate the higher-order AC currents present in the DSM. The results from the second harmonic are shown in Fig. 3. For α~0.85 (Fig. 3a), the shape of the output current is very close, but the results are offset by a scaling factor ~3. This is likely due to a different impedance being used to simulate the results from that which was used to de-embed the experimental results. For α~3.0 (Fig. 3b), the simulation estimates the output power of the DSM much more accurately as well as the relationship to bias voltage.

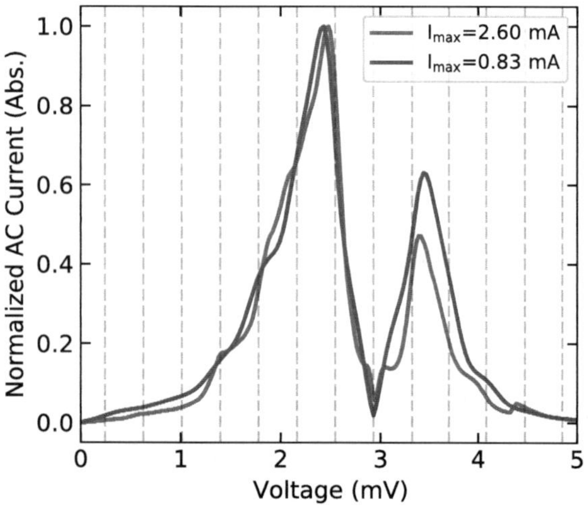

(a) AC tunneling current at the second harmonic for $\alpha \sim 0.85$. Experimental data is in blue, and simulated data is in red.

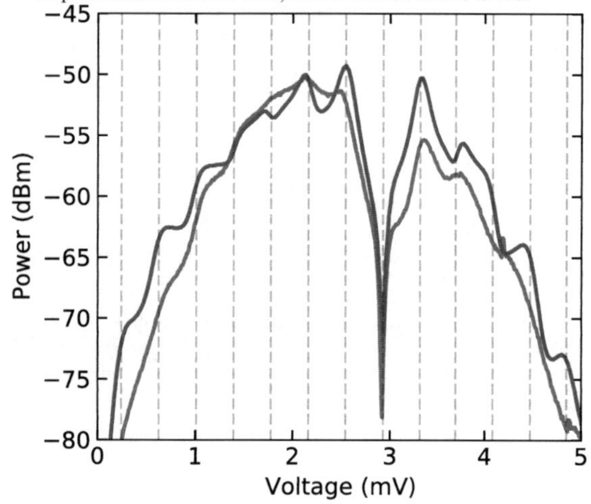

(b) AC power at the second harmonic for $\alpha \sim 3.0$. Experimental data is in blue, and simulated data is in red.

Fig. 3 Simulated AC tunneling current at the second harmonic compared to experimental results.

The match between simulation and experiment could potentially be improved by simulating the DSM in electromagnetic simulation software (e.g., Ansys HFSS). This would allow the embedding impedances to be found through simulations instead of de-embedding the experimental pumped I-V curves. Furthermore, the impedance of higher-order harmonics (p>2) could be found which might affect the performance of the second harmonic. The match could also be improved by having more information about the experimental setup. For example, local-oscillators often emit many other frequencies apart from their fundamental tone. This is due to the LO multipliers generating high-order harmonics and allowing low frequency signals to leak through. Low frequency leakage can affect the shape of the I-V curve, and could be the reason the I-V curve needed to be rounded in Fig. 2.

Simulations of SIS Multiplier Efficiency

Since the power delivered to the junction and the power delivered to the load are known, conversion efficiency in this section will be defined as

$$\eta = \frac{P_{L,p=2}}{P_{J,p=1}} \times 100\ \%$$

where $P_{L,p=2}$ is the output power delivered to the load at $f = 2f_{LO}$, and $P_{J,p=1}$ is the input power delivered to the junction at $f = f_{LO}$. In order to have a good understanding of where the maximum efficiency is found, the efficiency was calculated for a range of bias voltages and a range of input AC voltage amplitudes V_{thev}. The efficiency results using the embedding impedance values from the previous section are shown in Fig. 4a, while the efficiency using a different set of impedances ($Z_1 = 0.1$ and $Z_2 = 0.5$) is shown in Fig. 4b.

In both cases, the peak efficiency occurs at $\tilde{V} \sim 1 - \tilde{V}_{ph}$, similar to the simple simulation from Fig. 1. From Fig. 4a, we can see that the DSM was pumped near its maximum efficiency at α~3.0 (Fig. 3b); although, more output power could be found above this level. Fig. 4b is included to show that conversion efficiency can be optimized by altering the embedding impedance. The embedding impedances used in Fig. 4b were chosen by hand. Higher efficiency can be found by optimizing these simulations. This will allow optimal performance SIS multipliers to be designed in the future.

Conclusions

We have compared spectral-domain simulations of an SIS frequency multiplier to experimental results. The simulation software was designed specifically to simulate higher-order harmonics, while the experimental results were taken from a distributed SIS multiplier. Good agreement between simulations and experimental results was found, suggesting that the simulation software is able to adequately capture the experimental system. The simulation software can now be used to design new SIS multipliers with high output power and high conversion efficiency.

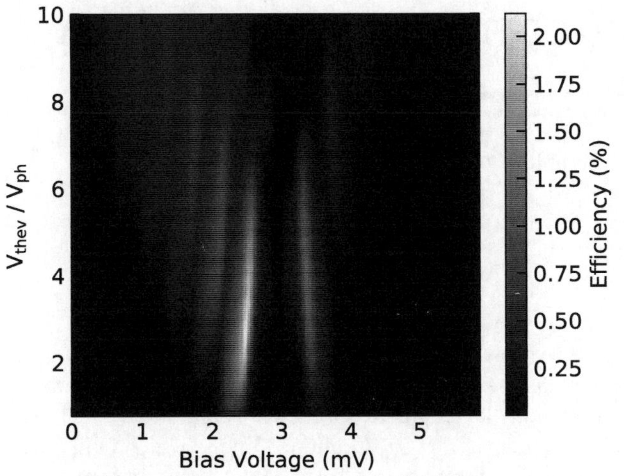

(a) Simulated conversion efficiency for $Z_1 = 0.7 - j0.3$ and $Z_2 = 0.015$

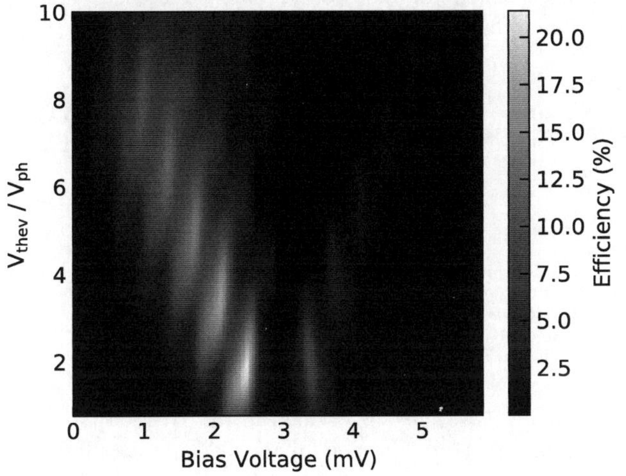

(b) Simulated conversion efficiency for $Z_1 = 0.1$ and $Z_2 = 0.5$

Fig. 4 Simulated conversion efficiency of an SIS frequency multiplier.

REFERENCES

[1] H. Rashid, V. Desmaris, A. Pavolotsky, and V. Belitsky. "Harmonic and reactive behavior of the quasiparticle tunnel current in SIS junctions," AIP Advances, 6(4), 045109, Apr 2016.

[2] H. Rashid, S. Krause, D. Meledin, V. Desmaris, A. Pavolotsky, and V. Belitsky. "Frequency Multiplier Based on Distributed Superconducting Tunnel Junctions: Theory, Design, and Characterization," IEEE Transactions on Terahertz Science and Technology, pp. 1–13, 2016.

[3] V. Belitsky and E. L. Kollberg. "Superconductor–insulator–superconductor tunnel strip line: Features and applications," Journal of Applied Physics, 80(8), pp. 4741, 1996.

[4] J. Tucker and M. Feldman. "Quantum detection at millimeter wavelengths," Reviews of Modern Physics, 57(4), pp. 1055–1113, 1985.

[5] S. Withington and E. L. Kollberg. "Spectral-domain analysis of harmonic effects in superconducting quasiparticle mixers," IEEE Transactions on Microwave Theory and Techniques, 37(1), pp. 231–238, 1989.

Wideband waveguide power combiner for ALMA Band 7+8 (275-500 GHz) Local Oscillator

A. Gonzalez, T. Kojima, K. Kaneko and S. Asayama
National Astronomical Observatory of Japan (NAOJ), Mikata, Tokyo, Japan
*Contact: Alvaro.Gonzalez@nao.ac.jp

Abstract— This paper describes the concept and design of a waveguide diplexer to combine the power of two Local Oscillator (LO) sources for different frequency bands. This diplexer is based on a novel design idea which uses two hybrid couplers combined with different waveguide filters. The proposed concept allows to input signals in two different frequency bands in two different input ports, and to have them combined in a common wideband port. At the same time, a signal at certain frequency is properly rejected at the input port which does not accept that frequency. This waveguide component allows the generation of a single high-power wideband LO signal with high spectral purity at sub-millimeter wavelengths for the first time. In particular, we are targeting the generation of a single LO signal for the full ALMA band 7+8 (275-500 GHz) by the combination of the LO sources used for ALMA band 7 (275-373 GHz) and band 8 (385-500 GHz).

INTRODUCTION

At NAOJ, we are developing high critical current density SIS mixers to cover two sub-millimeter ALMA [1] bands. In order to benefit from all equipment and know-how acquired during the design and production of ALMA band 8 receivers, we are aiming at covering ALMA band 7 (275-373 GHz) and band 8 (385-500 GHz) simultaneously. The fractional bandwidth of this target band is 60.7% at a central frequency of 370.8 GHz. For a typical 4-8 GHz 2SB IF signal, the required LO frequency coverage is therefore from 283 to 492 GHz. This represents a fractional bandwidth of 56%, which is too wide for typical LO sources at sub-mm wavelengths [2-3]. A possible solution to this problem is to combine the power of two different LO sources, one for each ALMA band, by means of a waveguide component. If a waveguide diplexer is used to combine the output power of two narrower band LO sources instead of a single wideband LO source, it is possible to achieve a single wideband high-power high-spectral-purity single LO signal at sub-mm wavelengths.

CONCEPTUAL IDEA

The 275-500 GHz diplexer is based on the conceptual design presented in Fig. 1. It uses two different hybrid couplers to divide and combine the LO signals coming from the different frequency band inputs. The two pass-bands of the diplexer have been defined by high-pass filters (HPF) between the two hybrids. The cross-over between the different pass-bands of the waveguide diplexer has been chosen to be at the 12 GHz frequency gap between the RF bands, where there is a strong atmospheric absorption line which prevents astronomical observations. The actual gap between LO sources will actually be 28 GHz for a 4-8 GHz 2SB IF bandwidth, which allows for certain tolerance in the fabrication of the diplexer.

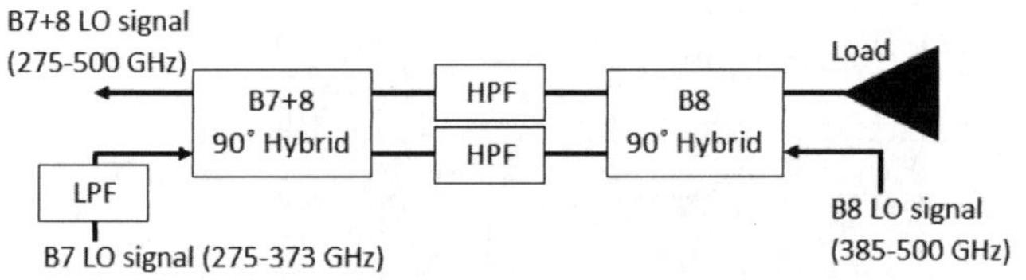

Fig. 1. Block diagram of the proposed wideband diplexer to combine two different LO sources at different frequencies. Band 7, 8 and 7+8 refer to the ALMA telescope bands, which are 275-373 GHz, 385-500 GHz and 275-500 GHz, respectively

The band 7 LO signal is divided in the B7+8 hybrid and reflected at the HPF between hybrids. The reflections are recombined in phase in the B7+8 LO signal port and 180-degrees out of phase in the B7 input port. The band 8 LO signal is divided in the B8 hybrid, passes the HPF and is divided again in the B7+8 hybrid. The combination of band 8 power will be in-phase in the B7+8 LO signal port and 180-degrees out of phase in the B7 input. A low-pass filter (LPF) has been placed at the B7 input port for extra protection of the B7 LO source from non-idealities in the design and fabrication. The hybrid close to the B7+8 port must be a wideband component, since it must work for both the B7 and B8 LO signals. However, the hybrid on the right-hand side, including the load connected to one of its ports, is only "seen" by the 385-500 GHz signal and can be therefore a simpler component with that reduced frequency coverage.

WAVEGUIDE DESIGN

The proposed LO combiner has been designed step by step. Firstly, the different building blocks in Fig. 1 have been designed independently, and then, connected together with

A FULL VERSION OF THIS PROCEEDINGS ARTICLE HAS BEEN SUBMITTED TO THE IEEE TRANSACTIONS OF TERAHERTZ SCIENCE AND TECHNOLOGY (TST) FOR PEER-REVIEW, PLEASE CHECK THE WEBSITE: HTTPS://WWW.MTT.ORG/TERAHERTZ

appropriate waveguide lengths, and bends. Finally, extra waveguides have been added to finalize the mechanical design of a waveguide block with flanges to connect to other components. The diplexer has three different waveguide ports for each of the bands involved: WR-3.0 (760 x 380 μm) for band 7, WR-2.2 (558 x 279 μm) for band 8, and WR-2.3 (580 x 290 μm) for band 7+8, and it is therefore highly asymmetric, which represent an additional design challenge. The design has been performed with the hybrid Mode-Matching/Finite-Elements software WaspNET [4], and the performance has been verified with HFSS [5] Finite Element simulations. The designed waveguide structure and the corresponding simulation results are presented in Fig. 2 and 3, respectively.

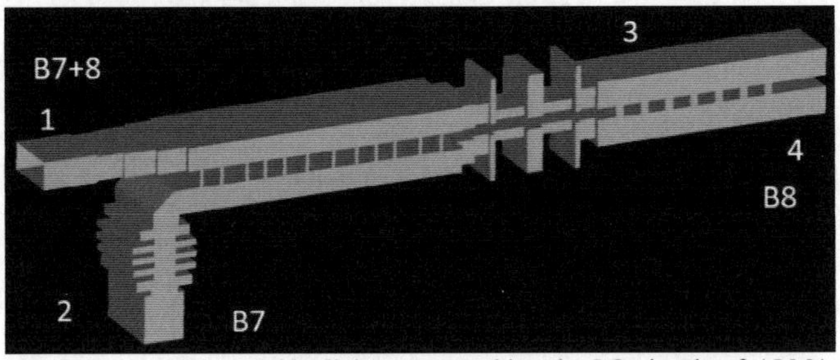

Fig. 2. Designed waveguide diplexer to combine the LO signals of ALMA band 7 and 8 individual LO sources. The band 7 and 8 inputs, together with the B7+8 output are clearly labeled, together with port numbers. Port 3 is to be terminated with a matched load.

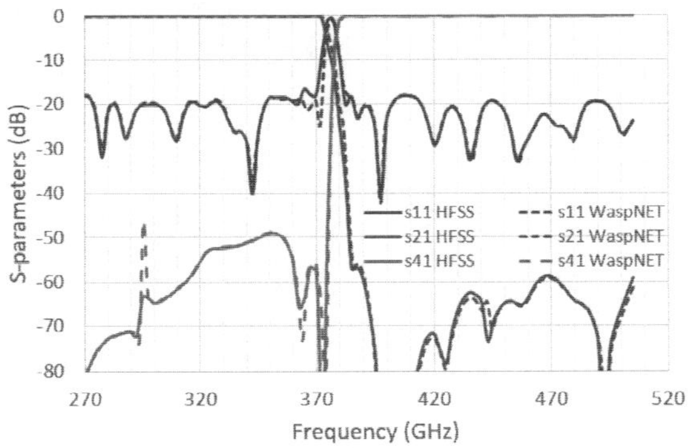

Fig. 3. S-parameter performance simulated by HFSS and WaspNET

For practical implementation, port 3 in Fig. 2 must be terminated with a matched load. This load has been implemented with an absorber of appropriate size and shape to minimize the s11 at the absorber input. The design and material of the absorber are the same as used for ALMA band 8 components.

For the simulation of effects related to loss, we have used the same loss model as derived from measurements of a waveguide multiplexer recently fabricated and for use in this frequency range [6]. Simulations considering loss have been performed with HFSS and are very demanding in terms of computer resources. Due to computer limitations, Microwave Office [7] has been used to add the effects of the extra loss in input and output waveguides in the actual waveguide block and the HFSS simulation results of the band 8 termination to the HFSS simulation of the diplexer considering loss. Rounded corners associated to fabrication by direct machining have also been added in these simulations. Results of the S-parameters of this 3-port simulation are presented in Fig. 5. The loss seen by the band 7 and 8 LO sources in the center of each band are -1.7 and -1.5 dB, respectively. The loss increases to around -2.0 dB at the lower end of band 7 due to the proximity of the waveguide cut-off. Frequencies close to the cross-over frequency range also see increased loss. The loss at 365 GHz and at 393 GHz are -2.3 and -2.2 dB, respectively. This loss is acceptable for the available input power levels. In terms of reflection loss, values are better than 19.2 dB at all frequencies of interest. Rejection of band 7 and 8 signals at band 8 and 7 input ports is better than -50 dB.

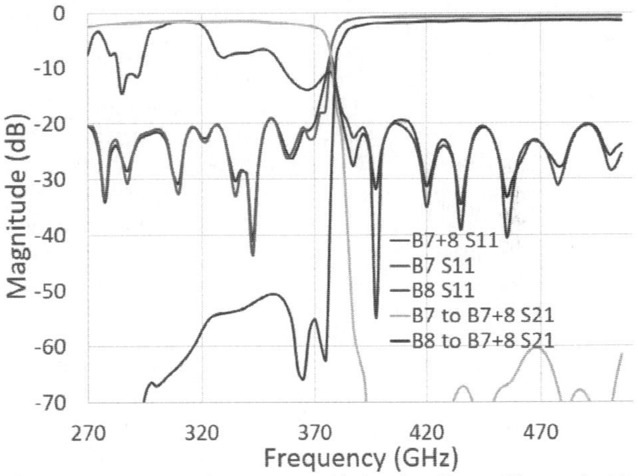

Fig. 5. S-parameter performance considering the effects of loss and with a load in the unused port of the band 8 hybrid. Simulations performed by HFSS and Microwave Office

FABRICATION AND MEASUREMENTS

Two prototype diplexers have been fabricated by direct machining and S-parameters have been measured with good agreement with measurements. These results have been included in an extended paper submitted to the IEEE Transactions on Terahertz Science and Technology. One of the diplexers has been used in an ALMA band 7+8 SIS mixer noise measurement setup with good results. This will be the subject of a future publication.

REFERENCES

[1] ALMA observatory. [Online] Available: www.almaobservatory.org

[2] J.L. Hesler, T.W. Crowe, "High-Power Solid-State Terahertz Sources", SPIE Newsroom, 28 July 2015. DOI: 10.1117/2.1201507.005859

[3] T. W. Crowe, J. L. Hesler, S. A. Retzloff and D. S. Kurtz, "Higher power multipliers for terahertz sources", 41st International Conference on Infrared, Millimeter, and Terahertz waves (IRMMW-THz), Copenhagen, 2016, pp. 1

[4] MIG WaspNET, Software Description. [Online]. Available: http://www.mig-germany.com/

[5] High Frequency Structure Simulator (HFSS). Sofware Description. [Online] Available: http://www.ansys.com/.

[6] T. Kojima, A. Gonzalez, S. Asayama, Y. Uzawa, "Design and Development of a Hybrid-Coupled Waveguide Multiplexer for a Multiband Receiver", IEEE Trans. THz Sci. Technol., vol. 7, no. 1, pp. 10-19, Jan. 2017

[7] AWR Microwave Office. Software description. [Online] Available: http://www.awrcorp.com/

Modelling proximity effects in x-ray Transition Edge Sensors (TESs) for space-based applications

Rebecca C. Harwin*, David J. Goldie and Stafford Withington

Cavendish Laboratory, University of Cambridge, CB3 0HE, UK

*Contact: rch66@cam.ac.uk

Abstract—Central to the operation of current and future space instruments such as I-XFU and SPICA are transition edge sensors (TESs) where the superconducting properties of the TES are spatially modulated due to the presence of additional normal metal structures such as the X-ray photon absorber itself. Due to the long-range, lateral superconducting proximity effect the superconducting properties of the TES are modified so that a detailed microscopic model is needed to account for its effective superconducting transition temperature T_c, its magnetic field dependencies and even perhaps its intrinsic noise properties.

The best way to account for this proximity effect in a microscopic model is to use the Usadel equations, modelling the TES as a weak link. So far, this has been done for standard TES geometries, i.e. S-S'-S weak links, and for particular values of the phase difference between the two electrodes which allow the Usadel equations to be simplified and solved using analytical methods. We have extended the existing models to model more complex TES geometries, for example a S-S'-S"-S'-S weak link, and to allow for arbitrary values of phase across the sensor.

The results of these models are used in a Resistively Shunted Josephson Junction phenomenological model to find the resistance of the sensor as a function of temperature and current, R(T,I). From this the experimentally measurable small signal electrothermal parameters α and β can be determined, which will allow the results from this work to be compared with future experimental observations. Additionally, we can use this model to obtain the variation of the supercurrent as a function of the magnetic field experienced by the TES, which will allow us to examine the effects of stray magnetic fields, such as those found in space, on TES performance.

A Smooth Walled Four Pixel Feed Horn Array Operating at 1.4 THz

Andre Hector[1,2,*], Andrey V. Trifonov[3], Edward Tong[3], Paul Grimes[3], and Ghassan Yassin[1]
[1]*Department of Physics, University of Oxford, Keble Road, OX1 3RH, Oxford, United Kingdom*
[2]*STFC, Rutherford Appleton Laboratory Space Department, OX11 0QX, Didcot, United Kingdom*
[3]*Harvard-Smithsonian Center for Astrophysics, MA 02138, Cambridge, USA*
[*]Contact: andre.hector@physics.ox.ac.uk

Abstract— We have designed and fabricated a feed horn array to operate between 1300 – 1500 GHz. It consists of four smooth walled, two section feeds with a single flare angle discontinuity near the throat of each horn. The design was carried out using a modelling package that employs modal matching together with genetic algorithm and downhill simplex optimisation techniques. This determines the optimum horn dimensions given a set of stringent requirements on beam circularity and cross polarisation levels. The simulated far-field radiation patterns exhibit very good beam circularity, low sidelobes and low cross polarisation levels across the frequency range of operation.
The array has been fabricated out of brass by the workshop of Oxford University using a direct machining technique that employs drill tools shaped to the profile of the designed horn. Far-field beam pattern measurements presented in this paper were carried out at the Harvard-Smithsonian Center for Astrophysics and suggest good agreement with the theoretical predictions.

INTRODUCTION

Multiple pixel feed horn arrays are a popular choice to efficiently couple astronomical and local oscillator signals into Superconductor-Insulator-Superconductor (SIS) or Hot-Electron-Bolometer (HEB) mixers. Such arrays consist of a number of horns and offer high aperture efficiencies, low sidelobes, low stray light sensitivities and low cross polarisations required by many astronomical applications [2]. The style and properties of the individual feeds that make up an array determine how well astronomical signals couple to the receiver. Therefore, the performance of a feed horn array will largely depend on the horn design and the accuracy of fabrication.

Traditionally, corrugated horns have been the preferred choice for use as high quality astronomical feeds because they exhibit circularly symmetric antenna beam patterns with low levels of cross-polarisation and sidelobes over a wide frequency bandwidth [3]. The problem with such feeds however, is that they require many corrugations per wavelength to function which makes them increasingly more difficult, time consuming and expensive to fabricate at higher frequencies, particularly in the Terahertz (THz) range. This becomes especially important when fabricating arrays with large numbers of horns.

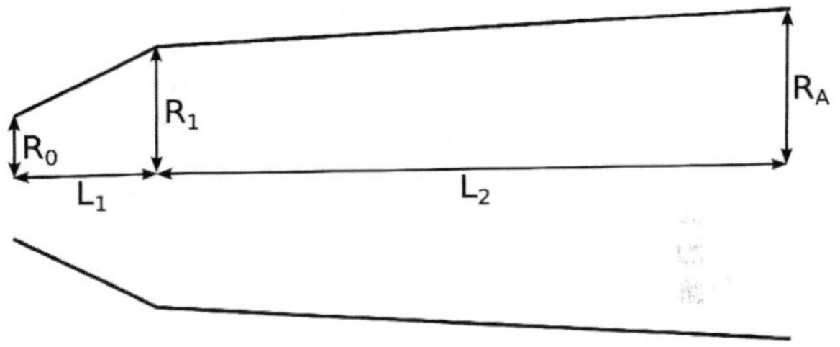

Fig. 1 Schematic diagram showing the inner profile of a smooth walled, two section feed horn.

A solution to the problem is to use a generalised version of the Potter horn with the phasing section removed [1], [4]. This type of horn has the advantage of being smooth walled and therefore easier to fabricate than a corrugated feed. On the other hand, it suffers from a relatively narrow bandwidth of ~ 10%. Nonetheless, such feeds have circular beams as it is possible to excite the TM11 mode within them by creating a sharp, step like discontinuity between the waveguide and the conical section of the horn. Provided that the incident TE11 and the excited TM11 modes arrive in phase at the horn aperture, this will result in sidelobe cancellation and low cross polarisation levels in the radiation pattern of the horn. Furthermore, by using a flare, rather than step, discontinuity it is possible to increase the operational bandwidth to ~ 15% and an even wider bandwidth can be obtained by adding more discontinuities at the throat of the horn [2].

In this paper, we describe the design, fabrication and beam pattern measurements of a smooth walled, four pixel feed horn array that operates between 1.3 and 1.5 THz which was designed and tested in collaboration with the Harvard-Smithsonian Center for Astrophysics. Our approach presents a fast and relatively simple solution to the issue of fabricating multiple feed horns at high frequencies and could, in principle, greatly speed up the fabrication of feed horn arrays in future instruments.

HORN DESIGN

A schematic diagram showing the inner profile of the horns in our array is presented in Fig. 1. This two section design, employing a single flare angle discontinuity, was obtained using a modelling package that determines the optimum horn dimensions given particular requirements on the far-field radiation pattern properties such as beam circularity and cross polarisation levels [5].

The simulated radiation patterns from our final design are shown in Fig. 2. These results have excellent beam circularity, low sidelobes and low cross polarisation levels between 1300 and 1500 GHz.

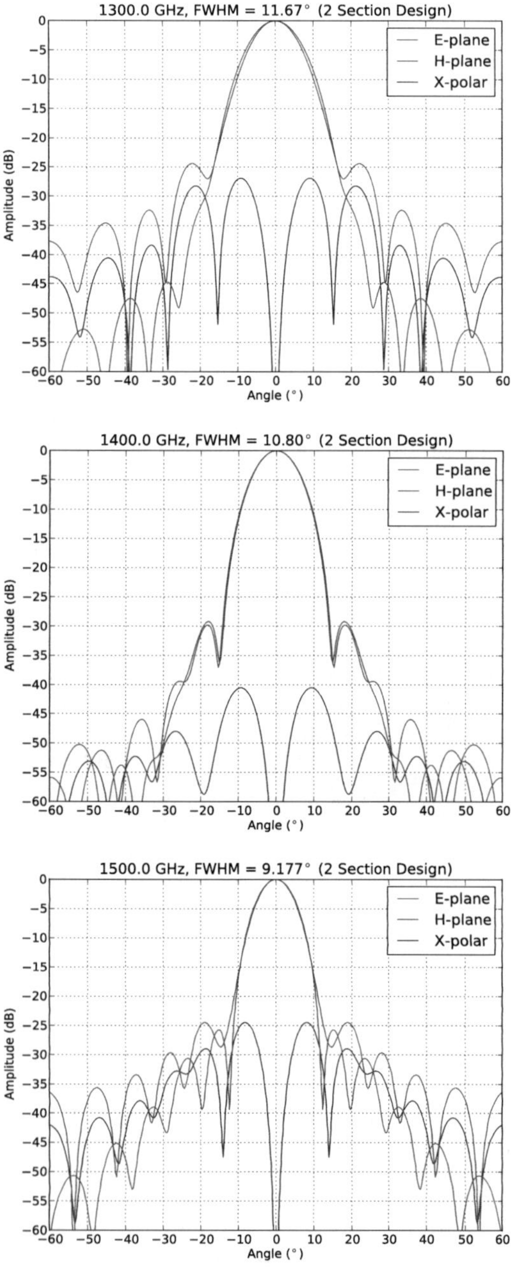

Fig. 2 Simulated far-field radiation patterns at 1300, 1400 and 1500 GHz.

Array Fabrication

The array was fabricated out of a brass block by the mechanical workshop of Oxford University. In order to create each of the horns in the array, a direct machining technique was used that employs a drill tool shaped to the horn's profile [6], [7]. A zoomed in, microscope image of such a tool is shown in Fig. 3.

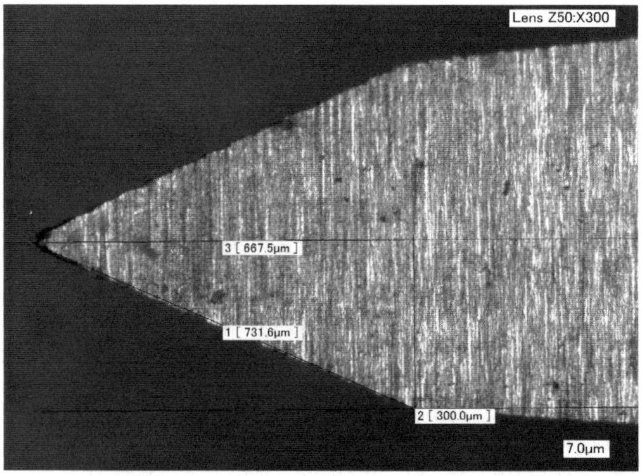

Fig. 3 High frequency (1.4 THz) feed horn fabrication tool.

The fabrication process itself begins with the creation of the circular input waveguide by direct drilling from one side of the block. The size of the drill required depends on the size of the waveguide. In our case, a 0.22 mm diameter drill was used.

Once the waveguide is complete, a high speed fabrication tool, such as the one shown in Fig. 3, is used to drill the required horn profile within the brass block.

The dimensions of our optimised horn design are given in Table 1 while the fabricated feed horn array is shown in Fig. 4 and 5.

TABLE I
FEED HORN DESIGN DIMENSIONS

Parameter	Length (mm)
R_0	0.112
R_1	0.304
R_A	0.764
L_1	0.420
L_2	5.110

When drilling feed horns from two different sides of a metal block, great care must be taken as any misalignment between the input waveguide and the horn itself will lead to deviations from theory in the measured radiation patterns.

Having said that, this method of fabrication has the advantages that it is more robust and reliable than electroforming. As well as this, once the machine tool and the metal block have been aligned, it is possible to repeat the process and quickly manufacture any number of identical feed horns.

Following repeated tests, we found that in order to get good experimental agreement with the simulations at these frequencies, the tolerances on the dimensions of the drill tool must be approximately 5 microns.

Fig. 4 The completed 1.4 THz array, shown beside a coin for scale.

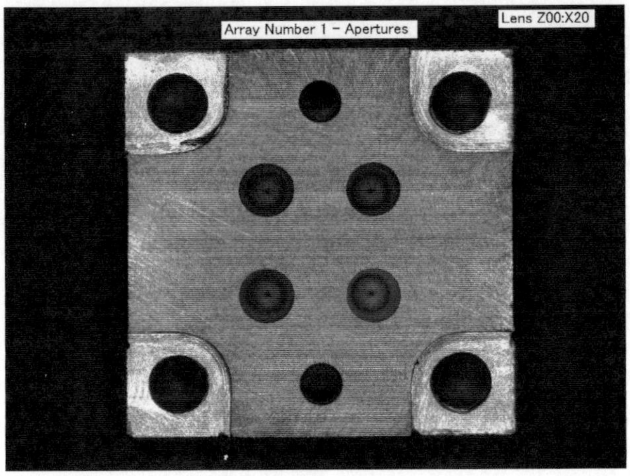

Fig. 5 Close up view of the completed 1.4 THz feed horn array.

MEASUREMENT AND RESULTS

The beam pattern measurements were carried out at the Harvard-Smithsonian Center for Astrophysics using an x-y stepper stage and a diagonal horn with an aperture diameter of 0.7 mm. The diagonal horn was used as the transmitter while our array was placed on the detector side. The separation between the two feeds was ~ 80 mm which ensured that the horn was in the far-field regime. A two dimensional beam scan at 1.35 THz is shown in Fig. 6. This result looks promising and it can be seen that the beam has good circularity and symmetry.

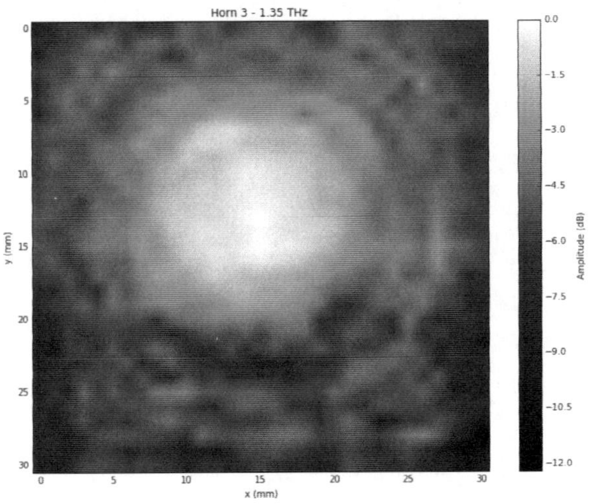

Fig. 6 Two dimensional beam scan at 1.35 THz.

As well as this, measurements were also carried out at 1.435 THz and a quadratic fit to the data is presented in Fig. 7 and 8. These results have been plotted over the theoretical predictions in the E and H-planes.

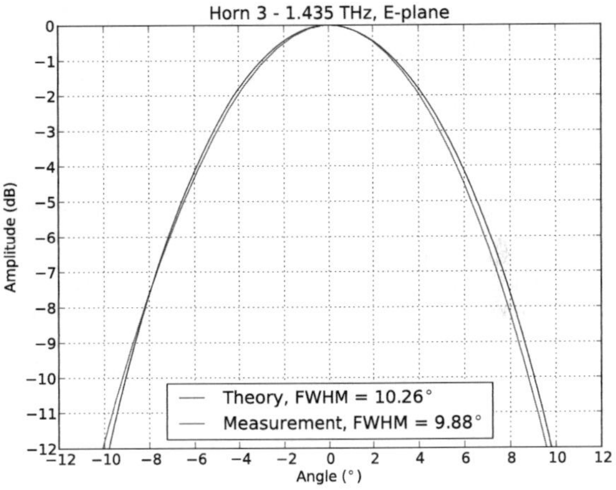

Fig. 7 E-plane theory and measurement at 1.435 THz.

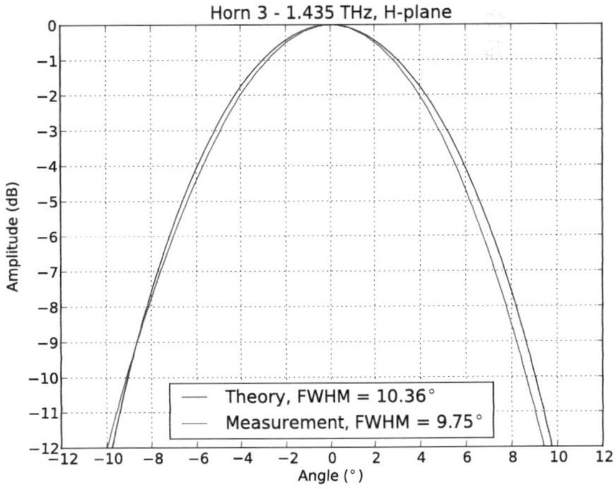

Fig. 8 H-plane theory and measurement at 1.435 THz.

From Fig. 7 and 8, it is evident that the measured beam patterns agree very well with the simulated predictions in both planes.

The Full Width at Half Maximum (FWHM) difference between the theory and result is 0.61 degrees in the H-plane and just 0.38 degrees in the E-plane. Furthermore, a FWHM difference of 0.13 degrees between the E and H-planes in the measured data suggests a highly circular beam. This is a very encouraging result which indicates that our method of horn design and fabrication is effective and viable for creating feed horn arrays that can operate at frequencies well into the THz regime.

Provided that careful attention is paid to the dimensions of the drill tools and that these can be accurately machined to the specified tolerances, there is no reason why this technology cannot be pushed to even higher frequencies.

Conclusions

We have designed, fabricated and measured a smooth walled, four pixel feed horn array operating between 1.3 and 1.5 THz.

The horns were designed using a modal matching technique that optimises the dimensions of the horn to produce excellent simulated far-field radiation patterns.

The array was fabricated at the University of Oxford using a highly accurate direct machining technique to match the dimensions of the fabricated horns to within 5 microns of the design specification.

Initial far-field radiation pattern measurements were carried out at the Harvard-Smithsonian Center for Astrophysics. They show encouraging agreement with theory suggesting that our method of horn design and array fabrication can offer a robust approach to making feed horn arrays operating in the THz regime.

In the future, we will aim to improve upon these preliminary results by extending the dynamic range to the sidelobe level and also increasing the frequency range of the measurements.

References

[1] P. D. Potter, "A New Horn Antenna with Suppressed Sidelobes and Equal Beamwidths,", *Microwave Journal*, vol. 6, pp. 71-78, 1963.

[2] J. Leech, B. K. Tan, G. Yassin, P. Kittara, S. Wangsuya, J. Treuttel, M. Henry, M. L. Oldfield, and P. G. Huggard, "Multiple flare-angle horn feeds for sub-mm astronomy and cosmic microwave background experiments," *Astronomy & Astrophysics*, vol. 532, A61, Jun. 2011.

[3] P. J. B. Clarricoats, Corrugated Horns for Microwave Antennas, Peter Peregrinus Ltd., 1984.

[4] H. M. Pickett, J. F. J. Hardy, and J. Farhoomand, "Characterization of a Dual-Mode Horn for Submillimeter Wavelengths,", *IEEE Trans. Microwave Theory Tech.*, vol. MTT-32, pp. 936-937, 1984.

[5] P. Kittara, A. Jiralucksanawong, G. Yassin, S. Wangsuya, and J. Leech, "The Design of Potter Horns for THz Applications Using a Genetic Algorithm,", *Int. J. Inf. Millim. Waves*, vol. 28, pp. 1103-1114, 2007.

[6] J. Leech, B. K. Tan, G. Yassin, P. Kittara, A. Jiralucksanawong, and S. Wangsuya, "Measured performance of a 230 GHz prototype focal-plane feedhorn array made by direct drilling of smooth-walled horns," in *Proc. 21st Int. Symposium Space Terahertz Technology*, 2010, p. 114.

[7] J. Leech, G. Yassin, B. K. Tan, M. Tacon, P. Kittara, A. Jiralucksanawong, and S. Wangsuya, "A New, Simple Method for Fabricating High Performance Sub-mm Focal Plane Arrays by Direct Machining using Shaped Drill Bits," in *Proc. 20th Int. Symposium Space Terahertz Technology*, 2009, p. 214.

Evaluation of aperture efficiency by using ray-tracing software in designing a wide field-of-view telescope

Hiroaki Imada[1,*], Makoto Nagai[2]
[1]Institute of Space and Astronautical Science, Japan Aerospace Exploration Agency, 3-1-1 Yoshinodai, Sagamihara, Kanagawa, 252-5210, Japan
[2]Advanced Technology Center, National Astronomical Observatory of Japan, 2-21-1, Osawa, Mitaka, Tokyo, 181-8588, Japan
*Contact: imada@astro.isas.jaxa.jp

Abstract— A wide field-of-view telescope is essential for large area observations which need as high mapping speed as possible. The size of field of view is limited by aberrations in most cases. One of the figure of merit of a radio telescope is aperture efficiency. Analytical expression of aperture efficiency affected by aberrations has been obtained recently. In this paper, we have implemented the analytical expression to commercial ray tracing software so that we can use aperture efficiency as well as Strehl ratio in designing a radio telescope.

INTRODUCTION

Various plans of astronomical observation are proposed in a radio band recently, e.g. LiteBIRD, 10-m THz telescope in Antarctica, CCAT, LST, and so on. The target bands of these telescopes are sub-mm or THz bands, which are at higher frequencies than ever. In addition, these plans demand very wide field-of-view (FOV) covering 1 square degree or more.

A radio telescope with a single beam or multi-beam has been designed so far. For a feed located at the center of FOV, we can calculate performance both analytically and numerically by using a fundamental Gaussian beam and physical optics simulation. For a feed located at the other area of FOV than the center, physical optics simulation is only available. Physical optics can provide a figure-of-merit such as aperture efficiency but not provide an optical design. An optical design must be given by using another way and also we have difficulty in looking for an optimal design. In other words, we do not have a systematic method to design a wide FOV telescope for multi-beam.

Ray tracing based on geometrical optics can design a wide FOV telescope for optical and infrared astronomy. Geometrical optics helps us achieve a wide FOV radio telescope but cannot provide a figure-of-merit which is need in evaluating a radio telescope. References [1], [2], and [3] revealed the relation between geometrical optics and aperture efficiency. Therefore we made a tool based on ray tracing but it can provide aperture efficiency.

FORMULATION

In this section, formulation of aperture efficiency affected by Seidel aberrations is summarized according to [1], [2], and [3]. [1] revealed that aperture efficiency is intrinsically equal to the product of two spillover efficiencies and coupling efficiency, i.e.,

$$\eta_{\mathrm{A}} = \eta_{\mathrm{sp,en}}\eta_{\mathrm{coup}}\eta_{\mathrm{sp,ex}},$$

where $\eta_{\mathrm{sp,en}}$ is spillover efficiency at entrance pupil and $\eta_{\mathrm{sp,ex}}$ is spillover efficiency at exit pupil. Coupling efficiency η_{coup} can be determined by the electric field of an incident wave and the virtual field from a feed at a pupil. In other words, coupling efficiency is given as a function of the direction of incident wave and the position of a feed because the electric fields are dependent on them.

References [2] and [3] demonstrated coupling efficiency as a function of the direction of incident wave and the position of a feed explicitly. The definition of coordinates is shown in Fig. 1. As a result, aperture efficiency is expressed as

$$\eta_{\mathrm{A}}(\hat{\boldsymbol{p}};\boldsymbol{r}_{\mathrm{det}}) = \eta_{\mathrm{sp,en}}\eta_{\mathrm{sp,ex}}\frac{\lambda^2 P_{\mathrm{n}}(\hat{\boldsymbol{p}};\boldsymbol{r}_{\mathrm{det}})}{A_{\mathrm{p,en}}\Omega_{\mathrm{A}}(\boldsymbol{r}_{\mathrm{det}})},$$

where $\hat{\boldsymbol{p}}$ is incident direction, $\boldsymbol{r}_{\mathrm{det}}$ feed position, λ wavelength, $A_{\mathrm{p,en}}$ the area of entrance pupil, Ω_{A} beam solid angle, and P_{n} normalized beam pattern. In order to observe effects of aberrations on aperture efficiency, wavefront error of the incident wave at the exit pupil is expressed as a form of summation of Zernike polynomials, i.e.,

$$\begin{aligned} W(\hat{\boldsymbol{p}};\boldsymbol{\rho};\boldsymbol{r}_{\mathrm{ref}}) = {} & A_1{}^1(\hat{\boldsymbol{p}};\boldsymbol{r}_{\mathrm{ref}})Z_1{}^1(\boldsymbol{\rho}) + B_1{}^1(\hat{\boldsymbol{p}};\boldsymbol{r}_{\mathrm{ref}})Z_1{}^{-1}(\boldsymbol{\rho}) \\ & + A_2{}^0(\hat{\boldsymbol{p}};\boldsymbol{r}_{\mathrm{ref}})Z_2{}^0(\boldsymbol{\rho}) + A_2{}^2(\hat{\boldsymbol{p}};\boldsymbol{r}_{\mathrm{ref}})Z_2{}^2(\boldsymbol{\rho}) \\ & + B_2{}^2(\hat{\boldsymbol{p}};\boldsymbol{r}_{\mathrm{ref}})Z_2{}^{-2}(\boldsymbol{\rho}) + A_3{}^1(\hat{\boldsymbol{p}};\boldsymbol{r}_{\mathrm{ref}})Z_3{}^1(\boldsymbol{\rho}) \\ & + B_3{}^1(\hat{\boldsymbol{p}};\boldsymbol{r}_{\mathrm{ref}})Z_3{}^{-1}(\boldsymbol{\rho}) + A_4{}^0(\hat{\boldsymbol{p}};\boldsymbol{r}_{\mathrm{ref}})Z_4{}^0(\boldsymbol{\rho}) \end{aligned}$$

,

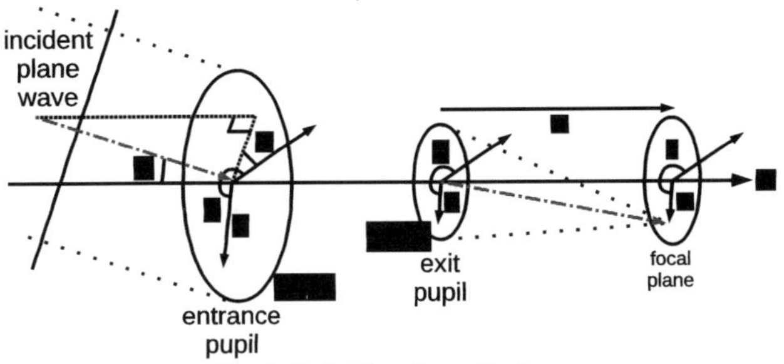

Fig. 1 Definition of coordinates.

A FULL VERSION OF THIS PROCEEDINGS ARTICLE HAS BEEN SUBMITTED TO THE IEEE TRANSACTIONS OF TERAHERTZ SCIENCE AND TECHNOLOGY (TST) FOR PEER-REVIEW, PLEASE CHECK THE WEBSITE: HTTPS://WWW.MTT.ORG/TERAHERTZ

where $A_n{}^m$ and $B_n{}^m$ are coefficients and $Z_n{}^m$ is Zernike polynomial. The meaning of the coefficients is summarized in Table I. Then, the electric field of the incident wave at the exit pupil is written as

$$E_{\rm ex}(\hat{\boldsymbol{p}};\boldsymbol{\rho}) = C\exp\left[\sqrt{-1}k\frac{f}{z_{\rm ref}}R_{\rm p,ex}\rho\sin\Theta\cos(\psi-\Phi)+\sqrt{-1}\frac{kR_{\rm p,ex}{}^2}{2z_{\rm ref}}\rho^2\right]\sum_{m,n}C_n{}^m(\hat{\boldsymbol{p}};\boldsymbol{r}_{\rm ref})Z_n{}^m(\boldsymbol{\rho})$$

$C_n{}^m$ is a function of $A_n{}^m$ and $B_n{}^m$ which are determined by using ray tracing software. The virtual field of a feed is also expanded into the summation of Zernike polynomials as well as the incident wave.

$$E_{\rm det}(\boldsymbol{\rho};\boldsymbol{r}_0) = C'\exp\left[\sqrt{-1}k\frac{f}{z_{\rm ref}}R_{\rm p,ex}\rho\sin\Theta\cos(\psi-\Phi)+\sqrt{-1}\frac{kR_{\rm p,ex}{}^2}{2z_{\rm ref}}\rho^2\right]\sum_{p,q}D_p{}^q(\boldsymbol{r}_{\rm ref};\boldsymbol{r}_0;w_0)Z_p{}^q(\boldsymbol{\rho})$$

$D_p{}^q$ is determined by beam pattern measurement or simulation such as HFSS. Using coefficients $C_n{}^m$ and $D_p{}^q$, we obtain coupling efficiency as follows:

$$\eta_{\rm coup} = \frac{\left|\sum_{m,n}\sum_{p,q}\delta_{np}\delta_{mq}C_n{}^m(\hat{\boldsymbol{p}};\boldsymbol{r}_{\rm ref})\left(D_p{}^q(\boldsymbol{r}_{\rm ref};\boldsymbol{r}_0;w_0)\right)^*\right|^2}{\sum_{m,n}\left|C_n{}^m(\hat{\boldsymbol{p}};\boldsymbol{r}_{\rm ref})\right|^2\sum_{p,q}\left|D_p{}^q(\boldsymbol{r}_{\rm ref};\boldsymbol{r}_0;w_0)\right|^2}$$

.

If a feed pattern is assumed to be an axially symmetric fundamental Gaussian beam, then coefficients $D_p{}^q$ is determined by

$$\begin{aligned}E_{\rm det}(\boldsymbol{\rho};\boldsymbol{r}_0) &= C'\exp\left[\sqrt{-1}k\frac{f}{z_{\rm ref}}R_{\rm p,ex}\rho\sin\Theta\cos(\psi-\Phi)+\sqrt{-1}\frac{kR_{\rm p,ex}{}^2}{2z_{\rm ref}}\rho^2\right]\sum_{p,q}D_p{}^q(\boldsymbol{r}_{\rm ref};\boldsymbol{r}_0;w_0)Z_p{}^q(\boldsymbol{\rho})\\ &= C'\exp\left[\sqrt{-1}k\frac{f}{z_{\rm ref}}R_{\rm p,ex}\rho\sin\Theta\cos(\psi-\Phi)\right]\exp\left[-\frac{R_{\rm p,ex}{}^2\rho^2}{w^2}+\sqrt{-1}\frac{kR_{\rm p,ex}{}^2\rho^2}{2R}\right]\end{aligned}$$

Finally, aperture efficiency for a fundamental Gaussian feed is expressed as

$$\begin{aligned}\eta_{\rm A} &= \eta_{\rm sp,en}\eta_{\rm coup}\eta_{\rm sp,ex}\\ &= \frac{2R_{\rm p,ex}{}^2}{w^2}\left|D_0{}^0+\sqrt{-1}k\left(A_2{}^0D_2{}^0+A_4{}^0D_4{}^0\right)-\frac{k^2}{2}\left[\left(D_0{}^0+\frac{D_2{}^0}{\sqrt{3}}\right)\left(A_1{}^1\right)^2+2\sqrt{2}\left(\frac{D_2{}^0}{\sqrt{3}}+\frac{D_4{}^0}{\sqrt{5}}\right)A_1{}^1A_3{}^1\right.\right.\\ &\quad+\left(D_0{}^0+\frac{D_2{}^0}{5\sqrt{3}}+\frac{D_4{}^0}{\sqrt{5}}+\frac{9D_6{}^0}{5\sqrt{7}}\right)\left(A_3{}^1\right)^2+\left(D_0{}^0+\frac{2D_4{}^0}{\sqrt{5}}\right)\left(A_2{}^0\right)^2+\left(\frac{4D_2{}^0}{\sqrt{5}}+\frac{6\sqrt{15}D_6{}^0}{5\sqrt{7}}\right)A_2{}^0A_4{}^0\\ &\quad\left.\left.+\left(D_0{}^0+\frac{2\sqrt{5}D_4{}^0}{7}+\frac{6D_8{}^0}{7}\right)\left(A_4{}^0\right)^2+\left(D_0{}^0+\frac{\sqrt{3}D_2{}^0}{2}+\frac{D_4{}^0}{2\sqrt{5}}\right)\left(B_2{}^2\right)^2\right]\right|^2\end{aligned}$$

IMPLEMENTATION

We implemented a script to calculate aperture efficiency expressed in the above equation to Zemax 13. The script can calculate aperture efficiency from the coefficients of a Zernike polynomial series. Also the root sum square of the difference between the analytical expression and a theoretical maximum determined by the edge taper. Constrains can be taken into account, e.g., aperture efficiency to be achieved, weighted by the position on the focal plane, and so on.

TABLE I ABERRATION COEFFICIENTS.

$A_1{}^1(\hat{\boldsymbol{p}};\boldsymbol{r}_{\rm ref})$, $B_1{}^1(\hat{\boldsymbol{p}};\boldsymbol{r}_{\rm ref})$	Tip, tilt, distortion
$A_2{}^0(\hat{\boldsymbol{p}};\boldsymbol{r}_{\rm ref})$	Defocus, curvature of field
$A_2{}^2(\hat{\boldsymbol{p}};\boldsymbol{r}_{\rm ref})$, $B_2{}^2(\hat{\boldsymbol{p}};\boldsymbol{r}_{\rm ref})$	Astigmatism
$A_3{}^1(\hat{\boldsymbol{p}};\boldsymbol{r}_{\rm ref})$, $B_3{}^1(\hat{\boldsymbol{p}};\boldsymbol{r}_{\rm ref})$	Coma
$A_4{}^0(\hat{\boldsymbol{p}};\boldsymbol{r}_{\rm ref})$	spherical

DESIGNING A WIDE FOV TELESCOPE WITH SCRIPT

A wide FOV telescope was designed by using the implemented script. The telescope consisted of two axially symmetric conic mirrors. Initial values for their parameters are shown in TABLE II and the system before optimization is shown in Fig. 2. The parameters to be changed during optimization were the conic constants of both of the mirrors and the radius of curvature of the secondary mirror. A stop was located at the secondary mirror. The center of FOV was weighted more than the edge in order to attain higher aperture efficiency at the center.

Two types of optimization were carried out. One adopted aperture efficiency as a figure of merit and the other adopted Strehl ratio. Figs. 3 and 4 show the systems optimized by aperture efficiency and Strehl ratio, respectively. The parameters obtained after optimization are shown in Tables III and IV. The parameters after optimization by aperture efficiency are different from those by Strehl ratio. Figs. 5 and 6 shows the maps of aperture efficiency as a function of an incident angle. In the case where aperture efficiency is adopted as a figure of merit, the aperture efficiency for the center of FOV is lower than those for the other incident angles, which arises from failure to weigh the center of FOV more than the others in evaluating aperture efficiencies.

TABLE III PARAMETERS AND INITIAL VALUES.

	conic constant	Radius of curvature	diameter	Thickness
Primary	-1	-1000 mm	-600 mm	-400 mm
secondary	-1.8	-230 mm	-100 mm	600 mm

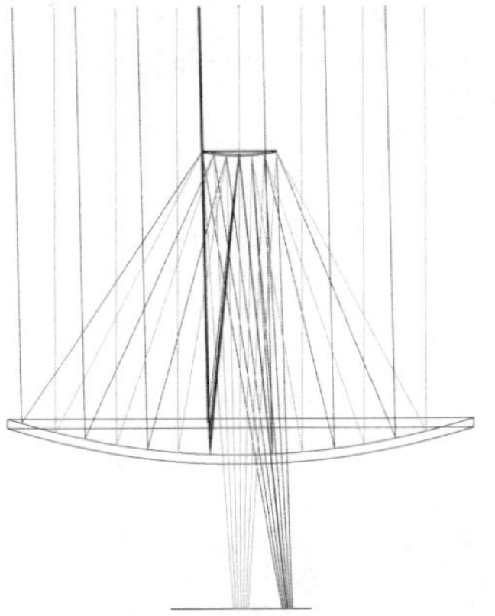

Fig. 2 An optical system composed of two axially symmetric conic mirrors before optimization.

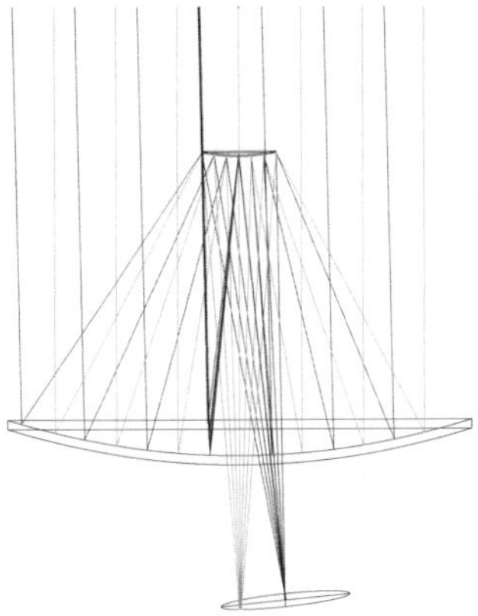

Fig. 3 After optimization by evaluating aperture efficiency.

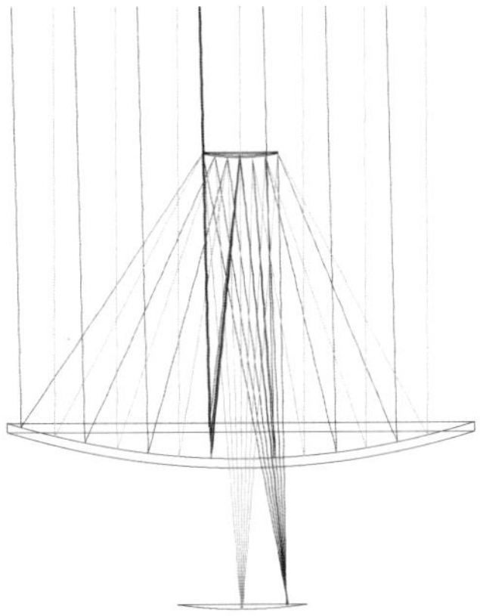

Fig. 4 After optimization by evaluating Strehl ratio.

TABLE IIIII PARAMETERS OPTIMIZED BY APERTURE EFFICIENCY.

	conic constant	Radius of curvature
Primary	-1.042528	-1000 mm
secondary	-3.018659	-237.151 mm

TABLE IVV PARAMETERS OPTIMIZED BY STREHL RATIO.

	conic constant	Radius of curvature
Primary	-1.162993	-1000 mm
secondary	-3.256167	-239.866 mm

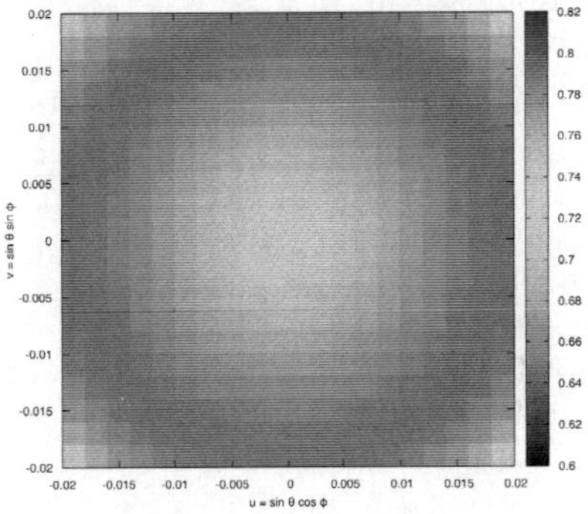

Fig. 5 Map of aperture efficiency optimized by evaluating aperture efficiency.

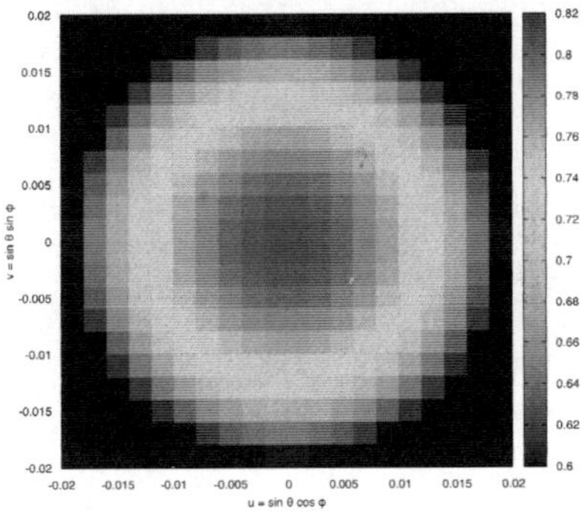

Fig. 6 Map of aperture efficiency optimized by evaluating Strehl ratio.

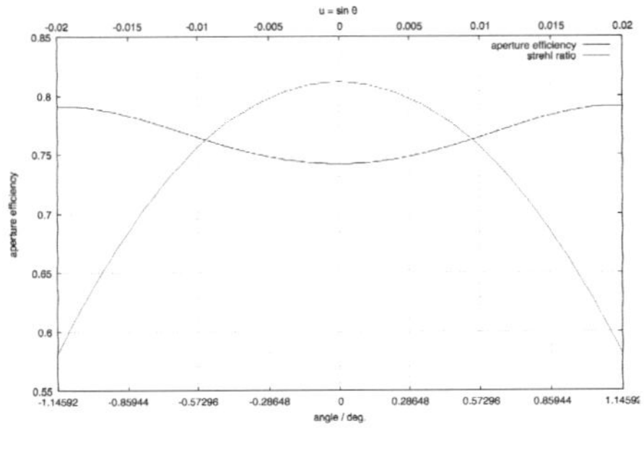

A FULL VERSION OF THIS PROCEEDINGS ARTICLE HAS BEEN SUBMITTED TO THE IEEE TRANSACTIONS OF TERAHERTZ SCIENCE AND TECHNOLOGY (TST) FOR PEER-REVIEW, PLEASE CHECK THE WEBSITE: HTTPS://WWW.MTT.ORG/TERAHERTZ

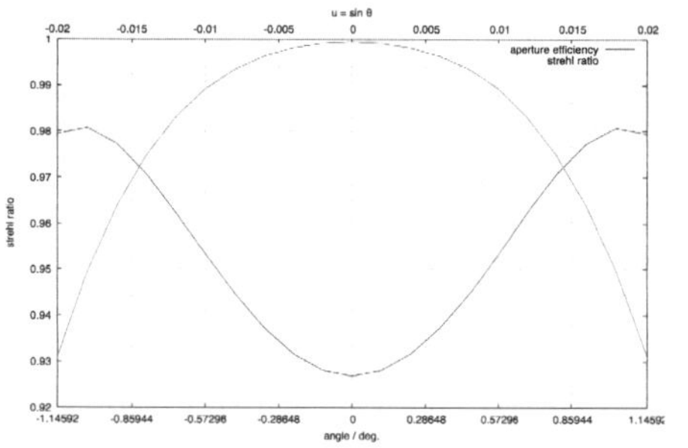

Fig. 7 Upper panel: Aperture efficiency vs. incident angle. Lower panel: Strehl ratio vs. incident angle. Purple lines represent optimization by aperture efficiency and green ones by Strehl ratio.

Fig. 7 shows the aperture efficiency and Strehl ratio obtained through the optimization as a function of incident angle. The angle zero corresponds to the center of FOV. When focusing on the angle of 0 degrees and 1.1 degrees, the Strehl ratios are about 0.93 but the aperture efficiencies are 0.74 for 0 degrees and 0.59 for 1.1 degrees, respectively. The fact that aperture efficiency depends on the types of aberrations is responsible for this discrepancy between aperture efficiency and Strehl ratio. The results in Fig. 7 imply that it is possible to fail to obtain target aperture efficiency when Strehl ratio is evaluated only.

CONCLUSIONS

We have developed a tools to calculate aperture efficiency based on the formulation in [1], [2], and [3] and implemented to the commercial ray tracing software, Zemax. Designing a radio telescope can be carried out by adopting aperture efficiency as a figure of merit. It has been revealed that aperture efficiency can be a different value for the same Strehl ratio. It arises from the dependence of aperture efficiency on the types of aberrations.

ACKNOWLEDGMENT

We are grateful to Professor Naomasa Nakai and Professor Nario Kuno at University of Tsukuba for supporting simulation softwarre.

REFERENCES

[1] M. Nagai and H. Imada, "Formalism of Beam Coupling and Aperture Efficiency of Antennas," in prep., 2017.
[2] H. Imada, "Theoretical study of Radio Telescope Optics with Wide Field of View," PhD thesis, University of Tsukuba, Tsukuba, Japan, 2016.
[3] H. Imada and M. Nagai, "Analytical expression of aperture efficiency affected by Seidel aberrations," in prep., 2017.

Simultaneous phase-locking of two THz-QCLs using an HEBM and a comb generator

Y. Irimajiri*, I. Morohashi, M. Kumagai, S. Ochiai, A. Kawakami, S. Nagano, , Y. Uzawa, and I. Hosako
National Institute of Information and Communications Technology, Koganei, Tokyo, Japan
* Contact: irimaji@nict.go.jp, phone +81-42-327-6089

***Abstract*—** We have demonstrated simultaneous phase-locking of two THz-QCLs using an HEB mixer and a comb generator. We have also demonstrated the simultaneous phase-locking using an AMC (Amplifier/Multiplier Chain) as a THz reference. This system can be applicable for a THz radar system etc.

I. Introduction

Phase-locking of THz-QCL [1] is important for some applications such as astronomical/atmospheric observations and communications. We have demonstrated simultaneous phase-locking of two THz-QCLs using an HEBM [2] and a comb generator [3]. Two THz-QCLs with different lasing frequencies at the 3 THz band were fabricated in our laboratory, and they were cooler by different coolers, a 45 K mechanical cooler and a LHe flow-type cryostat, respectively. A Mach-Zehnder type flat optical (1.55 μm)-comb with a UTC-PD (Uni-Traveling Carrier Photodiode) was used as a THz reference. In order to generate two THz reference signals with different frequencies for each THz-QCL, three modes of the comb were selected using a programmable bandpass filters. THz waves from the THz reference and the THz-QCLs were fed into an HEBM using beam splitters. Two beat notes were detected simultaneously by an HEBM and they were put into an each PLL system for the phase-locking of the THz-QCLs.

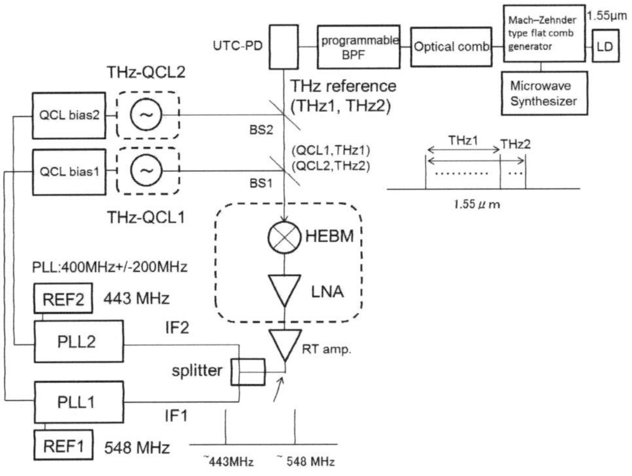

Fig. 1. A system block diagram of simultaneous phase-locking of two THz-QCLs using an HEBM and an optical comb generator followed by a UTC-PD.

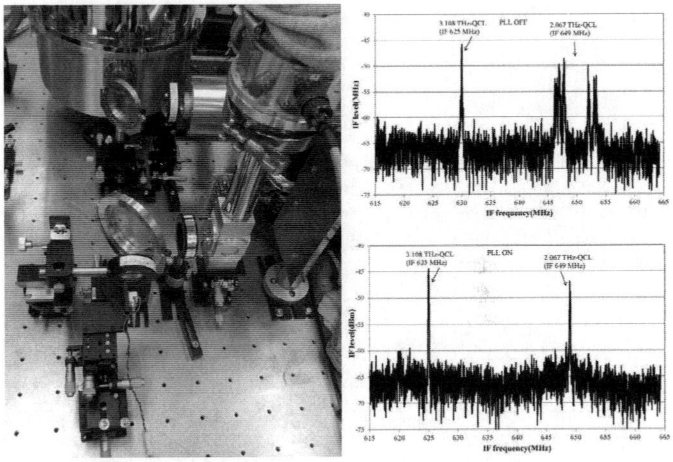

Fig. 2. A photograph of setup for the simultaneous phase-locking of two THz-QCLs (left). Each THz-QCL was cooled by different cooler, 45 K mechanical cooer and a LHe flow-type cryostat. THz reference was generated by a comb generator followed by a UTC-PD. Mylar beam splitters were used for LO injection to an HEB mixer. The detected spectrum show beat signals of PLL OFF(upper right)/ON(lower right) for 2 THz and 3 THz-QCLs.

TABLE I
FREQUENCY CONFIGULATION FOR SIMULTANEOUS PHASE-LOCKING OF TWO THz-QCLs AT 3-THz BAND USING A COMB GENERATOR AS A THz REFERENCE

Synth. Frq.(GHz)	Mode no.	UTC-PD(GHz)
17.991	172	THz1:3094.452
17.991	173	THz2:3112.443
THz-QCL device no.	QCL frq.(GHz)	IF(MHz)
QCL11dM(NICT)	QCL1:~3095	IF1:~548(THz1-QCL1)
QCL11kI(NICT)	QCL2:~3112	IF2:~443(THz2-QCL2)

TABLE II
FREQUENCY CONFIGULATION FOR SIMULTANEOUS PHASE-LOCKING OF TWO THz-QCLs AT 2-THz AND 3-THz BAND USING A COMB GENERATOR AS A THz REFERENCE

Synth. Frq.(GHz)	Mode no.	UTC-PD(GHz)
17.97241	115	THz1:2066.82715
17.97241	173	THz2:3109.22693
THz-QCL dvice no.	QCL frq.(GHz)	IF(MHz)
2THzQCL(Longwave)	QCL1:~2067.5	IF1:~649(THz1-QCL1)
QCL11lk(NICT)	QCL2:~3108.6	IF2:~625(THz2-QCL2)

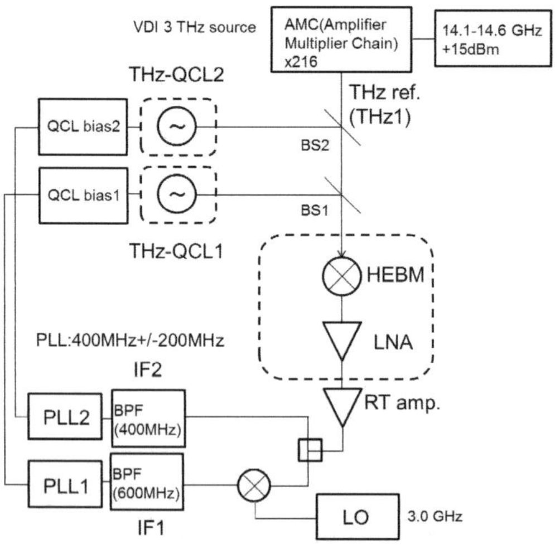

Fig. 3. Phase-locking system of two THz-QCLs using an AMC as a THz reference. In this case, we have a single THz reference signal. Therefore, the lasing frequency of two THz-QCLs should be within an IF bandwidth of an HEBM. Although, the gain IF bandwidth of the HEBM is around 3 GHz, the beat signal can be detected even at ~14 GHz if we use an appropriate LNA.

TABLE III
FREQUENCY CONFIGULATION FOR SIMULTANEOUS PHASE-LOCKING OF TWO THz-QCLs AT 3-THz BAND USING AN AMC AS A THz RFERENCE

THz reference	THz ref. frq.(GHz)	
AMC	THz1:3108.4	
THz-QCL device no.	QCL frq. (GHz)	IF(GHz)
QCL1:QCL11kI(NICT)	QCL1:~3112	IF1:~3.6(THz1-QCL1)
QCL2:QCL11lk(NICT)	QCL2:~3108	IF2:~0.4(THz1-QCL2)
		IF3:~4.0(QCL1-QCL2)

II. MEASUREMENT SETUP AND RESULT

Fig. 1 shows a setup for the simultaneous phase-locking of two THz-QCLs at 3 THz-band. Table I shows frequency configuration of two THz-QCLs at 3-THz band using a comb generator as a THz reference. The synthesizer frequency was set to 17.991 GHz. The mode number of 172 and 173 were selected from optical comb using a programmable filter and were put into a UTC-PD. Two THz reference signals of 3094.453 GHz (THz1) and 3112.443 GHz (THz2) were generated by a UTC-PD. THz waves from THz-QCL with lasing frequency of ~3095 GHz (THz-QCL1) and ~3112 GHz (THz-QCL2) were injected as a local oscillator for an HEB mixer. Both the THz waves from the THz-QCL1 and the THz-QCL2 were injected as a local signal for an HEBM, therefore, we should care about saturation of LO signal. The beat signal between THz1 and THz-QCL1 was ~548 MHz, and that between THz2 and THz-QCL2 was ~443 MHz. These signals were put into an each PLL circuit for phase-locking. We have also demonstrated simultaneous phase-locking of THz-QCLs at 2.067 THz (provided by Longwave photonics) and 3.108 THz (made in NICT) as shown in Fig. 2 and Table II.

We have also demonstrated the simultaneous phase-locking using an AMC (Amplifier/Multiplier Chain) as a THz reference. In this case, we have a single THz reference signal. Therefore, the lasing frequency of two THz-QCLs should be within an IF bandwidth of an HEBM. Although, the gain IF bandwidth of the HEBM is around 3 GHz, the beat signal can be detected even at ~14 GHz if we use an appropriate LNA. Two beat notes between THz-QCLs and the THz reference were used for phase-locking. A beat note between two THz-QCLs was also observed. We can use this beat note the phase-locking of the 2nd THz-QCL. The measurement setup and photograph are shown in Fig. 3 and Table III.

The simultaneous phase-locking of THz-QCLs could be achieved using a superlattice harmonic mixer. We will also try gas cell measurement using an HEBM with two phase-locked THz-QCLs as LOs.

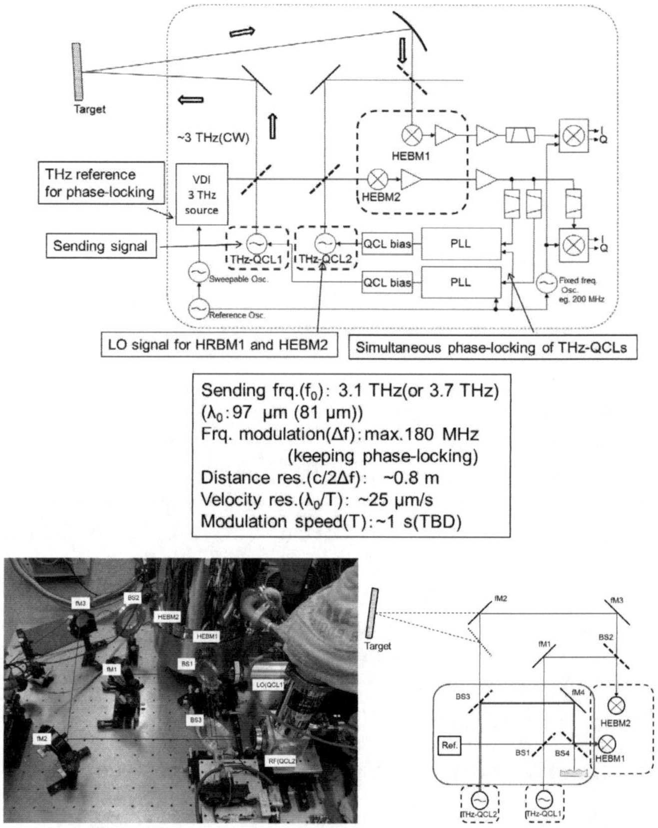

Fig. 4. System block diagram of a THz radar system (above), a photograph of an experimental setup (bottom left), and modified optics (bottom right).

One of an application of the simultaneous phase-locked THz-QCL is a THz radar system. Fig. 4 shows a system block diagram and an experimental setup of the THz radar (TeDar). We have successfully detected beat signals. However, we found the optics should be changed to avoid unnecessary beat signal. The modified optics is shown in Fig. 4 (bottom right).

We will make some demonstration experiments using this system by CW measurement which measures I, Q data of the signals.

1, Measure a metal target position in a few tens of micron from the phase measurement.
2, Measure another type of material such as a dielectric material.
3, Measure how far the target can be distant (ex. more than 1 m).
4, Measure through a dielectric material put in the path.
5, Measure environmental temperature and humidity change with fixed target (propagation delay, atmospheric attenuation).

III. CONCLUSIONS

We have demonstrated simultaneous phase-locking of two THz-QCLs using an HEB mixer at 2-3 THz band. An optical comb generator followed by a UTC-PD or an AMC were used as a THz reference. This system can be applicable for a THz radar system etc.

REFERENCES

[1] Y. Irimajiri, M. Kumagai, I. Morohashi, A. Kawakami, N. Sekine, S. Nagano, S. Ochiai, S. Tanaka, Y. Hanado, Y. Uzawa, and I. Hosako, "Precise Evaluation of a Phase-Locked THz Quantum Cascade Laser," *IEEE Trans. THz Sci. Technol.,* vol. 6, Issue 1, pp.115-120, 2016.

[2] Y. Irimajiri, M. Kumagai, I. Morohashi, A. Kawakami, N. Sekine, S. Nagano, S. Ochiai, S. Tanaka, Y. Hanado, Y. Uzawa, and I. Hosako, "Development of a Superconducting Low-Noise 3.1-THz Hot Electron Bolometer Receiver," *IEEE Trans. THz Sci. Technol.,* vol. 5, no. 6, pp.1154-1159, Nov. 2015.

[3] I. Morohashi, Y. Irimajiri, T. Sakamoto, T. Kawanishi, M. Yasui, and I. Hosako, "Generation of millimeter waves with fine frequency tunability using Mach-Zehnder-modulator-based flat comb generator," *IEICE Trans. Electron.*, vol. E96-C, no. 2, pp. 192-196, 2013.

Development of Quantum Cascade Lasers at 2.7 THz for Heterodyne detection

François Joint[1,2], Thibaut Vacelet[1], Gregory Gay[1], Yan Delorme[1], Raffaele Colombelli[2]

[1]*LERMA, Laboratoire d'Études du Rayonnement et de la Matière en Astrophysique et Atmosphères*

[2] *Centre de Nanosciences et de Nanotechnologies, CNRS, Univ. Paris-Sud, Université Paris-Saclay, 91405 Orsay cedex, France*
Email: francois.joint@u-psud.fr

***Abstract*-We have developed a single mode emission, low power consumption THz QCL operating at a specified target frequency for sensitive heterodyne detections. The Fabry-Pérot THz QCL is of limited use since its output power is distributed over many spectral modes. The approach we have chosen is the distributed-feedback (DFB) architecture, in particular the 3rd-order DFB grating that can provide single mode emission as well as small beam divergence. To obtain single mode operation at the desired frequency we have fabricated several devices with different grating periods and/or grating duty cycle. This strategy allows to thinly cover a relatively broad range of emission frequencies. Upon electro-optical characterisation of the lasers, the devices that best suit the application can be selected. The laser design is based on judicious electromagnetic modelling. Particular care has been taken to reduce the overall size as much as possible, in order to minimise the power dissipation. We obtained devices with very low electrical dissipation, which is suitable for embedded THz detection systems.**

I. INTRODUCTION

There is a particular interest in astronomy for the detection of radiation from cold interstellar gases. These emissions typically fall in the THz range of the electromagnetic spectrum: for instance, an important transition of deuterated hydrogen falls at 2.7 THz (90 cm^{-1}). Heterodyne detection is ideally suited to capture these signals with high spectral resolution.

Heterodyne detection requires local oscillator sources that operate a few GHz away from the frequency of interest. THz quantum cascade lasers [*1*] (QCL) emerged recently as suitable sources for the detection of signals above 2 THz. The combination of a THz QCL with an ultra-sensitive hot electron bolometer (HEB), cooled at 4K for mixing [*2*], is a very attractive solution to achieve heterodyne detections with very high sensitivity and at frequencies around or far beyond 2 THz.

II. 3RD ORDER DFB LASER: FABRICATION AND CHARACTERISATION

The third order DFB QCL has been developed at C2N Orsay. The QCL is based on a metal-metal waveguide with deeply etched lateral corrugation (fig. 1). The active region is a 14 µm thick $GaAs/Al_{0,15}GaAs$ quantum cascade structure which has been grown by molecular beam epitaxy on a semi-insulated GaAs wafer. The design of the active region is based on a four-well structure with a longitudinal optical phonon resonant depopulation mechanism. After the growth, the wafer is thermo-bonded with gold to a GaAs wafer. The grating and wire-bonding pads were defined by optical lithography then by gold deposition (Ti/Au, 5/250 nm). Laser ridges and lateral corrugation were defined by inductively coupled plasma reactive ion etching.

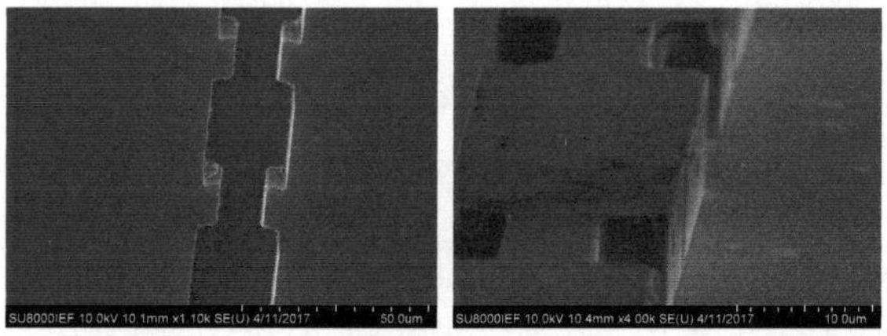

Fig. 1: SEM images of the third-order DFB QCL with deeply etched lateral corrugation

The emission of such QCL is mainly determined by the periodicity of the grating and also by the filling factor between the narrow and the wide part of the waveguide.

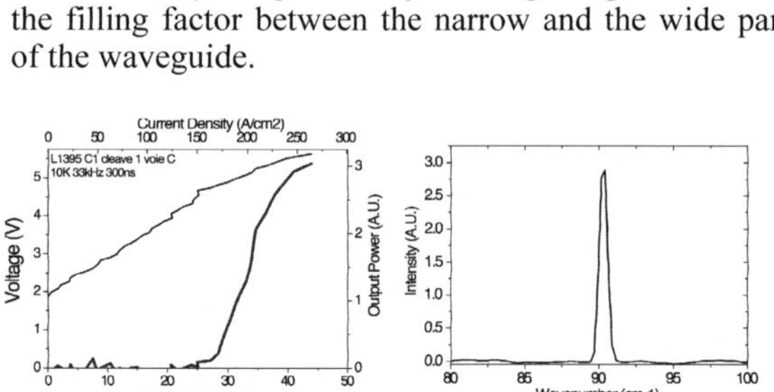

Fig. 2: I-V curve and I-L curve (left) of the third-order DFB QCL and measured spectrum of the QCL (right) at 10K in pulsed mode

We have characterised the lasing spectrum with a Fourier Transform Spectrometer using a deuterated triglycine sulfate (DTGS) Terahertz detector. The important aspect of this work is that we achieved single mode emission at the target frequency ($90cm^{-1}$). By finely adjusting the grating periodicity and filling factor during the design of the lithography mask, we managed to obtain devices with the emission exactly at our target frequency. This is of interest when we consider heterodyne detections of molecular transitions as one needs a local oscillator emission very close to the specific frequencies.

The measured laser emission stays single mode from the lasing threshold to the limit of applied current on the device and up to 75 K. In In CW (Continuous Wave) mode, the current-voltage-power (I-V-L) curves of the third-order DFB QCL (Figure 2) at 10 K show a very low threshold driving current (< 30 mA) while the DC dissipation of the device stays below 250 mW over the whole operation range. These characteristics make the component compatible for compact integration. The maximum output power measured with an absolute power meter in CW at 10 K was 800 μW, which is among the state-of-art results.

As shown in the modeling and simulations [3] of a third order DFB waveguide, the third-order diffracted mode is used for the distributed feedback while the first and the second-order diffracted modes for the output coupling. If the effective index of the cavity is set at $n_{eff} = 3$, and the grating periodicity is equal to half of the free space wavelength, then the radiations coming from each of the apertures are adding up constructively at both ends of the waveguide. One can compare the functioning of the 3rd order DFB QCL light extraction process to a one dimensional end-fire antenna array.

We have performed the far-field beam pattern measurements using a room temperature Golay cell with a 2D rotation test setup. The laser operating in pulsed mode was cooled down to 4K and was at a distance of 90 mm from the detector. We can see that the QCL has a single lobed emission and the FWHM of the beam is roughly 15x20 degree.

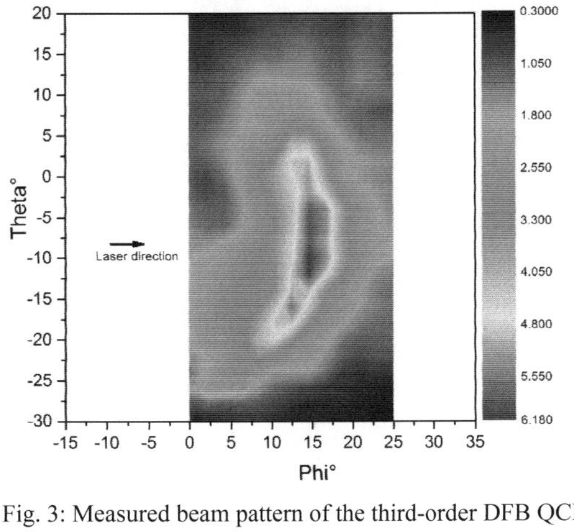

Fig. 3: Measured beam pattern of the third-order DFB QCL.

III. CONCLUSION

We have designed and fabricated the 3rd Order DFB lasers and have demonstrated by measurements that it is possible to get a single mode emission at a specified frequency. The single lobed beam has a FWHM around 15 x 20 degree, which is rather small compared to standard metal-metal waveguide. The low driving currents and low power dissipation of those devices make them suitable for ultra-compact systems, such as THz heterodyne detectors for space applications.

In order to increasing the coupling efficiency between the QCL and our mixers, we are currently working on several solutions to improve or re-shape the QCL's beam pattern, such as hollow dielectric waveguides, mirrors and dielectric lenses.

We thank Edmund H. Linfield and Lianhe Li from Leeds University for providing us with the QCL epitaxy structure (L1395). We acknowledge financial support from the Centre National d'Etudes Spatiales (CNES). The device fabrication has been performed at the nano-center CTU-C2N.

IV REFERENCES

1. C. Sirtori, S. Barbieri, R. Colombelli, *Wave engineering with THz quantum cascade lasers. Nat. Photonics. 7, 691–701 (2013).*
2. G. Gay, *Mélangeurs à bolomètres à électrons chauds sur membranes fonctionnant au-delà du THz pour l'instrument post-Herschel (2013) (available at https://tel.archives-ouvertes.fr/ tel-00976996/).*
3. M. I. Amanti, G. Scalari, F. Castellano, M. Beck, J. Faist, *Low divergence Terahertz photonic-wire laser. Opt. Express. **18**, 6390–5 (2010).*

A 4.745 THz Local Oscillator for the upGREAT receiver

M. Justen[1*], U.U. Graf[1], L. Bosco[2], K. Otani[2], J. Faist[2], and J. Stutzki[1]

[1] *1. Physikalisches Institut, Universität zu Köln, 50937 Köln, Germany*
[2] *Institute for Quantum Electronics, ETH-Zürich, CH-8093 Zürich, Switzerland*
[*] Contact: justen@ph1.uni-koeln.de

Abstract—We present the development of a local oscillator for the H channel of upGREAT, a 7pixel heterodyne receiver on board of the SOFIA aircraft which is used to observe the OI line at 4.745THz.

The local oscillator has to provide the pump power for all 7 HEB mixer pixels. Further requirements are mechanical stiffness, optical stability, frequency stability and a certain tunability.

We include details of the QCL development, refocusing optics as well as cryogenics and vibration cancellation.

Superconducting diamond films as perspective material for direct THz detectors

N. Titova[1], A. Kardakova[1,*], N. Tovpeko[1], S. Ryabchun[1], S. Mandal[2], D. Morozov[2], G.M. Klemencic[2], S.R. Giblin[2], O.A. Williams[2], G.N. Goltsman[1]

[1]*Physics Department, Moscow State University of Education (MSPU), 119435, Moscow, Russia*
[2]*School of Physics and Astronomy, Cardiff University, CF24 3AA Cardiff, United Kingdom.*
*Contact: kardakova@rplab.ru

Abstract— Superconducting films with a high resistivity in the normal state have established themselves as the best materials for direct THz radiation sensors, such as kinetic inductance detectors (KIDs) [1] and hot electron bolometers (nano-HEBs) [2]. The primary characteristics of the future instrument such as the sensitivity and the response time are determined by the material parameters such as the electron-phonon (e-ph) interaction time, the electron density and the resistivity of the material. For direct detectors, such as KIDs and nano-HEBs, to provide a high sensitivity and low noise one prefer materials with long e-ph relaxation times and low values of the electron density.

As a potential material for THz radiation detection we have studied superconducting diamond films. A significant interest to diamond for the development of electronic devices is due to the evolution of its properties with the boron dopant concentration. At a high boron doping concentration, $n_B \sim 5 \cdot 10^{20}$ cm^{-3}, diamond has been reported to become a superconducting with T_c depending on the doping level. Our previous study of energy relaxation in single-crystalline boron-doped diamond films epitaxially grown on a diamond shows a remarkably slow energy-relaxation at low temperatures. The electron-phonon cooling time varies from 400 ns to 700 ns over the temperature range 2.2 K to 1.7 K [3]. In superconducting materials such as Al and TiN, traditionally used in KIDs, the e-ph cooling times at 1.7 K correspond to ~20 ns [4] and ~100 ns [5], correspondingly. Such a noticeable slow e-ph relaxation in boron-doped diamond, in combination with a low value of carrier density (~10^{21} cm^{-3}) in comparison with typical metals (~10^{23} cm^{-3}) and a high normal state resistivity (~1500 μΩ·cm) confirms a potential of superconducting diamond for superconducting bolometers and resonator detectors. However, the price and the small substrate growth are of single crystal diamond limit practical applications of homoepitaxial diamond films. As an alternative way with more convenient technology, one can employ heteroepitaxial diamond films grown on large-size Si substrates.

Here we report about measurements of e-ph cooling times in superconducting diamond grown on silicon substrate and discuss our expectations about the applicability of boron-doped diamond films to superconducting detectors. Our estimation of limit value of noise-equivalent power (NEP) and the energy resolution of bolometer made from superconducting diamond is order 10^{-17} W/Hz$^{1/2}$ at 2 K and the energy resolution is of 0.1 eV that corresponds to counting single-photon up to 15 um. The estimation was obtained by using the film thickness of 70 nm and $\rho \sim 1500$ μΩ·cm, and the planar dimensions that are chosen to couple bolometer with 75 Ω log-spiral antenna. Although the value of NEP is far yet from what might like to have for certain astronomical applications, we believe that it can be improved by a suitable fabrication process. Also the direct detectors, based on superconducting diamond, will offer low noise performance at about 2 K, a temperature provided by inexpensive close-cycle refrigerators, which provides another practical advantage of development and application of these devices.

[1] P.K. Day, et. al, Nature, 425, 817, 2003.
[2] J. Wei, et al, Nature Nanotech., 3, 496, 2008.
[3] A. Kardakova, et al, Phys. Rev. B, 93, 064506, 2016.
[4] P. Santhanam and D. Prober, Phys. Rev. B, 29, 3733, 1984
[5] A. Kardakova, et al, Appl. Phys. Lett, vol. 103, p. 252602, 2013.

Measurement of THz performance of plasmonic absorbers made of bulk aluminum

V. Janonis[1], S. Tumėnas[1], S. Indrišiūnas[1], B. Voisiat[1], G. Račiukaitis[1], and I. Kašalynas[1*]

[1] *Center for Physical Sciences and Technology, Saulėtekio Ave. 3, LT-10257 Vilnius, Lithuania*
[*]Contacts: irmantas.kasalynas@ftmc.lt

Abstract—Opaque metal films with periodically arranged subwavelength size holes have demonstrated strongly enhanced light transmission acting as optical filter for certain wavelengths [1]. The ability to engineer the spoof plasmons at almost any frequency in the THz regime were proposed recently [2]. It was demonstrated that the subwavelength holes in a perfect conductor give rise to similar anomalous transmission of electromagnetic waves via the spoof plasmons excitation effects. The change of dielectric medium at the surface of THz filter revealed the red-shift of the resonance peak demonstrating the enhanced surface plasmons excitation and interaction with localized plasmons modes [3-4].

In this work, the resonant apertures of different shape and size were processed periodically in a bulk aluminum using the direct laser writing (DLW) technique [5]. The reflection spectra were measured with a vacuumed far-infrared Fourier transform spectrometer at different incidence angle. To observe THz plasmons modes accurately, spectra were recorded with the spectral resolution of 2 cm^{-1}. It was found that the number of peaks, the peak position, and the line width were controlled in the THz regime via the change of initial groove parameters.

[1] T.W. Ebbesen, H.J. Lezec, H.F. Ghaemi, T. Thio, P.A. Wolff, Extraordinary optical transmission through sub-wavelength hole arrays, Nature 391, pp. 667-669 (1998).
[2] J.B. Pendry, L. Martin-Moreno, F. J. Garcia-Vidal, Mimicking surface plasmons with structured surfaces, Science 305, pp. 847-848 (2004).
[3] M. Tanaka, F. Miyamaru, M. Hangyo, Effect of thin dielectric layer on terahertz transmission characteristics for metal hole arrays, Optics Letters 30, pp. 1210–1212 (2005).
[4] V. Kaveckytė, R. Venckevičius, L. Minkevičius, G. Račiukaitis, G. Valušis, B. Voisiat, I. Kašalynas, Effects of thin dielectric layer on plasmons excitation in perforated metal films, Proc. of IRMMW-THz Conf., Mainz on the Rhine, Germany, Sept. 1–6, 2013, p. 167.
[5] B. Voisiat, I. Kašalynas, G. Račiukaitis, THz components and plasmonic structures fabricated by direct laser patterning of metals, Proc. of IRMMW-THz Conf., Tuscon, AZ, Sept. 14–19, 2014, p. 90.

Study of mid infrared hot electron bolometer mixers

A. Kawakami*[1], H. Shimakage[2], J. Horikawa[3], M. Hyodo[4], S. Saito[1], S. Tanaka[1], and Y. Uzawa[1]
[1]*National Institute of Information and Communications Technology, 651-2492, Kobe, Japan*
[2] *Ibaraki University, 316-8511, Ibaraki, Japan*
[3] *Fukui College, 916-8507, Fukui, Japan*
[3] *Kanazawa University, 920-1192, Ishikawa, Japan*
*Contact: kawakami@nict.go.jp

Abstract— To design antennas for mid-infrared (MIR) hot electron bolometers (HEBs), the surface impedance of Au thin films at cryogenic temperatures was evaluated using Fourier transform infrared spectroscopy (FTIR) with a sample cooling system. For the evaluation, resonator arrays that were constructed by gold (Au) thin film strips were fabricated and the corrected surface reactance was estimated. MIR HEBs with a twin-slot antenna for operation at 61 THz were designed and fabricated using the corrected surface impedance. For evaluating the mixer properties of the MIR HEBs, measurement setup without beam splitter was constructed. In this setup, IF output power characteristics of the MIR HEB mixer was observed. But the obvious difference of the IF output power between 1000 K and 30 K thermal loads have not been confirmed.

Introduction

Electromagnetic waves generally exhibit both particle and wave properties. Many conventional mid-infrared (MIR) detectors are designed with structures and mechanisms based on the particle nature of light. One of the reasons for this design choice is that photons possess high energy with short wavelengths. However, such detectors usually suffer from a tradeoff between sensitivity and response speed because sensitivity depends on the detection area, and the electrostatic capacitance of a detector tends to increase with increasing area. Therefore, there were several reports of IR detection utilizing the wave nature of light [1], [2]. We also propose the use of nano-antennas to improve the response performance of infrared detectors [3]-[6].

Significant research and development in superconducting hot-electron bolometer (HEB) mixer has led to the emergence of low noise electromagnetic wave receivers in the terahertz frequency range [7]-[9], and we also contributed to this research and development effort [10]. Such mixer typically consists of a planar antenna for efficiently capturing electromagnetic waves in the air, combined with a small detector made of a niobium nitride(NbN) thin film strip located at the feeding point of the antenna. The upper limit of the operating frequency of HEB is considered to be determined by the physical structure of the device. We therefore recognized the potential for achieving greater response speed and efficiency in infrared photo-detectors, by separating the mechanisms for receiving and detecting light as the optical antenna and micro detector, and optimizing their functions.

In this paper, we report on the evaluation of the surface impedance of gold (Au) thin films under cryogenic conditions. Using the corrected surface impedance, we designed and fabricated MIR HEBs. The fabricated MIR HEB was evaluated at 61.3 THz.

Design and Fabrication of MIR HEBs.

For the simulation of MIR circuits, the dielectric constant of the magnesium oxide (MgO) substrate [11] and the complex surface impedances of the Au films were needed. The impedances could be derived using the measured complex refractive indices, however, these derivations were insufficient for designing superconducting MIR devices operating at cryogenic temperature. Therefore, we corrected the surface impedance at cryogenic temperatures by using a FTIR with sample cooling system. For the evaluation, resonator arrays constructed of thin film strips of Au were fabricated. Their length and width were set at 1.3 μm and 0.2 μm, respectively. The Au thickness was 55 nm. In the transmittance spectrum, the resonant frequency was observed as absorption properties at around 61 THz. When the temperature changed from 300 K to 10 K, it was found that the resonant frequency shifted to the low

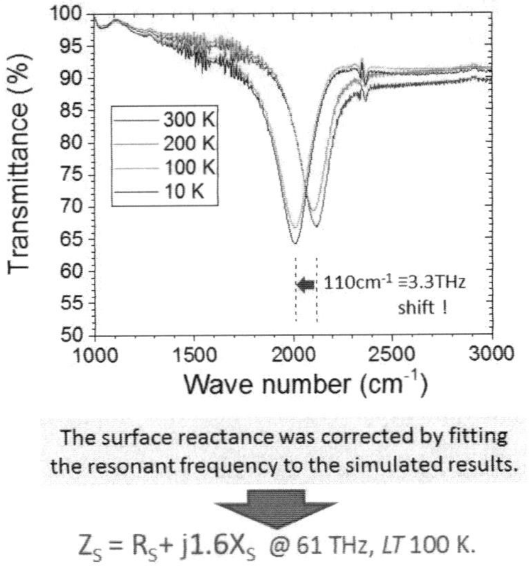

(b) The transmittance spectrums at various temperatures.

Fig. 1. Transmittance spectrums of the resonator array at various temperatures.

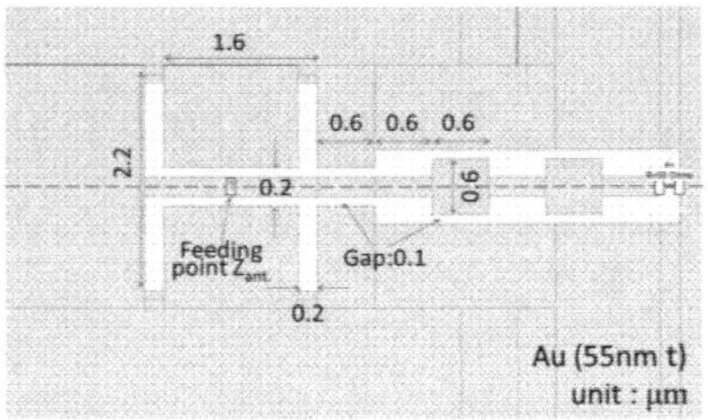

(a) Design of the twin-slot nano-antenna

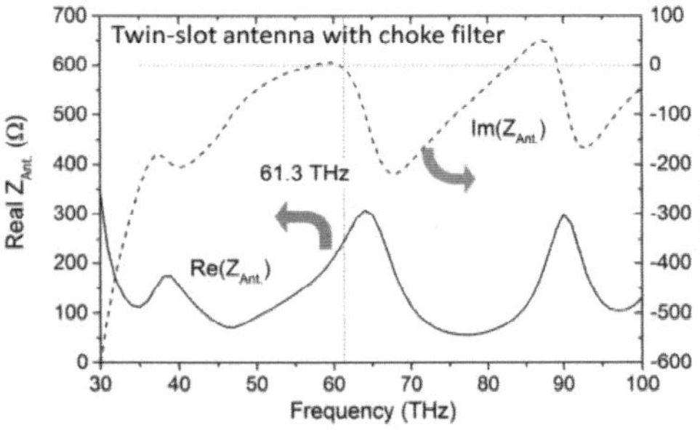

(b) Simulated impedance of the twin-slot nano-antenna at feeding point

Fig. 2. Design and impedance of the twin-slot nano-antenna.

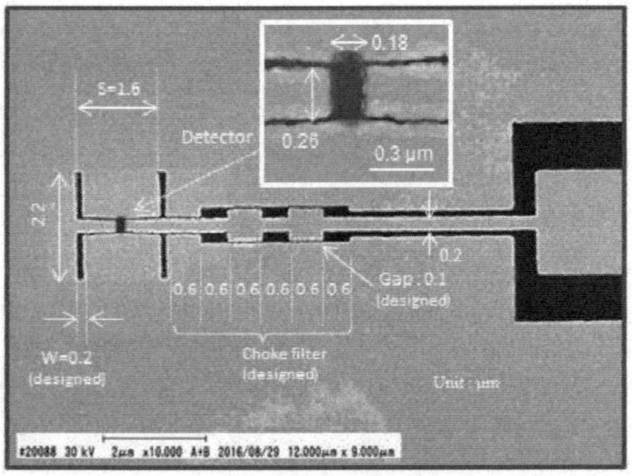

Fig. 3. SEM image of the MIR HEB,

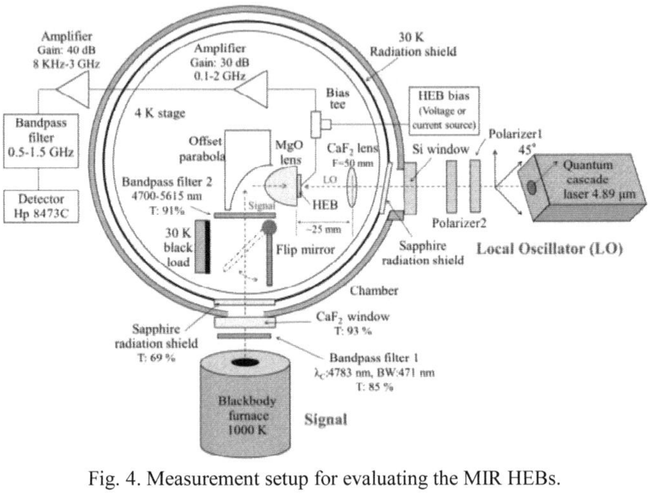

Fig. 4. Measurement setup for evaluating the MIR HEBs.

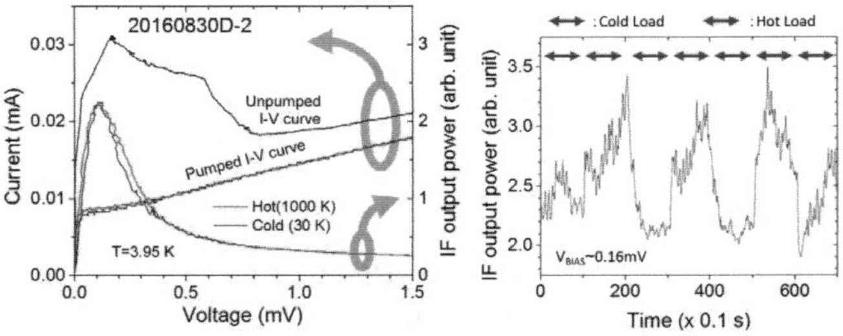

(a) I-V and IF output power characteristics (b) IF output power – Hot/Cold Loads

Fig. 5. I-V and IF output power characteristics of the MIR HEB as a function of bias voltage.

frequency side (See Fig. 1). By using resonant frequency under cryogenic, we corrected the surface reactance by fitting the resonant frequency to the simulated results. The correction value was found to be 1.6 times the surface reactance at room temperature.

MIR nano-antennas were designed for operation at 61.3 THz. Fig. 2. shows the design and the calculated impedance of the antenna. The antenna was a twin-slot antenna, and the slot length and width were decided by the simulated results and set at 2200 and 200 nm, respectively. The size of the choke filter was also decided. At the designed frequency, the antenna impedance was expected to be $Z_{Ant.} = 250 - j6\ \Omega$.

For impedance matching between the superconducting strip and the antenna, we used an Nb(2.5-δ nm)/NbN(5 nm) bilayer as the superconducting strip. Here, for electric contact, the surface of the bilayer was etched by Ar ion beam and the etched depth δ was estimated at less than 1 nm. Fabricating the MIR HEBs with an antenna structure requires the building of fine structures on nano scales. We developed a new fabrication process using electron beam lithography for all our lithography processes. The details of the fabrication process have already been described elsewhere [3], [4]. Fig. 3 shows a SEM image of the fabricated MIR HEB, the inset is a magnified view of the detector. The detector width and length were approximately 0.18 and 0.26 µm, respectively. The critical current and the normal resistance of the HEB were approximately 32 µA and 200 Ω, respectively.

EVALUATION OF MIR HEBS.

Fig. 4 shows the measurement setup for evaluating the mixer properties of the MIR HEBs. The HEB was cooled by a GM refrigerator. A mid-infrared quantum cascade laser (QCL) with a wavelength of 4.89 µm was used as the local oscillator (LO). In this case, the LO power was irradiated from the surface of the HEB. In order to concentrate the LO power, a CaF_2 lens was used. However, we had to slightly defocus the lens to reduce the influence of the vibration of the GM refrigerator. The signal from the thermal load was irradiated from the back of the HEB through an offset parabolic mirror and an anti-reflection coated MgO hyper-hemisphere. A black body furnace set at 1000 K was used as a reference signal. A carbon-nano-tube coated copper block was also used as a 30 K thermal load at which temperature was monitored. Band-pass filters and sapphire windows were used to exclude thermal radiation, except for the signal around LO-wavelength from the furnace. The total transmittance of the two band-pass filters and two windows in the signal input optical path was estimated to be about 50%. The IF signal was amplified by a 0.1–2 GHz cooled low-noise amplifier and an 8 kHz to 3 GHz room temperature amplifier. The signal was then filtered by a bandpass filter centered at 1

GHz with bandwidth of 1 GHz and the IF power was monitored using a Schottky diode detector.

Fig. 5 shows the pumped and unpumped I-V characteristics and IF output power characteristics of the MIR HEB mixer. The IF output showed a maximum at around 0.15 mV. However, the periodic noise caused by the vibration of the GM refrigerator was observed in the IF output characteristics, and the obvious difference of the IF output power between 1000 K and 30 K thermal loads have not been confirmed.

CONCLUSION

To design superconducting MIR devices, the surface reactance of Au thin films at cryogenic temperatures was evaluated using FTIR with a sample cooling system. Resonator arrays constructed with Au thin film strips were fabricated. The resonant frequency was set at about 61 THz. When the temperature was changed from 300 K to 10 K, it was found that the resonant frequency shifted to the low frequency side. By fitting the resonant frequency to the simulated results, the surface reactance at 10 K was corrected. MIR HEBs with a twin-slot antenna were designed and fabricated using the corrected surface impedance. For evaluating the mixer properties of the MIR HEBs, measurement setup without beam splitter was constructed. IF output power characteristics of the MIR HEB mixer was observed but the obvious difference of the IF output power between 1000 K and 30 K thermal loads have not been confirmed.

REFERENCES

[1] Y. Lobanov, M. Shcherbatenko, M. Finkel, S. Maslennikov, A. Semenov, B. M. Voronov, A. V. Rodin, T. M. Klapwijk, and G. N. Gol'tsman, "NbN Hot-Electron-Bolometer Mixer for Operation in the Near-IR Frequency Range," *IEEE Trans. Appl. Supercond.*, vol. 25, p. 2300704, 2015

[2] Y. Yao, R. Shankar, P. Rauter, Y. Song, J. Kong, M. Loncar, and F. Capasso, "High-Responsivity Mid-Infrared Graphene Detectors with Antenna-Enhanced Photocarrier Generation and Collection," *Nano Lett.* 14(7), pp 3749-3754, 2014

[3] A. Kawakami, S. Saito, M. Hyodo, "Fabrication of Nano-Antennas for Superconducting Infrared Detectors," *IEEE Trans. Appl. Supercond.*, vol. 17, pp. 632–635, 2011

[4] J. Horikawa, A. Kawakami, M. Hyodo, S. Tanaka, M. Takeda, H. Shimakage, "Study of Mid-Infrared Superconducting Detector with Phased Array Nano-Slot Antenna," *IEEE Trans. Appl. Supercond.*, 25, p. 2301005, 2015

[5] A. Kawakami, J. Horikawa, M. Hyodo, S. Tanaka, M. Takeda, H. Shimakage, "Fabrication of Superconducting Mid-infrared Photo-detectors with Dipole Nano-antennas," *IEEE Trans. Appl. Supercond.*, vol. 25, p. 2300904, 2015

[6] J. Horikawa, A. Kawakami, M. Hyodo, S. Tanaka, M. Takeda, H. Shimakage, "Evaluation of nano-slot antenna for mid-infrared detectors," *Infrared Phys. Technol.*, 67, pp. 21–24, 2014

[7] S. Krause, V. Mityashkin, S. Antipov, G. Gol'tsman, D. Meledin, V. Desmaris, and M. Rudzinski, "Reduction of Phonon Escape Time for NbN Hot Electron Bolometers by Using GaN Buffer Layers," *IEEE Trans. THz Sci. Technol.*, vol. 7, no. 1, pp. 53–59, 2017

[8] D. Cunnane, J. H. Kawamura, M. A. Wolak, N. Acharya, T. Tan, X. X. Xi, and B. S. Karasik, "Characterization of MgB_2 Superconducting Hot Electron Bolometers," *IEEE Trans. Appl. Supercond.*, vol. 25, no. 3, 2300206, 2015

[9] W. Zhang, J. R. Gao, M. Hajenius, W. Miao, P. Khosropanah, T. M. Klapwijk, and S. C. Shi, "Twin-Slot Antenna Coupled NbN Hot Electron Bolometer Mixer at 2.5 THz," *IEEE Trans. Appl. Supercond.*, vol. 1, no. 2, pp. 378-382, 2011

[10] Y. Irimajiri, A. Kawakami, I. Morohashi, M. Kumagai, N. Sekine, S. Nagano, S. Ochiai, S. Tanaka, Y. Hanado, Y. Uzawa, and I. Hosako, "Development of a Superconducting Low-Noise 3.1-THz Hot Electron Bolometer Receiver," *IEEE Trans. THz Sci. Technol.*, vol. 5, no. 6, pp. 1154–1158, 2015

[11] R. E. Stephens and I. H. Malitson, "Index of Refraction of Magnesium Oxide," *J. Res. Natl. Bur. Stand.* 49 pp. 249-252, 1952

Performance of SIS mixers for upgrade of CHAMP+ 7-pixel arrays

Andrey Khudchenko[1,*], Andrey M. Baryshev[1,2], Ronald Hesper[1], Marielle E. Bekema[1], Rob de Haan Stijkel[1], Jan Barkhof[1], Duc van Nguyen[2], Valery P. Koshelets[3], Pavel Dmitriev[3], Michael Fominsky[3], Kirill Rudakov[1,3,4], Stefan Heyminck[5] and Rolf Güsten[5]

[1]*NOVA/Kapteyn Astronomical Institute, University of Groningen, 9747 AD, Groningen, Netherlands*
[2]*Netherlands Institute for Space Research SRON, 9747 AD, Groningen, Netherlands*
[3]*Kotel'nikov Institute for Radio Engineering and Electronics, 125009, Moscow, Russia*
[4]*Moscow Institute of Physics and Technology, 141701, Dolgoprudny, Russia*
[5]*Max Planck Institute for Radio Astronomy, D-53121, Bonn, Germany*
*Contact: A.Khudchenko@sron.nl

Abstract— We present here the performance of SIS mixers for upgrade of CHAMP+ array instrument on APEX telescope. In total it includes 14 mixers: 7 for the low band (600-720 GHz) and 7 for the high band (790-950 GHz). The mixers are a replacement for the existing set, which was commissioned on APEX in 2006. The low band mixers are based on Nb/AlN/Nb single tunneling SIS junction and high band ones - on Nb/AlN/NbN SIS twin junctions. The corrected DSB noise temperature of the low band mixers is roughly between 60 K and 120 K for the entire frequency range, and the corrected DSB noise temperature of the high band mixers varies from about 200 K at low frequencies to 400 K at the high end.

INTRODUCTION

The CHAMP+ heterodyne array [1] was commissioned on the APEX (Atacama Pathfinder EXperiment) telescope [2] in 2006. The array consists of 14 pixels, divided into two sub-arrays of 7-pixels each, setup in a hexagonal configuration. The RF (radio frequency) tuning range is 600-720 GHz for the low band and 790-950 GHz for the high band sub-arrays, corresponding to atmospheric windows.

The low frequency array was based on the state of the art (at that time) superconductor-insulator-superconductor (SIS) mixers developed for the ALMA Band 9 receivers [3]. For the high band mixers, on the other hand, an one-off design was made, and the performance was not optimized as far as for the low band. Since then, the SIS mixer technology has been significantly improved and we have developed a new generation of detectors for both CHAMP+ bands. In this paper, we describe the performance of the new mixers installed in the CHAMP+ instrument in January this year (2017) and compare it with the old ones.

LOW BAND (600-720 GHZ) MIXERS

The original low band mixers installed in CHAMP+ instrument were fabricated using SIS junctions based on a Nb/AlO/Nb three-layer. The performance of these mixers is depicted in Fig. 1 by blue curves. Later, due to progress of SIS junction technology, it was found that SIS mixers with AlN barrier can have much higher current density providing significantly wider RF bandwidth and also better sensitivity [4]. This technology was successfully implemented to finalize production of the ALMA Band 9 receivers [3].

After the ALMA Band 9 cartridges were finished, additional mixers with AlN barrier were measured, and seven of them were selected to upgrade the CHAMP+ low band. The noise temperature of the new mixers is presented in Fig. 1 by red curves. One can see a significant improvement of performance in the entire RF range (from 30% in the middle of the band to several time at the edges). It should be mentioned, that the new mixers are operated in the first minimum of a critical current suppression curve.

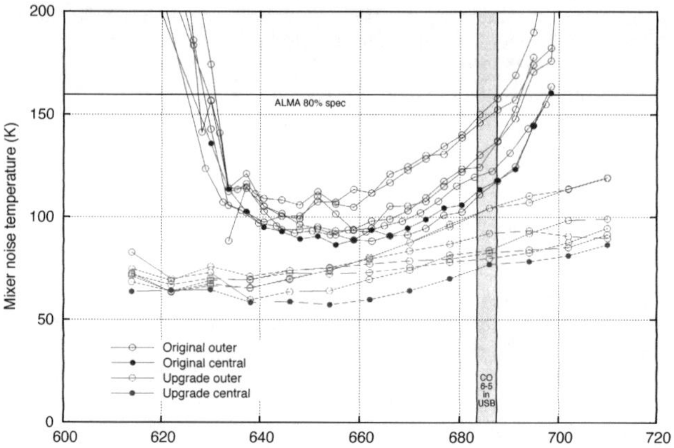

Fig. 1 DSB noise temperature of the new CHAMP+ low band mixers (red curves) compare to old ones (blue curves). Data is corrected for beam splitter insertion losses (5%). The curves with solid circles correspond to the central pixel of the array. The noise temperature in this figure is integrated over the entire 4–12 GHz IF band. Especially marked is the LO frequency range, where the CO j=6-5 line (one of the prime goals of the band) appears in upped sideband of 4-8 GHz IF band.

To describe other characteristics of the installed SIS mixers, we show as example the mixer installed in the central pixel. In Fig. 2 one can find the IV-characteristics, showing gap voltage of about 2.7 mV and normal resistance close to 10 Ohm. The IF power versus bias measurement, corresponding to the pumped IV-curve, is presented in Fig. 3. It includes classical hot/cold (300K/77K) curves and derived noise temperature dependence, showing a wide and stable area around 2 mV, which is fully suitable as mixer operating point.

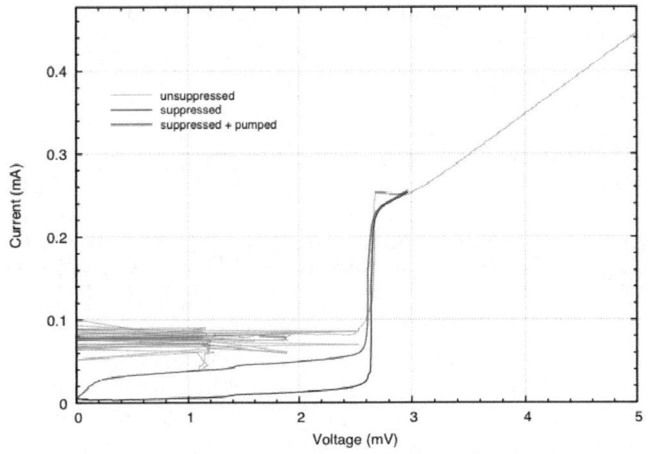

Fig. 2 IV-curves of the central pixel SIS mixer: unsuppressed (violet), with suppressed critical current (blue), with suppressed critical current and pumped by LO signal of 650 GHz (red).

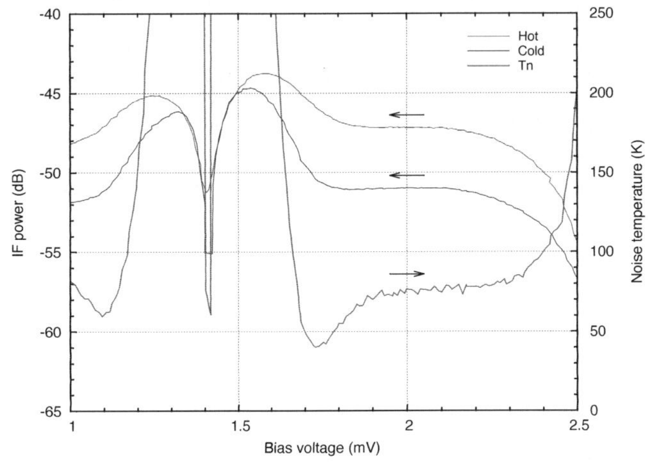

Fig. 3 Measured IF output power versus mixer bias voltage for hot (300K) and cold (77K) load – red and blue curves correspondingly. Using them, the noise temperature curve (green line) was calculated. Operating point is chosen close to 2 mV. The LO frequency is 650 GHz.

The flatness of the mixer sensitivity within the 4-12 GHz IF band is demonstrated by the curves in Fig. 4. Two LO points were chosen as an example.

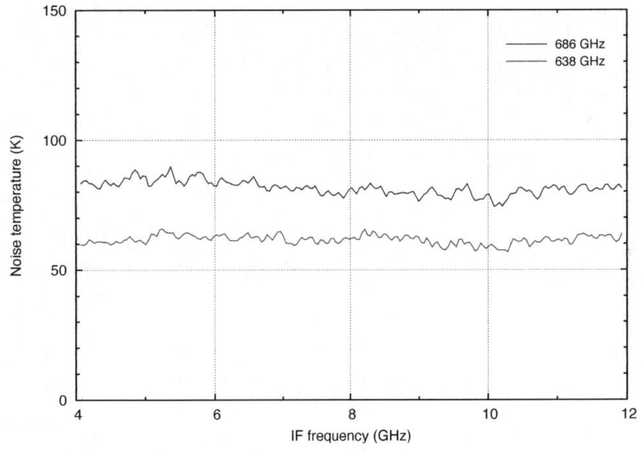

Fig. 4 IF noise temperature vs. frequency, measured for LO frequencies of 686 GHz (blue curve) and 638 GHz (red curve).

High Band (790-950 GHz) Mixers

For the original CHAMP+ high band mixers, a Nb/AlO/Nb SIS technology was used, the same as for the low band ones. However, the high band mixers were based on twin SIS junctions, in contrast to the single junction design for the low band, and the embedded circuit of the junctions consisted of a microstrip line with a NbTiN bottom and aluminum top wire [1], similar to the HIFI band 3 mixers [5]. A one-off design was made without further iterations. Because of this, the mixers performance was not optimized, leaving room for improvement. The corrected noise temperature of the original mixers, measured before the installation in the instrument, is shown in Fig. 5 by blue curves. The data is not covering the full band due to limited range of the lab LO available at the time (a BWO source). This performance was confirmed by a recent lab-measurement of the original central-pixel mixer after it was removed form the telescope (green curve). Roughly, the mixers noise temperature varies form 300..400 K at low frequencies to twice this number at the high end.

To upgrade the high band receivers, we have developed an SIS mixer based on high critical current density Nb/AlN/NbN tunnel junctions incorporated in a microstrip line consisting of a 300 nm thick bottom electrode made of NbTiN and a 500 nm thick top electrode made of Al. The microstrip electrodes are separated by a 250 nm SiO_2 isolator. The Nb layer of the SIS junction is deposited on the NbTiN film, while the NbN layer is contacting the Al top electrode. The nominal critical temperature of NbTiN film is about 14.5 K. More details of the design and fabrication were presented earlier in our paper [6]. This technology allows to reach a junction current density of about 30 kA/cm^2 keeping the quality (ratio of sub-gap resistance to a normal one) as high as 25-30. Using a twin junction design, we have reached a wideband response providing a good noise temperature in the required frequency range (see Fig. 5 red curves).

From the data on the Fig. 5 one can conclude that the upgrade of the mixers should improve CHAMP+ high band sensitivity by a factor of 1.3-1.5, roughly doubling the mapping speed.

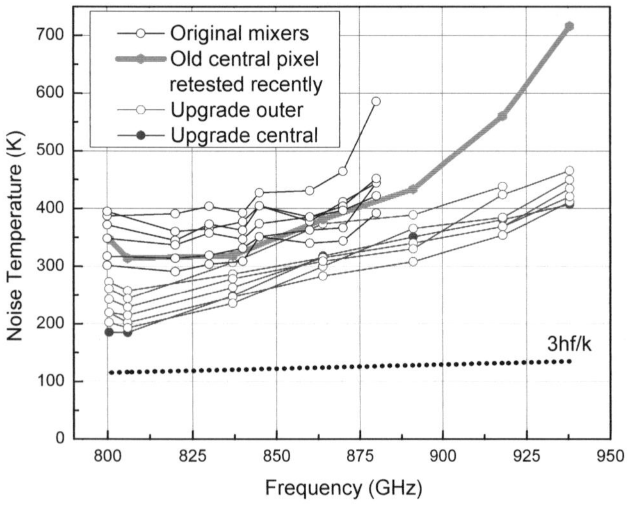

Fig. 5 DSB mixer noise temperature for the entire 4-12 GHz IF band vs. LO frequency. The noise temperature is corrected for about 10% insertion loss of a beamsplitter and for the hot LO (300K) contribution. The old mixers performances (blue curves), including additional recent verification for the central pixel (green curve), are compared here to results for the new mixers (red curves). For all the upgrade mixers the second minimum of the critical current suppression was used.

In addition to the integrated IF data in Fig. 5, we have measured the IF spectra for all LO points (see Fig. 6). Unfortunately, for some frequencies we had strong LO noise contribution, deteriorating IF sensitivity mainly around 4-6 GHz and of course influencing the results on Fig. 5.

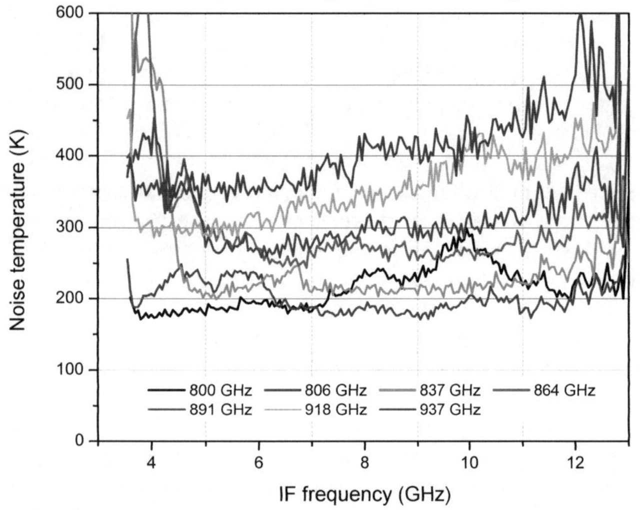

Fig. 6 Corrected noise temperature over the IF band for different LO points. Peaks are caused by LO noise and resonances in the testing horn. The presented lab data corresponds to one of the upgrade mixers installed later in an outer pixel of the array.

To estimate the mixers performance without this extra noise contribution we have plotted the best IF noise temperature within the 4-12 GHz band versus LO frequency (see Fig. 7). As expected, it is lower than the integrated results, but the sequence of curves does not change – the mixer with the lowest noise temperature in Fig. 5 stays the best in Fig. 7 as well. About the data in Fig. 7 one can say that it shows a performance for typical spectral lines observations.

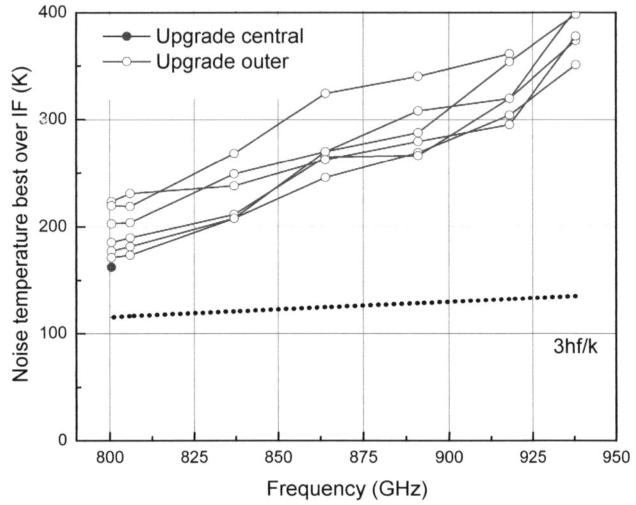

Fig. 7 Noise temperature at the best point in the IF band vs. for different LO frequencies. The data is corrected for the beam splitter as on Fig. 5. For the central pixel (solid dot) only the lowest frequency point was measured.

On Fig. 8 is shown an IV-curve of the central pixel mixer. One can see 1 Ohm serial resistance in a contact introduced by a silver epoxy paste and clearly exposed by unsuppressed IV-curve critical current. The pumped IV-curve shown by red curve corresponds to an optimal LO power and has a pumping level of about 16% (ratio of photon-assisted current to a gap current). The related Y-factor measurement is depicted by a dotted curve on the same figure, demonstrating a wide possible biasing range.

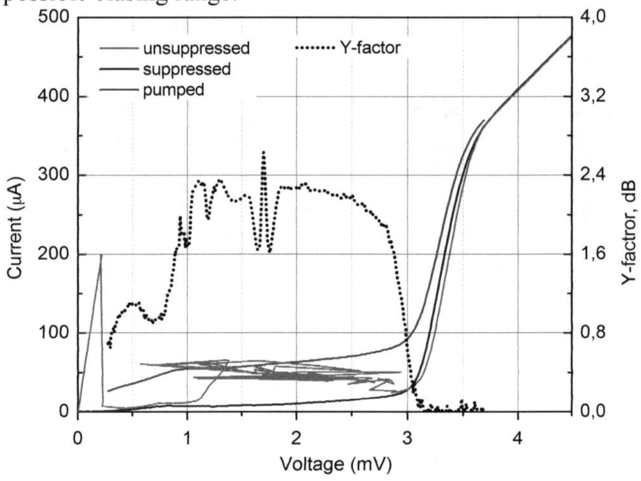

Fig. 8 IV-curves of one of the high band SIS junctions: autonomous unsuppressed (violet line), with suppressed critical current (blue line), pumped by LO signal providing an optimal noise temperature (red curve). The uncorrected Y-factor measured using 77 K/300 K loads is shown by dotted curve. The LO frequency is 800 GHz.

An essential advantage of the new SIS mixers is the higher gap voltage compared to the old ones with Nb electrodes: 3.15 mV against 2.8 mV. This plays a significant role for our band, because a frequency of 950 GHz gives a photon step of 3.9 mV [7], which exceeds the gap voltage. As a result, the voltage range available for SIS mixer biasing is wider by about 0.7 mV for the upgrade mixers, compared to the original ones, which is a big advantage for the mixers operation due to the presence of problematic Shapiro feature right in the middle of the photon step (see Fig. 8).

Conclusions

Upgrading the SIS mixers for the CHAMP+ instrument improved the high band sensitivity by about a factor of 1.4 and the low band sensitivity from 30% in the middle of the band to a few times at the edges. Moreover, the low band frequency coverage was significantly extended.

The low band mixers we can characterize as demonstrating state of the art performance. The high band ones can still be improved, as demonstrated by other groups [8][9].

Acknowledgment

We would like to recognize the efforts of the group led by T.M. Klapwijk from Delft University of Technology, fabricated the low band devices during the ALMA Band 9 production. Also, we thank J. Adema and A. Koops for their organizational and management support.

This work was supported by AETHER of RadioNet under FP7, Netherlands Research School for Astronomy (NOVA) NOVA-IV ALMA R&D grant, the grant No. 8168.2016.2 within the State Program for Support of Leading Scientific Schools and the Russian Foundation for Basic Research Grant No. 17-52-12051

References

[1] C. Kasemann, R. Güsten, S. Heyminck, B. Klein, T. Klein, S. D. Philipp, A. Korn, G. Schneider, A. Henseler, A. Baryshev, T. M. Klapwijk, "CHAMP+: a powerful array receiver for APEX", *Proc. of SPIE*, vol. 6275, 62750N, 2006.

[2] R. Güsten, R. Booth, C. Cesarsky, K. Menten, et al., "Ground-based and Airborne Telescopes", *Proc. of SPIE*, vol. 6267, 626714-1, 2006.

[3] A. M. Baryshev, R. Hesper, F. P. Mena, T. M. Klapwijk, T. A. van Kempen, M. R. Hogerheijde, B. D. Jackson, J. Adema, G. J. Gerlofsma, M. E. Bekema, J. Barkhof, L. H. R. de Haan-Stijkel, M. van den Bemt, A. Koops, K. Keizer, C. Pieters, J. Koops van het Jagt, H. H. A. Schaeffer, T. Zijlstra, M. Kroug, C. F. J. Lodewijk, K. Wielinga, W. Boland, M. W. M. de Graauw, E. F. van Dishoeck, H. Jager and W. Wild, "The ALMA Band 9 receiver. Design, construction, characterization, and first light", *A&A* 577, A129, 2015.

[4] T. Zijlstra, C. F. J. Lodewijk, N. Vercruyssen, F. D. Tichelaar, D. N. Loudkov, and T. M. Klapwijk, "Epitaxial aluminum nitride tunnel barriers grown by nitridation with a plasma source", 2007, Appl. Phys. Lett., 91, 233102.

[5] B. D. Jackson, G. de Lange, T. Zijlstra, M. Kroug, T. M. Klapwijk, J. A. Stern, "Niobium titanium nitride-based superconductor-insulator-superconductor mixers for low-noise terahertz receivers", *J. Appl. Phys.*, vol. 97, 113904, 2005.

[6] A. Khudchenko, A.M. Baryshev, K.I. Rudakov, P.M. Dmitriev, R. Hesper, L. de Jong, V.P. Koshelets, "High-Gap Nb-AlN-NbN SIS Junctions for Frequency Band 790–950 GHz," in IEEE Transactions on Terahertz Science and Technology, 2016. vol. 6. no. 1. pp. 127-132.

[7] J. R. Tucker, M. J., Feldman, "Quantum detection at millimeter wavelength", *Rev. Mod. Phys.*, vol. 57, no. 4, pp. 1055-1113, 1985.

[8] Y. Uzawa, Y. Fujii, A. Gonzalez, K. Kaneko, M. Kroug, T. Kojima, A. Miyachi, K. Makise, S. Saito, H. Terai, and Z. Wang, "Tuning Circuit Material for Mass-Produced Terahertz SIS Receivers", *IEEE Trans. on Appl. Supercond.*, vol. 25, no. 3, 2401005, 2015

[9] G. de Lange, B. D. Jackson, M. Jochemsen, W. M. Laauwen, L. De Jong, M. Kroug, T. Zijlstra, T. M. Klapwijk, "Performance of the flight model HIFI band 3 and 4 mixer units", *Proc. of SPIE*, vol. 6275, id. 627517, 2006.

The Advanced Microwave Radiometer – Climate Quality (AMR-C) Instrument for Sentinel-6

Jenna L. Kloosterman[1,*], Shannon Brown[1], Timothy Koch[1], Nho Vo[1], Fabien Nicaise[1], Erich Schlecht[1], Matthew Bloom[1], Lance Milligan[1], K. Scott Tripp[1], Pekka Kangaslahti[1], Isaac Ramos[1], Shannon Statham[1], Anders Skalare[1], Richard Cofield[1], Frank Maiwald[1]

[1]*Jet Propulsion Laboratory, California Institute of Technology, 91106 Pasadena, CA, USA*

*Contact: Jenna.L.Kloosterman@jpl.nasa.gov

Abstract— **The Advanced Microwave Radiometer – Climate Quality (AMR-C) is designed to measure the path delay due to atmospheric water vapor along Sentinel-6 altimeter path over one decade. The AMR-C receiver is based on heritage from previous AMR instruments with the addition of a THz-frequency radiometer, the High Resolution Microwave Radiometer (HRMR), for improved coastal zone accuracy and a Supplemental Calibration System (SCS) to meet level 3 requirements that the path delay error due to the altimeter-derived sea surface height be less at 0.8 cm and the path delay stability be maintained to 0.7 mm averaged over a 1-year time period.**

I. Introduction

The Sentinel-6 Mission will provide continuity to ocean topography measurements made from previous missions TOPEX-Poseidon (launched in 1992) [1], Jason-1 (2001) [2], OSTM/Jason-2 (2008) [3], and Jason-3 (2016) [4]. These measurements are important in determining ocean circulation, climate change, and sea level rise year-over-year. The results of these missions show a clear global mean sea rise of ~3+ mm/yr, shown in Figure 1 [5]. The Sentinel-6 mission consists of two satellites to be launched approximately 5 years apart. Each satellite is designed for a 5.5-year mission to extend measurements for at least another decade.

Sentinel-6 consists mainly of two instruments: an altimeter and a radiometer. These two instruments in combination will enable the mission to achieve its level 1 requirement to measure year-over-year global mean sea level stability to within 1 mm. These requirements flow to the payload level 3 requirements that the path delay error due to altimeter-derived sea surface height will be 0.8 cm. The level 3 requirements flow to the level 4 instrument requirements, which are derived from a global path delay retrieval algorithm. The level 4 instrument requirements are that microwave brightness temperature error will be less than 0.65 K at a 1 Hz sample rate and the radiometer brightness temperature will be stable to +/- 0.1K over a one year period.

The focus of this paper will be the radiometer instrument, the Advanced Microwave Radiometer – Climate Quality (AMR-C), which has completed the instrument preliminary design review (PDR) and is now finalizing the flight designs.

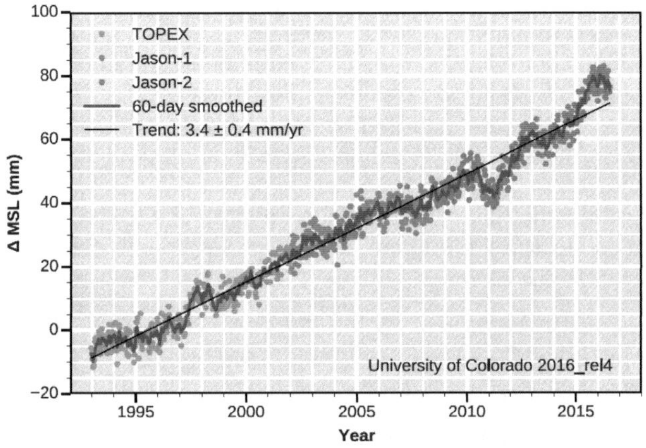

Figure 1. Global mean sea level rise from TOPEX, Jason-1, and Jason-2 after seasonal variations are removed [5]. The trend is a rise of ~3 mm/yr.

II. Instrument Description

The AMR-C receiver is based on heritage from the previous missions with addition of a High Resolution Microwave Radiometer (HRMR) [6] and a Supplemental Calibration System (SCS). The radiometer channels at 18.7 GHz, 23.8 GHz, and 34.0 GHz are inherited from previous AMRs and constitute the radio frequency subassembly (RFA). The 18.7 GHz channel estimates ocean surface components in observed brightness temperature, the 23.8 GHz channel estimates water vapor, and the 34.0 GHz channel estimates cloud liquid. HRMR consists of bands at 90 GHz, 130 GHz, and 168 GHz. The SCS is an additional calibration system in order to meet the level 3 payload requirement of long term radiometric stability. In addition to the RFA, HRMR, and SCS subassemblies, the AMR-C instrument also contains a parabolic mirror in the Reflector Subassembly (RSA), and the Electronics Unit (EU) in the Electronics Subassembly (ESA). An AMR-C instrument model with all five subassemblies is shown in Figure 2.

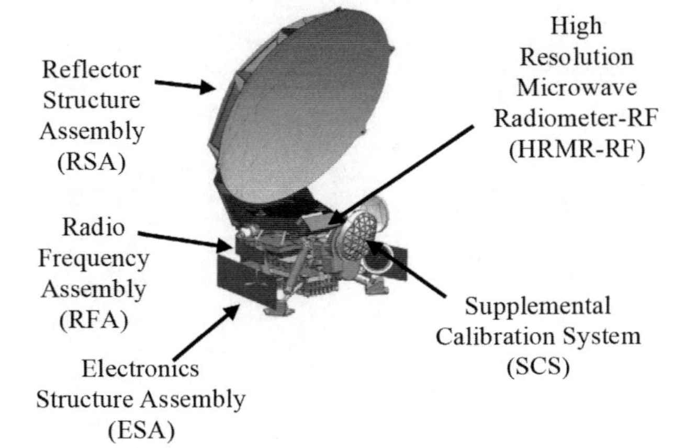

Figure 2. AMR-C instrument model with each of the major subsystems.

A block diagram of the AMR-C instrument is shown in Figure 3. HRMR sits at the focus of the primary reflector and the lower frequency channels in the RFA are offset. There are two identical lower frequency radiometer units in the AMR-C system, a nominal unit (H-polarization) and a redundant unit (V-polarization) shown in green. All three of these receivers have a separate EU containing the Power Converter Unit (PCU), a Data Acquisition and Control Unit (DAC), and a Housekeeping Unit (HKU). The DACs of the AMR-H and AMR-V units are crossed-strapped to the SCS shown in purple, which has fully redundant Control Mechanism Interface Electronics (CMIE) units, both of which can control either or both motors in the Standard Dual Drive Actuator (SDDA). Please note that cross-strapping in Figure 3 is only shown for the EU-H unit to reduce clutter in the figure. HRMR is in turquoise.

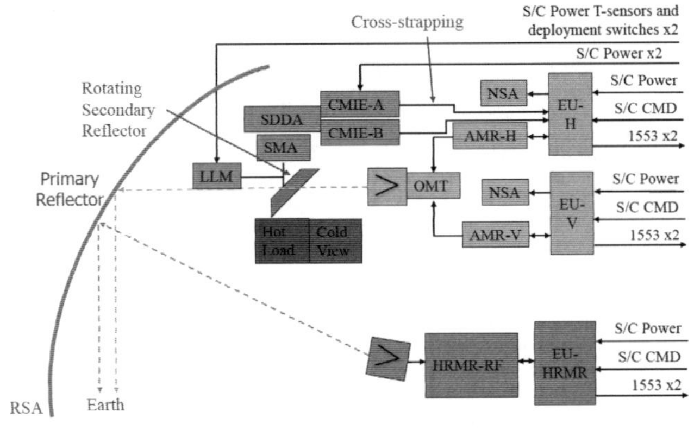

Figure 3. AMR-C block diagram. Each subsystem is color-coded (with its EU unit).

III. Instrument Design

A. AMR-H and AMR-V Receiver Design

Signal is relayed to the receiver through a circular feed horn. The signal is split by the Ortho-mode Transducer (OMT) into H and V units, nominal and redundant, respectively, although the polarization is arbitrary. The redundant unit will be used as a cold spare. From the OMT a diplexer divides the signal into 18/24 GHz and 34 GHz channels and the 18/24 GHz channel is then spilt into separate 18 and 24 GHz channels. A detector diode along with an ADC converts the signal to a digital signal, which is then relayed to the spacecraft and transmitted to the ground. A model of the receivers is shown in Figure 4 and an internal block diagram is shown in Figure 5. In operation, a Dicke switch at the receiver waveguide output toggles between the antenna signal and 50 Ω load for a differential measurement.

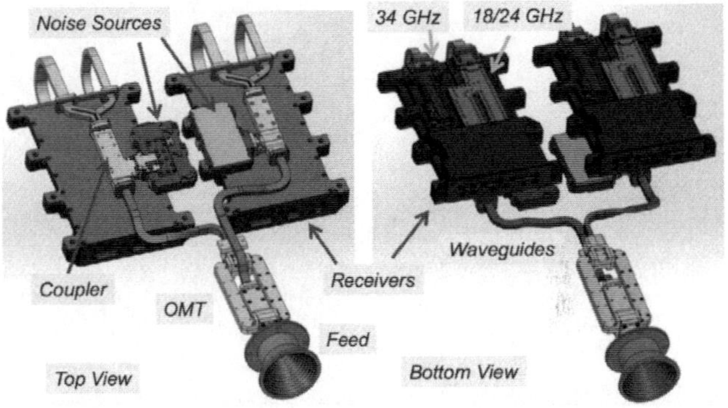

Figure 4. Top and bottom views of the AMR receivers for 18/24 and 34 GHz.

TABLE 1. LEVEL 6 AMR INSTRUMENT REQUIREMENTS.

Parameter	**Requirement**		
Input Return Loss (over channel passbands)	≥ 15 dB		
Dicke Switch Isolation	≥ 30 dB		
Channel Center Frequency	18.7 GHz	23.8 GHz	34.0 GHz
Center Frequency Tolerance	±50 MHz	±100 MHz	± 100 MHz
Center Frequency Knowledge	± 20 MHz	± 20 MHz	± 50 MHz
Channel Noise Bandwidth	200 MHz	400 MHz	700 MHz
Noise Bandwidth Tolerance	± 50 MHz	± 100 MHz	± 150 MHz
Passband Ripple	± 1 dB max	± 1 dB max	± 1 dB max
Stopband Rejection	> 50 dB	> 50 dB	> 50 dB
System Noise Figure	≤ 6.2 dB	≤ 6.5 dB	≤ 6.6 dB
System Gain/Temperature Coefficient	≤ 0.2 dB/℃	≤ 0.2 dB/℃	≤ 0.2 dB/℃
Post-detector Circuit Video (3 dB) Bandwidth	≥ 75 kHz	≥ 75 kHz	≥ 75 kHz
Backend Noise (relative to radiometric noise)	≤ 1/3	≤ 1/3	≤ 1/3
Input Dynamic Range	2.7 to 750 K		
Digitizer Sampling Rate	≥ 200 ksps		

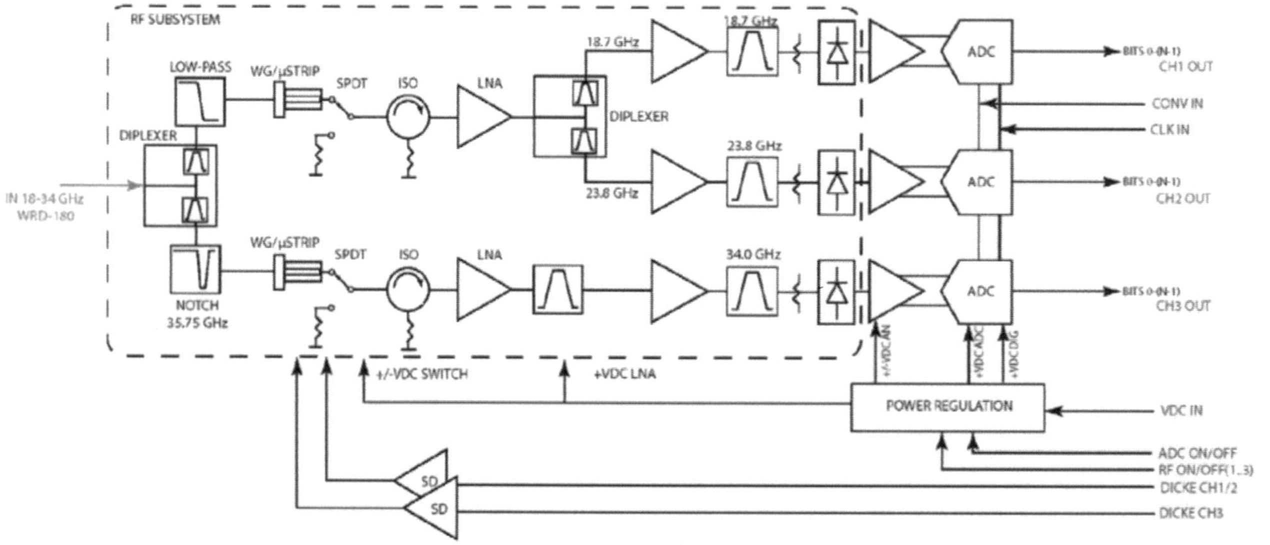

Figure 5. AMR receiver block diagram.

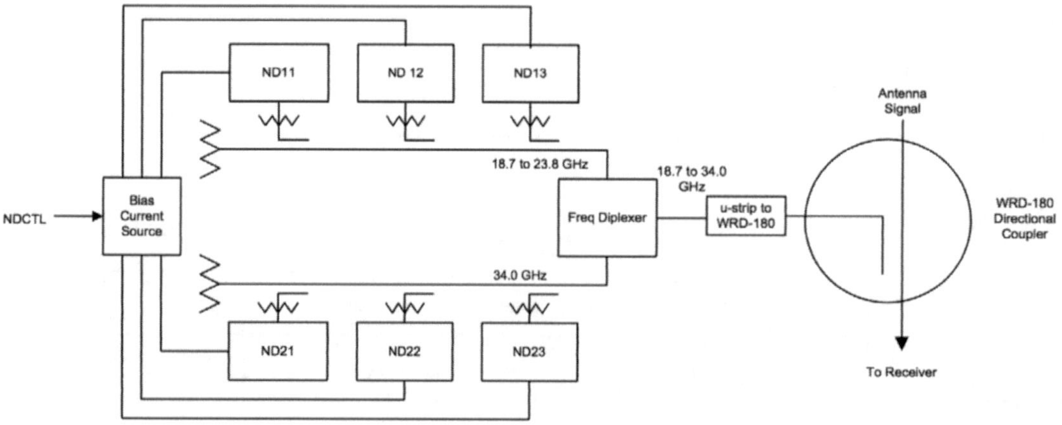

Figure 6. The AMR noise source block diagram.

A fully characterized noise source at the input of each receiver is used for internal gain stability calibration. Each noise source contains 3 sets of redundant diodes that can be used separately or together. A block diagram for the noise source is shown in Figure 6. The noise signals are coupled at the receiver input using a directional coupler.

The level 6 receiver requirements flow from the level 4 instrument requirements. These requirements are summarized in TABLE 1.

B. HRMR Receiver Design

Previous AMRs were limited to a 25 km diameter footprint on the ocean. In order to provide higher spatial resolution to improve the coastal zone measurement accuracy to a 3-5 km diameter footprint, a THz radiometer, HRMR, has been added to the AMR-C instrument. HRMR includes receiver bands at 90 GHz, 130 GHz, and 168 GHz and is based on radiometers designed for airborne and cubesat missions, the High-frequency Airborne Microwave and Millimeter-wave Radiometer (HAMMR) [7], and the Temporal Experiment for Storms and Tropical Systems (TEMPEST) [8], respectively. HRMR has been designed to attach to three mounting points at the focus of the RSA to minimize AMR beam blockage. The feedhorn and millimeter wave modules will be assembled and delivered on a radiatively-cooled plate, which will be enclosed for better thermal shielding.

HRMR will interface with EU hardware identical to the AMR units through its digitizer driver unit (DDU). This receiver utilizes low noise, high gain Indium Phosphide (InP) MMICs [9] to amplify incoming signal in order to detect it. Like the AMRs, HRMR signal is relayed through a feedhorn into diode detectors for each frequency. The calibration noise source is integrated in the multi-chip module (MCM). It has two noise diodes and directional couplers to provide stable calibration references. Additional calibration and stability is provided by the integrated Dicke switch that toggles between the antenna and reference load at 2 kHz rate to reduce NEDT, see Figure 13. A model of the HRMR receivers is shown in Figure 7 and design parameters are shown in TABLE 2.

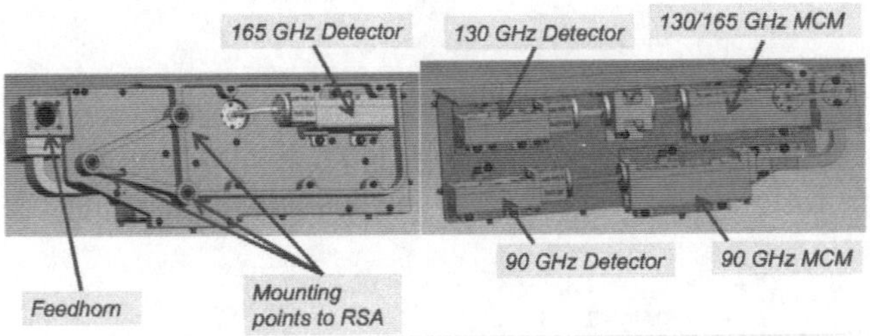

Figure 7. Top and bottom views of the HRMR receiver.

TABLE 2. HRMR RECEIVER DESIGN PARAMETERS

Parameter	**Requirement**		
Channel Center Frequency	90 GHz	130 GHz	168 GHz
Center Frequency Tolerance	±5 GHz	±5 GHz	± 5 GHz
Minimum Bandwidth	5 GHz	5 GHz	5 GHz
Noise Temperature	2000 K	2500 K	3500 K
Brightness Temperature Sensitivity	0.2 K	0.2 K	0.2 K
Deviation from White Noise Level Over 60 secs	< 0.2 K	< 0.2 K	< 0.2 K

C. The SCS

Due to long term fluctuations seen in the noise source from the Jason-3 mission [4] a Supplemental Calibration System (SCS) has been included on AMR-C. This subsystem is designed to turn the secondary mirror every 5-10 days so that the AMR receivers look at a warm load at ambient temperature (~200 K) and a cold load (cold sky, ~3 K), shown in Figure 8. As shown in Figure 3, the SCS only calibrates the AMR receivers, not HRMR, whose signal path is instead at the focus of the primary. These calibrations will be done over land in order to maximize observation times over the ocean.

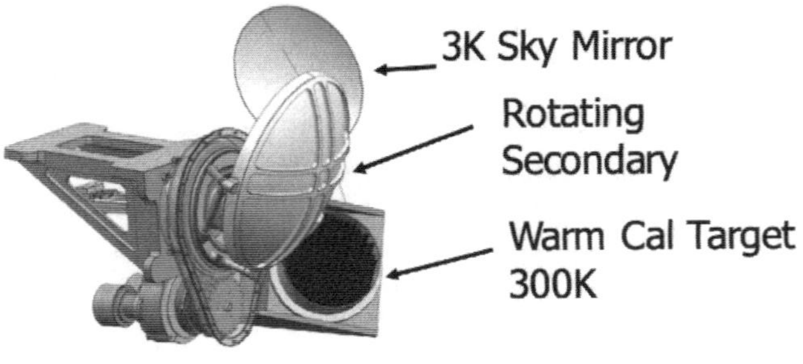

Figure 8. The SCS, which rotates a secondary mirror to look at ambient and cold calibration targets.

The SCS is driven by an SDDA motor, which is a block redundant, single fault tolerant mechanical/electronic assembly that provides a rotary output with fully characterized torque, speed, and current relationships. The gearbox couples dual spur gears for the first stage with dual harmonic gears in the final stage. The redundancy in the SDDA means that no single mechanism failure within the assembly will prevent the output from rotating. The SDDA power is supplied separately from the rest of the instrument. The mechanism control is cross-strapped to both the H and V flight computers. During launch the secondary mirror is held in place by the Launch Lock Mechanism (LLM).

IV. Initial Results

A. Thermal Modeling

The AMR-C instrument will have a PID-controlled thermal loop run by the spacecraft. The preliminary thermal design was simulated using a P-regulator and modeling shows that the receiver will meet its thermal requirements detailed below. The thermal analysis was done for three different cases: a hot winter, a hot summer, and a cold summer. Results are shown for several simulations lasting the duration of one orbit, which is 112 minutes long. Figure 9 models the AMR-H receiver thermal stability over one orbit showing that it can be kept to within ~0.04 °C/min. Similarly, Figure 10 shows the modeled thermal stability for HRMR. HRMR has no requirement, but the goal for this receiver is ≤ 0.1 °C of variation over an orbit. Peaks and minimums in these models are a result of the satellite's orbit as it transitions in and out of the sun. In Figure 11, models show the thermal variation within the AMR-H receiver will be ± 2.5 °C. Figure 12 shows the thermal variations between the feed horn assembly (FHA) and the AMR-H receiver. The requirement is these thermal variations not exceed 10 °C and models show that this difference is well within the model's margin.

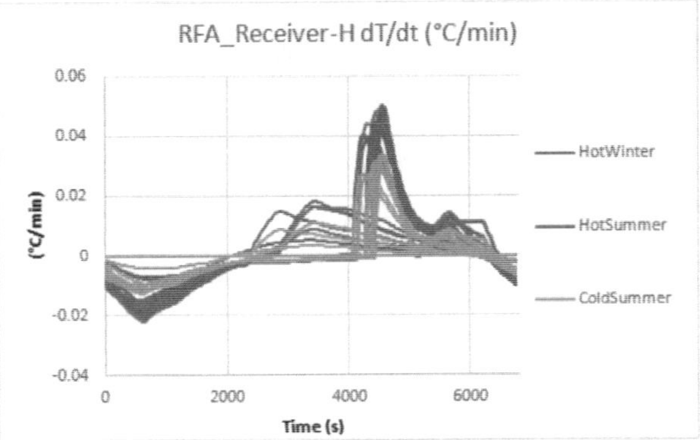

Figure 9. AMR-H receiver thermal stability can be kept to less than 0.04 °C/min during an orbit.

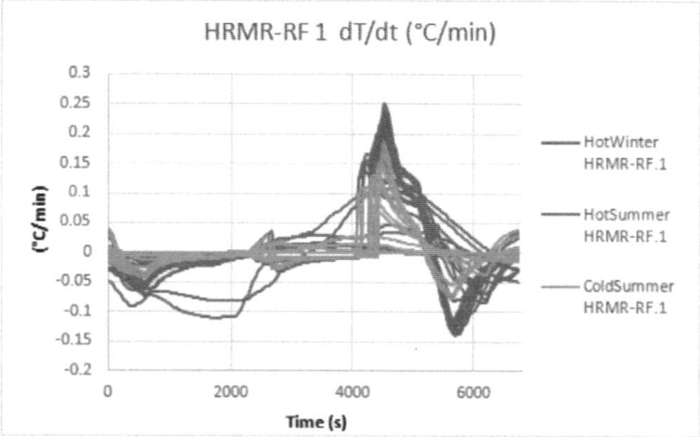

Figure 10. HRMR thermal stability models. The goal is ≤ 0.1 °C

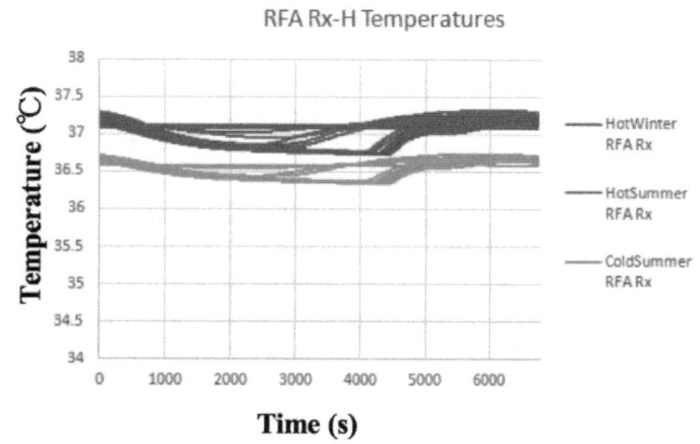

Figure 11. The AMR-H temperature range is ± 2.5 °C within the receiver.

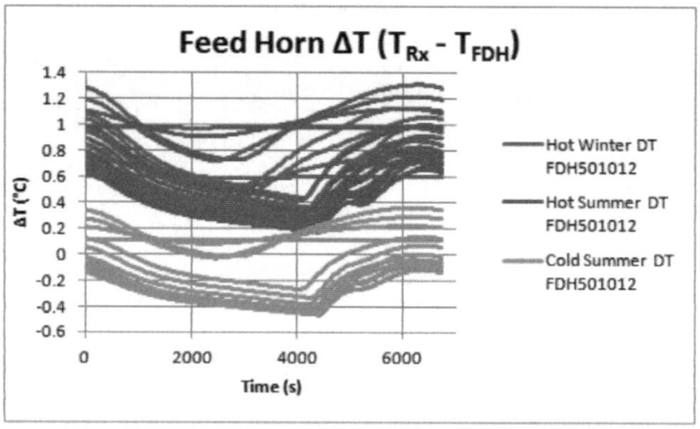

Figure 12. The thermal variations between the feed horn and the receiver over one orbit.

B. HRMR Prototype

The HRMR 90 GHz prototype's measured noise temperature is ~500 K. The noise equivalent differential measurement (NEDT) was measured for both 90 and 160 GHz. The NEDT is a measure of sensitivity that determines the threshold for the minimum differential temperature that the system can detect. This measurement is taken by looking at the difference between the receiver looking at a blackbody radiator and a 50 Ω reference load using a Dicke switch. The results of the NEDT measurements for 90 and 160 GHz prototype receivers are presented in Figure 13. The NEDT at 90 GHz is in green and the NEDT at 160 GHz is in blue. At the Dicke switch frequency of 2 kHz, the NEDTs ~0.1 K, which provides a 50% margin on the sensitivity requirement.

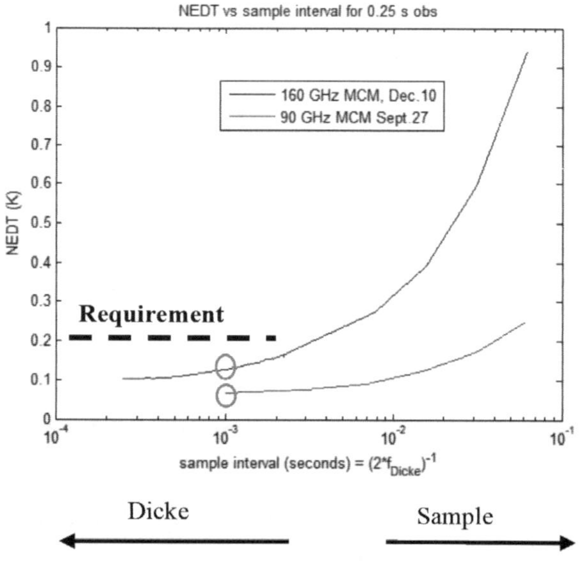

Figure 13. HRMR NEDT measurements for prototype HRMR receivers at 90 and 160 GHz.

Further measurements made on the prototype indicate that the power and mass are within the margins of their allotted budgets. These results are presented in TABLE 3.

TABLE 3. HRMR PROTOTYPE SPECIFICATIONS

Requirement	Prototype Measurement	Requirement	Margin
Power (W)	2.84	3.2	11%
Mass (kg)	1.98	2.2	10%
NEDT (K)	<0.1	0.2	50%
Deviation from white noise over 60 s [K]	0.05	<0.2	75%

V. Current Status and Future Work

The AMR-C team plans to deliver two flight instruments, one for each mission ~5 years apart. The instrument has passed the preliminary design review (PDR) and Phase C has begun. Hardware testing will begin in the summer of 2017 and the critical design review (CDR) will be in the fall of 2017. Instrument I&T for the first flight module will start in the Spring of 2018 for delivery to payload I&T in early 2019. Instrument I&T for the second flight module will begin in early 2019 after the delivery of the first flight module, and begin payload I&T in fall 2019. Sentinel-6 is expected to launch in 2020.

Acknowledgment

The research was carried out at the Jet Propulsion Laboratory, California Institute of Technology, under a contract with the National Aeronautics and Space Administration.

REFERENCES

[1] C. S. Ruf, S. J. Kienhm, and M. A. Janssen, "TOPEX/Poseidon Microwave Radiometer (TMR). I. Instrument description and antenna temperature calibration," in *IEEE Transactions on Geoscience and Remote sensing*, vol. 33, no. 1, pp. 125-137, Jan. 1995.

[2] Y. Ménard, L. L. Fu, P. Escudier, F. Parisot, J. Perbos, P. Vincent, S. Desai, B. Hanes, and G. Kunstmann, "The Jason-1 mission special issue: Jason-1 calibration/validation," in *Marine Geodesy,* vol. 26, no. 3-4, pp. 131-146, Jul. 2003.

[3] J. Lambin, R. Morrow, L. L. Fu, J. K. Willis, H. Bonekamp, J. Lillibridge, J. Perbos, G. Zaouche, P. Vaze, W. Bonnoura, F. Parisot, E. Thouvenot, S. Coutin-Faye, E. Lindstrom, and M. Mignogno, "The OSTM/Jason-2 mission," in *Marine Geodesy,* vol. 33, no. S1, pp. 4-25, Aug. 2010.

[4] F. Maiwald, O. Montes, S. Padmanabhan, D. Michaels, A. Kitiyakara, R. Jarnot, S. T. Brown, D. Dawson, A. Wu, W. Hatch, P. Stek, T. Gaier, "Reliable and Stable Radiometers for Jason-3," in *IEEE Journal of Selected Topics in Applied Earth Observations and Remote Sensing,* vol. 9, no. 6, pp. 2754-2762, June 2016.

[5] R. S. Nerem, D. P. Chambers, C. Choe, and G. T. Mitchumm, "Estimating mean sea level change from the TOPEX and Jason altimeter missions." in *Marine Geodesy,* vol. 33, no. S1, pp. 435-446, Aug. 2010.

[6] P. Kangaslahti, E. Schlecht, J. Jiang, W. R. Deal, A. Zamora, K. Leong, S. C. Reising, X. Bosch, and M. Ogut, "CubeSat scale receivers for measurement of ice in clouds," in *2016 14th Specialist Meeting on Microwave Radiometry and Remote Sensing of the Environment (MicroRad),* Espoo, 2016, pp. 42-47.

[7] S. Reising, P. Kangaslahti, A. B. Tanner, S. Padmanabhan, O. Montes, and C. Parashare, "Demonstration of High-frequency Airborne Microwave and Millimeter-Wave Radiometer (HAMMR) to Improve Spatial Resolution of Wet-Tropospheric Path Delay Measurements for Coastal and Inland Water Altimetry," in *NASA Earth Science Technology Forum 2014*.

[8] S. C. Reising, T. C. Gaier, C. D. Kummerow, V. Chandrasekar, S. T. Brown, S. Padmanabhan, B. H. Lim, S. C. van den Heveer, T. S. L'Ecuyer, C. S. Ruf, Z. S. Haddad, Z. J. Luo, S. J. Munchak, G. Berg, T. C. Koch, and S. A. Boukabara, "Overview of Temporal Experiment for Storms and Tropical Systems (TEMPEST) CubeSat constellation mission." In *2015 IEEE MTT-S International Microwave Symposium (IMS)*, 2015, pp. 1-4.

[9] X. Mei W. Yoshida, M. Lange, J. Lee, J. Zhou, P.-H. Liu, K. Leong, A. Zamora, J. Padilla, S. Sarkozy, R. Lai, and W. R. Deal, "First Demonstration of Amplification at 1 THz Using 25-nm InP High Electron Mobility Transistor Process," in *IEEE Electron Device Letters*, vol. 36, no. 4, pp. 327-329, April 2015.

Performance of a wide IF SIS-mixer-amplifier module for ALMA band 8 (385-500 GHz)

T. Kojima*[1], M. Kroug[1], K. Uemizu[1], Y. Niizeki[1], H. Takahashi[2], Y. Uzawa[3,1]
[1]National Astronomical Observatory of Japan, 181-8588, Mitaka, Japan
[2]The University of Electro-Communications, 182-8585, Chofu, Japan
[3]National Institute of Information and Communication Technology, 184-8795, Koganei, Japan
*Contact: t.kojima@nao.ac.jp

Abstract— This paper reports the receiver performance based on a low-noise heterodyne module with very wide intermediate frequency (IF) bandwidth in the radio frequency range 385–500 GHz. The module integrates a superconductor-insulator-superconductor (SIS) mixer with a 3–21 GHz low-noise preamplifier. We utilize high current density junctions for the SIS mixer to achieve good matching conditions between the SIS junction and the amplifier, and to maintain the IF performance over the designed local oscillator (LO) frequencies. The measurement results of the receiver using the heterodyne module demonstrate a typical noise temperature of 70-80 K over 3–18 GHz, at LO frequencies of 400–480 GHz.

INTRODUCTION

At NAOJ, we are carrying out a feasibility study for increasing the intermediate frequency (IF) bandwidth of SIS receivers at submillimeter wavelengths. In the Atacama large millimeter/submillimeter array (ALMA) project [1], enhanced instrument capability that would offer multiline observations without changing the local oscillator (LO) frequency, is one of the targets in future development programs [2], [3]. Most heterodyne receivers are equipped with a cryogenic isolator inserted between SIS mixer and low-noise amplifier. Using this configuration, excellent performance has been demonstrated but IF bandwidths are usually limited to 4-12 GHz. This is because for a passive component, e.g. the isolator, it is in general difficult to simultaneously obtain low reflection and insertion losses over a bandwidth exceeding one octave. In order to achieve a wider IF performance, a possible solution is to omit the isolator and to directly integrate SIS mixer and amplifier by making a short connection between them. However, there are several technical challenges such as impedance matching, thermal isolation and packaging.

SIS-MIXER-AMPLIFIER MODULE ASSEMBLY

Fig. 1 shows the mixer block and cryogenic low-noise amplifier. We used an SIS mixer chip similar to the one used for ALMA band-8 cartridges (radio frequency RF: 385-500 GHz) with only minor modifications in the tuning circuit to accommodate higher current density junctions with j_c=25-30 kA/cm^2 [4]. We note that the circuit was not optimized for wide IF bandwidth operation.

The 60 μm x 120 μm x 2.3 mm SIS mixer chip is mounted into the chip slot across a rectangular wave guide. Bonding wires are used to connect ground and IF pads to the microstrip line at the input of the cryogenic IF amplifier which itself is fixed by screws onto the mixer block, see Fig. 1. This configuration allows us to remove the amplifier and easily connect another component directly to the IF output of the SIS mixer. A superconducting magnet coil (NbTi) is placed behind the SIS mixer chip to suppress the Josephson current [5].

We used a 3–21 GHz cryogenic amplifier manufactured by Low Noise Factory with a typical noise temperature of 5 K and gain of 35 dB [6]. The amplifier input circuit incorporates a microstrip-based matching circuit for wideband noise and impedance matching to a source impedance of 50 Ω, with a typical return loss of −10 dB, and a bias-T circuit for 4-terminal sensing to apply bias voltages to the SIS junctions. A K-connector is implemented at the output of the module.

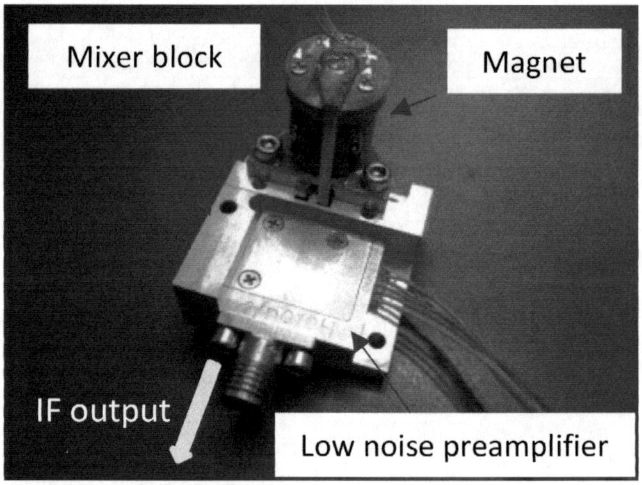

Fig. 1. Photograph of the SIS-mixer-preamplifier module. A drop-in type low-noise cryogenic amplifier is fixed by screws onto the mixer block. The RF signal and the LO power are injected through a waveguide from the back of the mixer block.

A FULL VERSION OF THIS PROCEEDINGS ARTICLE HAS BEEN SUBMITTED TO THE IEEE TRANSACTIONS OF TERAHERTZ SCIENCE AND TECHNOLOGY (TST) FOR PEER-REVIEW, PLEASE CHECK THE WEBSITE: HTTPS://WWW.MTT.ORG/TERAHERTZ

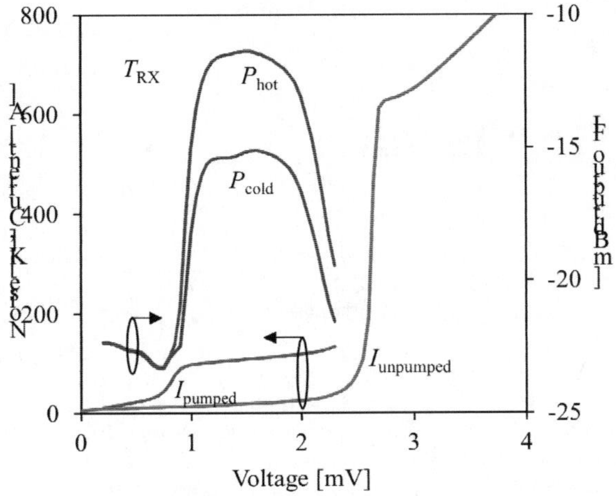

Fig. 2. Heterodyne response of the wide IF bandwidth receiver using the SIS-mixer amplifier module at an LO frequency of 440 GHz. The output power and double sideband (DSB) receiver noise temperature were measured with a power meter.

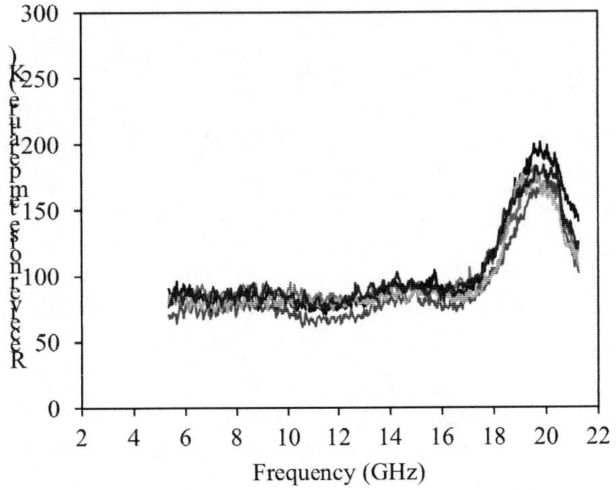

Fig. 3. DSB receiver noise temperature as a function of the IF at LO frequencies of 400 (green), 420 (black), 440 (red), 460 (yellow), and 480 GHz (blue).

Receiver Noise Temperature

The RF and IF receiver performances using the SIS-mixer-preamplifier module was evaluated with a measurement setup covering RF: 385–500 GHz and IF: 0.01-26.5 GHz. The receiver output power within the IF band, and power spectrum as a function of the IF were measured using a power meter and spectrum analyzer, respectively. The receiver noise temperature measurement was based on the standard *Y*-factor method with room temperature (295 K) and liquid-nitrogen-cooled (77 K) blackbody loads.

Fig. 2 shows the measured *I-V* characteristics of two parallel connected $Nb/AlN_x/Nb$ junctions, 0.8 μm in diameter, fabricated at the National Astronomical Observatory of Japan [7]. The critical current density of the SIS junctions was ~45 kA/cm^2. Even though this is considerably higher than the target value, the RF matching circuit still yields reasonably good coupling to the junctions. Fig. 2 also shows the heterodyne response of the wide IF receiver measured at an LO frequency of 440 GHz. The IF responses to hot and cold loads showed a *Y*-factor of 3.8 dB at a bias voltage of 1.4 mV, which corresponds to a double sideband (DSB) receiver noise temperature of 74 K. The dynamic resistance estimated from the LO-pumped *I-V* characteristic was 55 Ω at the chosen bias point, which is close to a 50-Ω standard impedance.

Fig. 3 summarizes measurement results of the DSB receiver noise temperature as a function of the IF at LO frequencies of 400–480 GHz. The measured noise temperature was typically 70-80 K and below 100 K, over 3–18 GHz. The variations are almost independent of the LO frequency. The noise temperature performance over 3–18 GHz is comparable to mass-produced ALMA band-8 receivers with an IF of 4–8 GHz [8]. However, it can be seen that the receiver performance deteriorates above IF 18 GHz at all LO frequencies. This is attributed to the IF matching conditions between SIS mixer and amplifier.

Detailed Characterization

In order to understand the degradation above 18 GHz and to further improve the performance, the module has to be characterized in detail. However, it is not straightforward to estimate how the components behave when they are integrated. The design of our module permits separate characterization of its constituents, and we have studied SIS mixer and IF amplifier using *S*–parameter measurements at microwave frequencies. The results and analysis have been included in an extended paper submitted to IEEE Transactions on Terahertz Science and Technology.

Acknowledgment

This work was supported in part by JSPS KAKENHI Grant Numbers 15K18057 and 25108005.

References

[1] ALMA Observatory. [Online]. Available: http://www.almaobservatory.org/
[2] S. Iguchi and D. Iono, "What are scientifically valuable developments for ALMA enhancement?" *Proc. SPIE* vol. 9145, Ground-based and Airborne Telescopes V, 914520, July 2014.
[3] D. Iono, S. Iguchi, and S. Asayama, "East Asian ALMA Future Development," in *Proc. 2016 URSI Asia-Pacific Radio Science Conference*, Seoul, Korea, August 2016.
[4] W. Shan, T. Noguchi, S. Shi, and Y. Sekimoto, "Design and development of SIS mixers for ALMA band 8," *IEEE Trans. Appl. Supercond.*, vol. 15, no. 2, pp. 503–506, June 2005.
[5] Y. Fujii, T. Kojima, A. Gonzalez, S. Asayama, M. Kroug, K. Kaneko, H. Ogawa, and Y. Uzawa, 2017 *Supercond. Sci. Technol.* 30 024001
[6] Low Noise Factory AB, Göteborg, Sweden. [Online]. Available: http://www.lownoisefactory.com/
[7] M. Kroug, S. Ezaki, T. Kojima, T. Noguchi "High Current Density SIS Junctions for Sub-mm Wave Mixers," presented in *Applied superconductivity Conference*, Denver, Colorado, USA, Sep. 2016
[8] T. Tamura *et al.*, "Performance and Uniformity of Mass-Produced SIS Mixers for ALMA Band 8 Receiver Cartridges," *IEEE Trans. Appl. Supercond.*, vol. 25, no. 3, pp. 1–5, June 2015.

A FULL VERSION OF THIS PROCEEDINGS ARTICLE HAS BEEN SUBMITTED TO THE IEEE TRANSACTIONS OF TERAHERTZ SCIENCE AND TECHNOLOGY (TST) FOR PEER-REVIEW, PLEASE CHECK THE WEBSITE: HTTPS://WWW.MTT.ORG/TERAHERTZ

Material Study for a THz SIS Mixer

M. Kroug[1,*], T. Kojima[1], K. Makise[2], Y. Uzawa[2]

[1]*National Astronomical Observatory, 181-8588, Mitaka, Japan*
[2]*National Institute of Information and Communication Technology, 184-8795, Koganei, Japan*
[*]Contact: matthias.kroug@nao.ac.jp

Abstract**—**The highest frequency reported for an SIS based mixer is 1.25 THz, as demonstrated with the Herrschel Band 5 receiver (Karpov, 2008). At NAOJ, we are in the early stage of developing the technology for a wave guide coupled SIS mixer that can be operated at frequencies up to 1.5 THz. Our target is to integrate high-gap junctions, e.g. $V_{gap} > 4$ mV, with low loss micro strip lines made from the layer stack Nb(epitaxial) / SiO_2 / Al which is in part similar to above mentioned Herrschel mixer. As for high-gap junctions, with high current density and low leakage, the most advanced technology is based on NbN(epi)/MgO/NbN tri-layers and has been reported by Dmitriev (ISSTT 2012). One of the challenges is that epitaxial quality Nb and NbN films can be grown only on certain substrates, e.g. sapphire for Nb and MgO for NbN, both of them difficult to thin down to ~10 μm thickness which is required when mounting devices into a THz wave guide. Recently, we started experimenting with Si/SiC as a substrate where the ~1 μm SiC buffer layer is grown by chemical vapor deposition. Nb films deposited by magnetron sputtering and at elevated substrate temperatures of 800 °C show electron diffraction patterns suggesting single crystal quality. We will present results on Nb film characterization through X-ray spectroscopy and electrical measurements and discuss the prospect of combining these Nb films with all-NbN SIS junctions.

A Terahertz Time-Domain Reflectometer

Bram Lap*, Andrey Khudchenko, Ronald Hesper, Andrey Baryshev,
Valery Koshelets, Pavel Dmitriev

Abstract— We have built a Terahertz Time-Domain Reflectometer based on a Michelson interferometer, where the device under test replaces the stationary mirror. This way, the reflectometer is sensitive to both the phase and amplitude properties of the device under test. Measuring in the time domain also makes it straight-forward to separate different contributions in the optical path, and for the same reason facilitates calibration.

This setup is built to characterize different loads at 4 K in the terahertz frequency range. This is important, since it was recently found that 2SB SIS receivers are extremely sensitive to standing waves in RF waveguide structure [1]. Part of which is caused by reflections from the loads. Understanding the behaviour of the loads is a prerequisite for optimizing the performance of the receivers.

In an additional investigation, the reflectometer is used to measure the RF properties of superconducting NbTiN films, particularly, the gap frequency, which is crucial for the development of SIS receivers at frequencies around 1THz.

The the latest results of these measurements will be presented at the conference.

[1] R. Hesper, A. Khudchenko, A.M. Baryshev, J. Barkhof and F.P. Mena, *A new high-performance sideband-separating mixer for 650GHz,* Proc. SPIE 9914, 99140G-99140G-11 (2016)

Feasibility Studies on Photon Counting Terahertz Interferometry

H. Matsuo[1*], H. Ezawa[1], H. Kiuchi[1], M. Honma[1], Y. Murata[2]

[1]*National Astronomical Observatory of Japan, Mitaka, Tokyo 181-8588, Japan*
[2]*Institute of Space and Astronautical Science, Sagmihara, Kanagawa 252-5210, Japan*
[*]Contact: h.matsuo@nao.ac.jp

Abstract—Feasibility studies on photon counting terahertz interferometry (PCTI) is presented for the future space-borne interferometer, including astronomical objectives, requirements on telescopes, photon counting detectors, array configuration, and imaging simulations.

In the past symposiums, we have discussed the intensity interferometer can be a powerful tool for imaging, and an experimental demonstration using Nobeyema Radioheliograph was presented. In terahertz frequency region, photon counting detector is useful for intensity interferometers, hence we named PCTI. In this symposium, we address some of the design issues and technical challenges, and discuss on feasibility of future space-borne interferometers.

From astronomical point of view, terahertz frequency region is important for observing exo-planet formation, star-formation in our galaxy and external galaxies, and active galactic nuclei. All of these observations require high angular resolution to identify the central activities of these sources. Within existing technologies, PCTI can provide both high angular resolution and high sensitivity, because photon counting detectors on each telescope can record photon signal referring to a precision onboard clock, and realize long baseline interferometry. Appropriate arrangement of satellite orbit and precision measurement of relative position of telescopes realize aperture synthesis imaging. Under low background observing condition from space, cryogenic telescopes with moderate diameter (< 1 m) will provide enough sensitivity for many far-infrared sources catalogued by IRAS and AKARI satellites. With maximum baseline of 20 km at 3 THz, angular resolution of 1 milli-arcsecond will resolve proto-planetary disks, active galactic nuclei and star-forming regions in external galaxies. Formation flight of multiple telescopes is one of the key technology and precision measurement of telescope position and time is needed, some of these are common to VLBI and gravitational wave telescopes in space.

The concept of PCTI is based on the delay time measurement using terahertz photon bunches and complex visibilities are defined by intensity cross correlations only. Since the intensity cross correlation drops more rapidly than amplitude correlation, telescope arrangements should be optimized according to the source structure, in other words the spatial filter of PCTI have relatively narrow bandwidth. Also, a large number of photon is needed to measure the delay time with enough accuracy, and a bright calibration source is needed within the field of view. Some of the imaging simulations will be presented and discussed.

Photon counting detectors are being developed using SIS photon detectors targeting NEP of less than 10^{-18} W/Hz$^{0.5}$ with low leakage junction of less than 1 pA. Assuming a post detection bandwidth of 1 GHz, astronomical terahertz photon at a rate of 100 M photons/sec can be countable. The photon counting capability of detectors are to be demonstrated.

Dielectric deposition for tuning the frequency of THz quantum cascade lasers

B. Mirzaei[1,*], D. J. Hayton[2], D. Thoen[1], J. R. Gao[1,2], T. Y. Kao[3], Q. Hu[3] and J. L. Reno[4]
[1]*Kavli Institute of NanoScience, Delft University of Technology, Delft, the Netherlands*
[2]*SRON Netherlands Institute for Space Research, Groningen/Utrecht, the Netherlands*
[3]*Massachusetts Institute of Technology(MIT), Cambridge, Massachusetts, USA*
[4]*Center for Integrated Nanotechnologies, Sandia National Laboratories, Albuquerque, USA*
*Contact: b.mirzaei@tudelft.nl

Abstract— We report an extensive study of the effect of an additional dielectric layer on the frequency of terahertz quantum cascade lasers (QCLs). QCLs with third-order distributed feedback structure at frequencies of 3.5 and 4.7 THz are used in our experiment. The dielectric layer applied is either Silicon-dioxide (SiO_2) or Polymethylmethacrylaat (PMMA). We find that both dielectric layers can down shift the lasing frequency up to 6GHz on a 3.5THz QCL, and 13GHz for a 4.7THz QCL. Full 3D FEM simulations suggest that the effect is dominated by the effective thickness of the dielectric on the vertical walls of the laser structure, and also confirm that for a given dielectric layer the effect is stronger in the 4.7THz QCL due to its larger out-distribution of electric-magnetic field. The knowledge provides guideline to shift the frequency of an existing QCL used as a local oscillator in practical applications.

INTRODUCTION

3'rd order DFB THz QCLs [1], which can be operated above 50K using a compact, low power stirling cooler, with a single mode emission in combination with reasonable output power (~0.5mW) and single spot far-field beam pattern, are good candidates for high resolution spectroscopy, especially in an astronomic instrument. However, due to the limited accuracy of the lithography, the limited electric tuning range, and the limited bandwidth of a practical mixer, additional fine frequency tuning possibility using an external dielectric layer would be extremely useful.

In this work we extensively study the effect of the dielectric layers of SiO2 and PMMA (both materials having reasonable refractive indices and low absorption coefficients) on the shift of the lasing frequency of a working QCL. The advantage of the latter is its reversibility, namely the dielectric is removable. We identify the importance of the thickness of the dielectric layer on the walls of the laser mesa-structure. Furthermore, we observe stronger effect in a 4.7THz than what in a 3.5THz QCL, which is supported by a 3D simulation.

QCLS AND DIELECTRIC LAYERS USED

We use 3'rd order DFB THz QCLs at two frequencies of 3.5 and 4.7THz. Since the resonance frequency is dependent on the effective index, it can be lowered by adding a dielectric layer around the laser where the electric-magnetic field exists. This is the essence of the technique studied in this work. The out extending electric field is shown in Fig.1, which is different for each QCL. For the 4.7THz QCL, it is more localized along the period edges, easy to be contacted by the additional dielectric layer whilst for 3.5THz QCL, it is more concentrated inside and around the air gaps, somewhere harder to be reached, suggesting a more effective frequency shift for 4.7THz QCL.

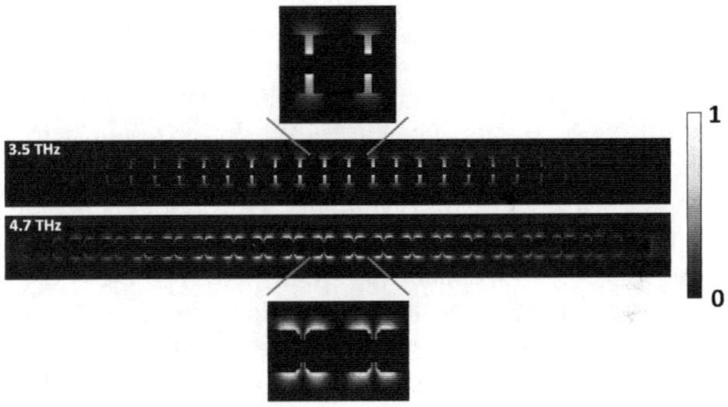

Fig. 2 Top view of the normalized magnitude of the extending electric field to the free space of two QCLs at 3.5 THz and 4.7 THz used for our experiments, showing different field distributions.

MEASUREMENT AND SIMULATION RESULTS

We measure the laser spectrum by a Fourier Transform Spectrometer (FTS) with a resolution of ~0.6GHz, and find the central frequency by fitting the data (the inset of the Fig.2.a). Fig.2.a shows the measured data of the 3.5THz QCL with the thicknesses of sputtered SiO2, varying between 0, 60, 240 and 480nm, corresponding to frequency down shifts of 2, 3 and 3.9GHz, respectively. We observe almost no effect on the output power.

Since we only measure the thickness on the flat surfaces of top and substrate (surface layer thickness), not on the side walls, we find the latter (lateral layer thickness) by 3D simulation using COMSOL 5.1.The results are plotted in Fig.2.b, where we get the lateral layer thicknesses of ~2, 5, and 20 nm for the measured surface layer thicknesses of 60,

240, 480nm respectively. We find that the shift is mainly due to the lateral layer thickness whilst the effect of the surface layer thickness in the same scale is so smaller. The Fig.2.b shows that the shift has almost a linear dependence to the lateral layer thickness with a rate of ~40MHz/nm for a fixed surface layer thickness.

We apply PMMA on the 3.5THz QCL in the mentioned way and assume that it sits on the structure uniformly since the liquid covers the entire laser and evaporates slowly. We observe a frequency down shift of ~6.3GHz corresponding to 340nm of the layer thickness found by simulation and a linear relation of the frequency shift to the PMMA thickness by a rate of ~10MHz/nm. The smaller rate than what is found in the SiO2, is due to the lower refractive index of PMMA (~1.54) [2] comparing to SiO2 (~2) [3] and also its higher absorption coefficient that reduces the influence on the neff. The latter is also responsible for a drop of ~18% in the output power.

Now we describe the results of the same experiments of 4.7 THz QCLs, which are motivated directly by our instrument applications (such as NASA balloon borne telescope STO2) due to existence of an astronomic important oxygen [OI] line at 4.745THz. By sputtering of 410nm of SiO2 on the surface we find a down shift of ~13.3 GHz, nearly 3 times larger than the case of 3.5THz QCL with a comparable thickness. A drop in power of ~20% is observed which is due to the higher absorption in 4.7THz and also stronger interaction of the field with SiO2. We couldn't find the lateral layer thickness based on simulation, since it shows a larger shift than the experiment even with zero lateral layer thickness. We attribute it to the problematic surface condition of this laser which may partly absorb the EM field on one hand, and have a poor adhesion to SiO2 or PMMA on the other hand. The simulation results of the frequency shift dependence of the 4.7THz QCL on the lateral layer thickness of SiO2 and uniform layer thickness of PMMA, show much larger effect, by rates of ~200 and ~88 MHz/nm respectively, which actually confirms the intuitional estimation of a stronger effect on 4.7THz laser due to its different out extended field.

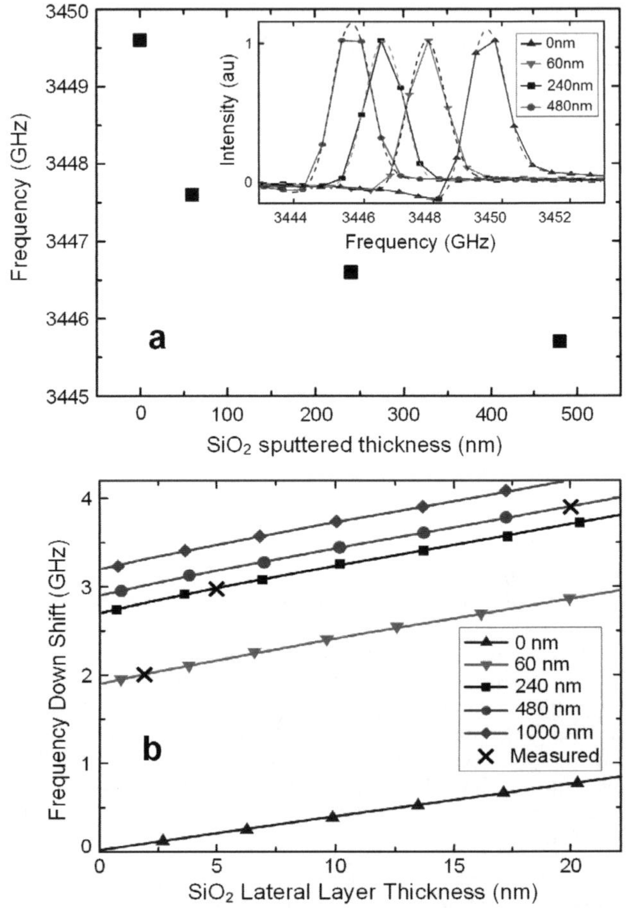

Fig. 2 a) The measured frequencies of the 3.5THz QCL before and after of the SiO2 sputtering in thicknesses of 60, 240 and 480nm. Inset shows the raw FTS spectrums (solid lines) together with the fitted curves (dashed lines). b) The simulated dependence of the frequency downshift on the lateral layer thickness, in five surface layer thicknesses of 0, 60, 240, 480 and 1000nm. The crosses show the experimental data points.

REFERENCES

[1] M. I. Amanti, G. Scalari, F. Castellano, M. Beck, and J. Faist, "Low divergence Terahertz photonic-wire laser," *Opt. Express* 18, 6390, 2010.

[2] R. Kitamura, L. Pilon and M. Jonasz, "Optical constants of silica glass from extreme ultraviolet to far infrared at near room temperature," *Appl. Opt.* 46, 8118, 2007.

[3] P. D. Cunningham et al, "Broadband terahertz characterization of the refractive index and absorption of some important polymeric and organic electro-optic materials", *J of Appl. Phys.* 109, 043505, 2011.

4 and 8-pixel THz Fourier phase gratings

B. Mirzaei[1,*], J. R. G. Silva[2], Y. C. Luo[1], X. X. Liu[3], L. Wei[4], D. J. Hayton[2], J. R. Gao[1,2], and C. Groppi[5]
[1]*Kavli Institute of NanoScience, Delft University of Technology, Delft, the Netherlands*
[2]*SRON Netherlands Institute for Space Research, Groningen/Utrecht, the Netherlands*
[3]*Microelectronics Department, Delft University of Technology, Delft, the Netherlands*
[4]*Imaging Physics Department, Delft University of Technology, Delft, the Netherlands*
[5]*School of Earth and Space Exploration, Arizona State University, AZ, USA*
[*]Contact: b.mirzaei@tudelft.nl

Abstract— Modelling, manufacturing and characterization of two 4 and 8-pixel Fourier phase gratings operated at 1.4 THz are reported, mainly applicable as local oscillator multiplexers for heterodyne receivers. Comparing the measurements with full 3D simulations shows good agreements and provides good understanding. Power efficiency of around 70% is experimentally derived for both gratings. We demonstrate the application of both, as multiple beam local oscillators to simultaneously pump (or operate) an array of 4-pixel superconducting mixers.

INTRODUCTION

Since terahertz heterodyne receivers are approaching quantum-noise limited sensitivity [1], to further improve spatial observing efficiency, multi-pixel heterodyne receivers [2] are necessary, which improve the mapping speed of the telescope significantly. Considering the challenges in providing suitable THz sources and complexities to synchronize the phase of many individual sources, generating multiple beams from a single source e.g. by a Fourier phase grating is more favorable.

Besides this primarily application in THz frequencies, phase gratings can be implemented in various fields, including imaging, coherent communication and optical computing.

Although several milestones have been reached towards developing THz phase gratings, little work can be found, beyond 1 THz.

In this work we present two phase gratings of 4 and 8-pixels, designed to operate at 1.4 THz by employing 13 Fourier coefficients. The designed surface structures, shown in Fig. 1(a,b), were manufactured on aluminum plates as shown in Fig. 1(c), by a computer numerical control (CNC) micro-milling machine.

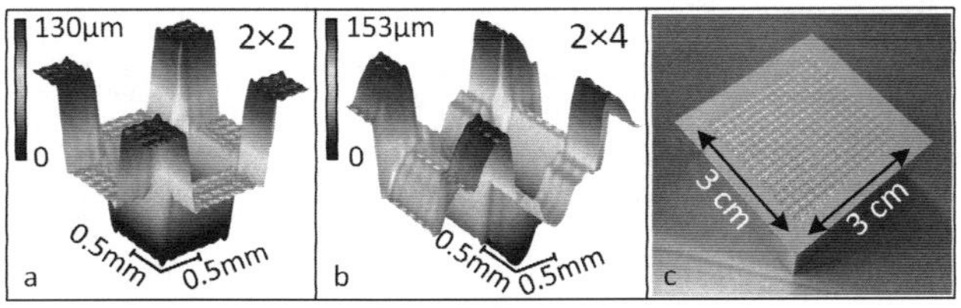

Fig. 1 3D surface profiles of unit cells of 4 (a) and 8-pixel (b) gratings with exaggerated z axis and a photograph of the manufactured 8-pixel grating (c).

MEASUREMENTS AND SIMULATION RESULTS

For grating's characterization, a high power 1.4 THz FIR gas laser in a proper setup is used. Despite small differences between the designed and fabricated gratings, we measured power efficiencies of both gratings to be around 70%, which is in a good agreement with the simulated values based on the ideal design. Fig. 2 shows the 3D simulated diffraction orders of the 8-pixel grating (a), and the good agreement between the simulated and measured diffracted beam pattern size and spatial distribution (b).

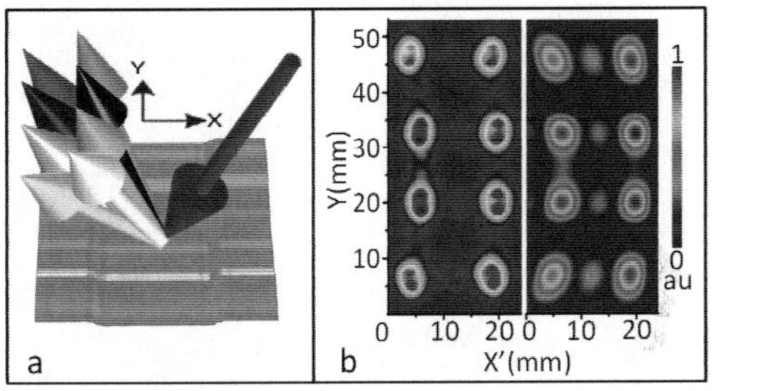

Fig. 2 a) Arrow schematics of the desired diffraction orders of the 8-pixel grating, calculated by 3D simulations, b) measured (left) and simulated (right) diffraction beam pattern of the 8-pixel grating.

Motivated by such results, we evaluate the real manufactured structure using full 3D simulation in COMSOL, where we find for the first time that, even with non-ideal machining, one can reach nearly the theoretically predicted efficiency and beam pattern. Moreover we find that by decreasing the number of Fourier coefficients from 13 to 6 in design, the efficiency drops about 4% while the minimum radius of curvature of the surface gets 4 times larger which ease the manufacturing by the same scale. These findings have a high impact on the manufacturability of a grating, making its fabrication less critical.

We also study the effect of the incident angle on the grating performance, where we find that by changing from 15° to 30° in four steps, the power distribution among diffraction orders becomes less uniform whilst the power efficiency decreases negligibly by less than 2%.

ARRAY DEMONSTRATION

We implement both gratings in a heterodyne experiment, for which we have built a 2x2 HEB array, using antenna coupled, quasi-optical NbN HEB mixers. We apply the 4-pixel grating together with the 1.4 THz gas laser to optically pump the array. Since we have no 8-pixel mixer array available, we examine the 8-pixel grating using the same source by coupling three groups of four left, four centre or four right beams. For all these configurations, we succeeded in fully pumping all the mixers and bring their current-voltage (IV) curves from superconductive to resistive state. Fig. 3 shows the results for the 4-pixel grating.

Applying the isothermal technique [3] to estimate the beam power at the detector and having estimated all the losses in the optical path from the laser to the mixer array and also taking the efficiency of the grating into account, we end up with about 25 μW for the total required input power, and about 50 μW for an 8-pixel mixer array. This order of power is reachable at lower frequencies (< 2 THz) by multiplier based solid state sources [4] and at higher frequencies by quantum cascade lasers [5].

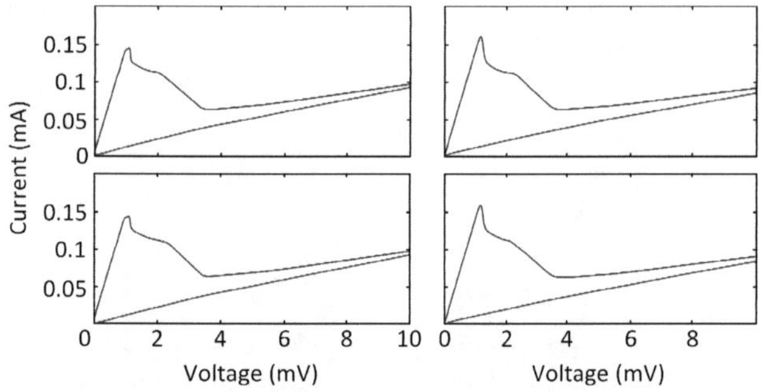

Fig. 3 I-V curves of the HEB array in superconductive (red) state with no incoming radiation, and resistive state (blue) when they are fully pumped by the grating diffracted beams.

This experiment establishes the fact that the grating is an extremely promising technology for generating a multiple beam local oscillator at the supra-THz, required for future astronomic observatories, such as NASA Galactic-extragalactic ultra-long-duration spectroscopic stratospheric observatory (GUSTO).

REFERENCES

[1] W. Zhang, P. Khosropanah, J. R. Gao, E. L. Kollberg, K. S. Yngvesson, T. Bansal, R. Barends, and T. M. Klapwijk, "Quantum noise in a terahertz hot electron bolometer mixer," *Appl. Phys. Lett.* 96(11), 111113, 2010.

[2] C. Walker, Terahertz Astronomy, CRC Press, Taylor & Francis Group, 2016.

[3] H. Ekström, B. S. Karasik, E. Kollberg and K. S. Yngvesson, "Conversion gain and noise of niobium superconducting hot-electron-mixers," *IEEE Trans Microw. Theory Tech.* 43(4), pp. 938–947, 1995.

[4] J. V. Siles, R. H. Lin, C. Lee, E. Schlecht, A. Maestrini, P. Bruneau, A. Peralta, J. Kloosterman, J. Kawamura and I. Mehdi, "Development of High-Power Multi-Pixel LO Sources at 1.47 THz and 1.9 THz for Astrophysics: Present and Future," in *Proc ISSTT.* 2015, pp. 40–42.

[5] B. S. Williams, "Terahertz quantum-cascade lasers," *Nat. Photon.* 1, pp. 517–525, 2007).

Design of an Optical Beam Combiner for Dual-Frequency Observation with ALMA

D. Montofré[1,2]*, A. Baryshev[1], R. Hesper[1] and F.P Mena[3]
[1]*Kapteyn Astronomical Institute, University of Groningen, Postbus 800, 9700 Av Groningen, The Netherlands*
[2]*Department of Electrical Engineering, Universidad de Chile, Av Tupper 2007, Santiago, Chile*
*Contact: montofre@astro.rug.nl

Abstract— The aim of this work is to improve high-frequency calibration data on long baseline observations for the ALMA antennas. A dual-frequency atmospheric phase error calibration method is proposed and will be implemented by the simultaneous observation in two ALMA Bands, specifically 6 and 9, coupled by means external optics in a few baselines. This method is envisioned to demonstrate the advantage of receiving signals simultaneously at different frequencies from the same point of the sky. It will permit an increase of accuracy in determining the phase correction needed to reduce the effects of the atmosphere and, therefore, it will enable higher resolutions when imaging at high frequencies using the ALMA interferometer. While maintaining the existing receiver optics, an optical layout that couples Bands 6 and 9 is proposed. Here we demonstrate that very limited impact on the existing ALMA system is needed. Furthermore, we present in detail the optical layout, made within the formalism of ray optics. The initial results demonstrate the feasibility of the proposed system.

INTRODUCTION

At the present, the Atacama Large Millimeter/Submillimeter Array (ALMA) [1] represents the largest astronomical project in existence, which is composed by 66 high precision antennas with advanced technology [2]. Despite being located in one of the best sites for performing mm and submm radio astronomy (Chajnantor plateau located at the Atacama Desert in northern Chile) [3], ALMA is not exempt from image degradation caused by atmospheric effects or instabilities in the instruments. To ensure that it operates at its full potential during the next years and decades it is necessary to maintain a continuous development program.

As a part of this program we propose to develop a dual-frequency atmospheric phase-error calibration method using ALMA Bands 6 (211-275 GHz) and 9 (602-720 GHz) by means of external optics on a few ALMA baselines employing existing ALMA receivers. This method is devised to demonstrate the advantages of receiving signal simultaneously at different frequencies from the same area of the sky. When implemented, it will permit an increase of accuracy in determining phase correction due to atmosphere, improve UV coverage, double instantaneous frequency coverage for line searches and improve cross calibration between different bands. This project involves designing an optical system for combining the beams of existing Bands 6 and 9, implementing limited modifications of signal transfer at few ALMA antennas and making dual-frequency observations with ALMA. Here we demonstrate that very limited impact on the existing ALMA system is needed and present the optical layout that it will be used.

In this work, we present the results obtained so far in the development of this dual band optical calibration device. We start with a general background about elliptical mirrors and how these can be used to reduce losses in the system. Then, the space constraints to mount the system are presented along with the proposed layout that allows splitting the beam and redirect it into Bands 6 and 9. Finally, the mechanical design of the proposed structure to make dual-frequency observation possible, using ALMA telescope, is presented.

THEORETICAL BACKGROUND

Elliptical mirrors are widely used in the high-frequency ALMA bands, because they allow the focal plane to fit the incident beam coming from the secondary reflector to their respective receiver. An illustration of a general two elliptical mirror set-up is shown in figure 1. When off-axis mirrors are used, it is unavoidable to introduce distortion to the beam. However, the parameters of this kind of system are chosen to reduce this effect. Among those parameters, the bending angle (α_1, α_2) must be kept as small as possible in order to guarantee minimal distortion losses (L_{dist}) and cross-polarization losses (L_{dist}). The total distortion loss (or polarization loss, since both formulas differ just by a factor 2) can be calculated subtracting the effect introduced by each mirror [4],

$$L_{dist} = \left| \frac{{w_1}^2 tan^2(\alpha_1)}{8\,{f_1}^2} - \frac{{w_2}^2 tan^2(\alpha_2)}{8\,{f_2}^2} \right| \quad (1)$$

where w corresponds to beam radius at the mirror surface, α to the bending angle and f to the focal distance of the mirror.

A FULL VERSION OF THIS PROCEEDINGS ARTICLE HAS BEEN SUBMITTED TO THE IEEE TRANSACTIONS OF TERAHERTZ SCIENCE AND TECHNOLOGY (TST) FOR PEER-REVIEW, PLEASE CHECK THE WEBSITE: HTTPS://WWW.MTT.ORG/TERAHERTZ

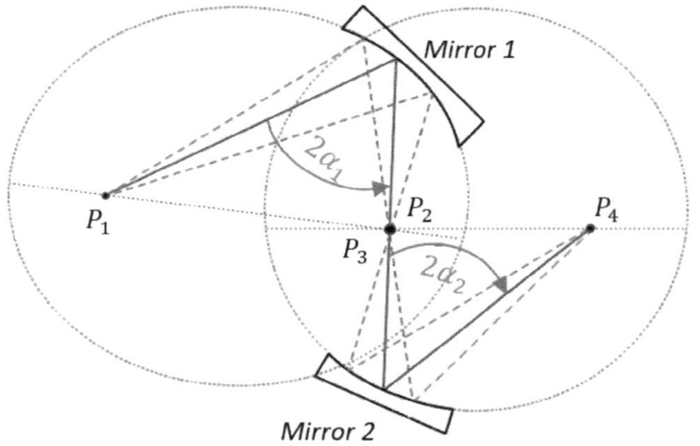

Figure.1 - Optical layout to minimize distortion using two elliptical mirrors. The main ray is represented by the blue solid line and the marginal rays are represented by the blue dashed lines. The point P_1, P_2, P_3, P_4 correspond to focal points of the ellipse defined by the mirror 1 and 2 respectively.

Symbol	Quantity	Value in (mm)
Φ	Maximum Radius	73
L	Maximum Length	188
H	Maximum Height	185

Table I Dimensions of the Safe Work Zone

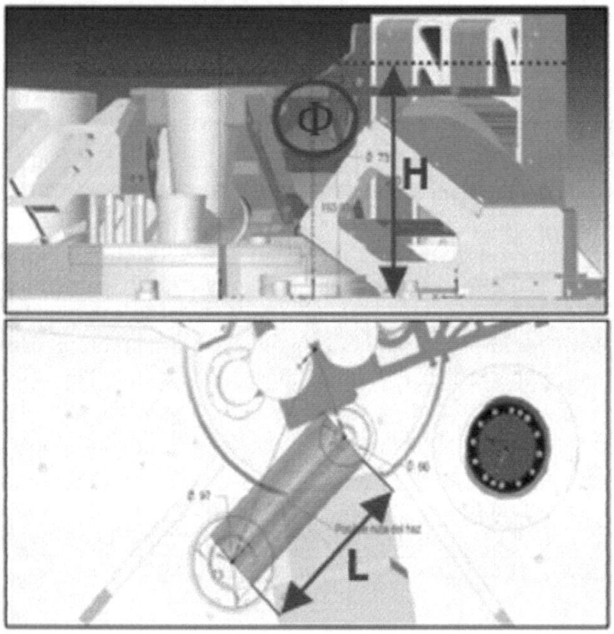

Fig. 2 The top and bottom pictures show, respectively, a side and top view of the Safe Work Zone.

SAFE WORK ZONE and OPTICAL DESIGN

As starting point, it is important to highlight that not all space inside the cryogenic chamber is available to be used in this optical system. There already exist many optical, electronic and calibration devices for every one of the bands. For this reason,
is important to define a Safe Work Zone (SWZ), which guarantees that the beam will not crash with any object in its path, producing undesirable effects such as diffraction or interference if appropriate measures are not taken. Figure 2 (top) shows the straight path starting from Band 6 and ending in 9. This figure shows that there is a bracket (part of external optics of Band 4) blocking the path. To avoid this obstacle and any other that could appear in the optical path of the beam, the first goal is to clearly identify the SWZ by means of inspecting the CAD file of the cryogenic chamber, which is available in the ALMA database. Figure 2 (bottom) shows a top view of the SWZ, whose dimensions are presented in table I.

OPTICAL DESIGN

Since the different bands in ALMA are spatially separated in the focal plane, as shown in figure 2, the beams of the two band must be combined optically in order to perform simultaneous dual-band observations. An optical first order system has been designed following a geometrical optics approximation and using Zemax software. Using geometrical optics rather than Quasioptics is acceptable, because in the near field approximation geometrical optics describes a beam which envelops the Gaussian beam. This condition guarantees that both beams will follow the same optical path. However, it is important to be aware that the diameter of each reflecting element in the system must be chosen following Quasioptical approach, i.e., using at least a diameter of $4w$, where w corresponds to beam radius at any element surface. This condition guarantees that 99.97 % of the power of the beam will be transmitted after each reflection.

Figure 3 shows the optical layout of the proposed beam-combining system. The concept is to use a dichroic beam splitter surface ("dichroic" in short) to separate the Band 6 and 9 beams. The beam of Band 6 is reflected and then redirected to its respective window by an optical re-imaging system, consisting of five mirrors, two elliptical and three flat. The incidence angle upon the dichroic filter is kept as low as possible (~17.5 degree) to decrease the insertion losses [5]. The idea of using two elliptical mirrors in this system is that the beam distortion (produced by each mirror) can be compensated by properly matching their foci [6]. Other advantage of using elliptical mirrors is that the beam shape could be easily matched, in order to change the beam features as little as possible [7]. The beam of Band 9 is transmitted through the dichroic without any major changes. One of the main reasons that we chose for this arrangement is that dichroic tend to work better when the high frequency is transmitted.

All the optical components must be properly aligned, both in lateral offset and angle, for achieving high levels of efficiency. However, this situation is not realistic and the misalignments present in the system should be corrected. A tolerance analysis (not presented here) has been performed and its objective is to know how sensitive the proposed system

A FULL VERSION OF THIS PROCEEDINGS ARTICLE HAS BEEN SUBMITTED TO THE IEEE TRANSACTIONS OF TERAHERTZ SCIENCE AND TECHNOLOGY (TST) FOR PEER-REVIEW, PLEASE CHECK THE WEBSITE: HTTPS://WWW.MTT.ORG/TERAHERTZ

is to any angle misalignment and how the focus matching between the bands 6 and 9 will change. This analysis shows that the best elements in the system to correct any misalignment are the flat mirror 1 and

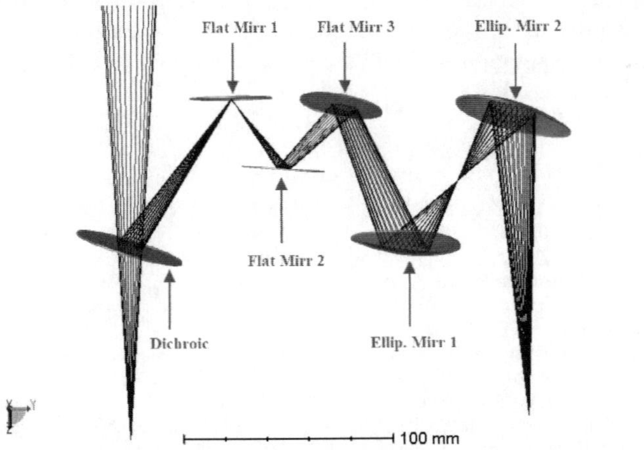

Fig. 3 Optical design of the beam combiner system made using Zemax software. The design has been made using a geometrical optical approximation by means of the Zemax sequential system option.

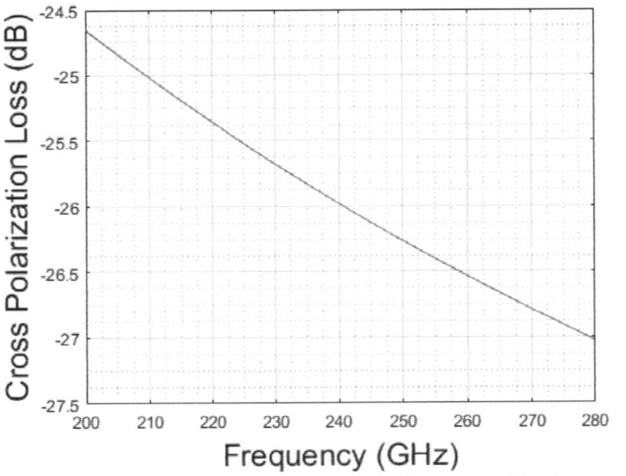

Fig. 4 Cross polarization loss as function of the frequency. The frequency range spans that cover by Band 6. The loss has a value of -26 dB at the central point.

the elliptical mirror 2, since they provide a maximally decoupled adjustment in either offset or angle. These mirrors will be equipped with micrometer screws for manual adjustment.

For the Band 9 beam, any deterioration is purely due to the dichroic. Typical transmission of a modern mesh-type dichroic is of the order of 92% [8], giving a noise temperature increase of about 25K for Band 9. The polarization purity is expected to be reduced no more than a few dB.

For Band 6, the situation is more complicated. Reflection of the dichroic, typically of the order of 96% [8], will increase the noise temperature by at most 12 K. We believe that subsequent loss in the beam can be quite low when high-quality mirrors are used in the re-imaging optics. Figure 4 shows the cross-polarization losses calculations as described by equation 7. The plot shows a cross-polar deterioration of about -26 dB at central frequency. For an eventual full deployment, this may be considered too high, but for a proof of concept demonstrator we deem it acceptable.

Element	Beam Radius (mm)	Element Diameter (mm)
Dichroic	10.03	55.0
Flat Mirror 1	8.65	40.0
Flat Mirror 2	8.52	40.0
Flat Mirror 3	8.91	40.0
Elliptical Mirror 1	10.79	52.0
Elliptical Mirror 2	11.66	55.0

Table II Beam radius at each element of the beam combiner system and actual size of each element. The beam radius has been calculated according to Quasioptical beam formula and the element diameter correspond to 4 times the beam radius plus an extra margin to compensate any possible misalignment.

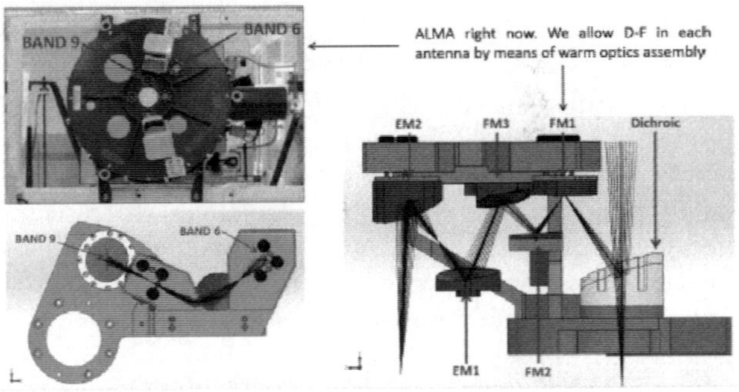

Fig. 5 Top view of ALMA cryostat (left top) showing the actual layout of the Bands. The left bottom image shows a top view of the support structure for the beam combiner assembly. A side view of the system is shown on the right side.

MECHANICAL DESIGN

In order to make dual-band observations possible, we propose to insert an optics assembly directly on top of the front-end cryostat, containing the dichroic and the re-imaging optics.

Figure 5 shows the current design of the support structure for this assembly. The dichroic will be located above the Band 9 vacuum window, allowing the beam to pass almost unmodified. On the other hand, the new Band 6 beam, now coaxial with the Band 9 beam, will be reflected by the dichroic and re-imaged by the set of mirrors into the original Band 6 focal point.

One of the most important requirement to accomplish by the whole support structure is that this should be linked to the cryostat without major modification, i.e., no drilling or using pieces that might endanger the optimal operation of other components previously installed. Furthermore, using any hole that might be used for the already set-up components is not allowed. This structure has been designed to operate exclusively in the 12- meter diameter antennas. Since the space restrictions in those with 7-meter diameter are higher [9], due to additional components aimed to compensate the difference in the optical path length of the beam for both antennas. As those components are not present in the 12-meter configuration, some extra holes (used to link those components) are available to be used. Those holes will be used to link the support structure to the cryostat. In order to fix the structure, the main support of the structure was designed thick enough to resist the weight of all the other components and with as little impact as possible on the alignment of the system due to either mechanical vibrations or elastic deformation of the parts.

Finally, an analysis involving a Quasioptical approach calculations, has been performed. This provides the real size that each element of the beam combiner system must have in order to reduce the spillover produced by the multiple reflections which the Band 6 beam is undergone along through its optical path. The values with the real diameter of each element are presented in table II. These values are minimal recommended to guarantee a negligible spillover of the beam.

CONCLUSIONS

An innovative and low distortion first order system, within the geometrical optics formalism, have been proposed. This system uses mirrors (two elliptical and three flat), which keeps the losses in a minimum level. Any misalignment presented in the system could be corrected, in principle, by tilting those mirrors equipped with micrometer screws. Additionally, a Quasi-Optical model has been used to corroborate the feasibility of this proposed design and calculate the suitable size of each system element.

This works demonstrate that the available space is enough to mount the support structure beam combiner system. This structure has been designed to do not interfere with other components of the cryostat and make its installation simple. Normal operation of the telescope will not be affected by this structure.

REFERENCES

[1] ALMA, Retrieved from http://www.almaobservatory.org. 2016.

[2] FEND-40.02.00-035-B-REP, ALMA *Front-end Optics Design Report*. 2007.

[3] R. Bustos, M. Rubio, A. Otárola, and N. Nagar, *"Parque Astronómico de Atacama: An Ideal Site for Millimeter, Submillimeter, and Mid-Infrared Astronomy,"*, vol. 126, p. 1126, Dec. 2014.

[4] A. Baryshev and W. Wild, ALMA *Band 9 Optical Layout*. ALMA Memo Series, ALMA, 9 2001.

[5] P. A. R. Ade, G. Pisano, C. Tucker, and S. Weaver, *"A review of metal mesh filters,"* in Society of Photo-Optical Instrumentation Engineers (SPIE) Conference Series, vol. 6275 of, p. 62750U, June 2006.

[6] J. A. Murphy, *"Distortion of a simple Gaussian beam on reflection from off-axis ellipsoid mirrors,"* Int. J. Infrared and Millimeter Waves, vol. 8, no. 9, pp. 1165–87, 1988.

[7] R. Ulrich, *"Far-infrared properties of metallic mesh and its complementary structure,"* Infrared Physics, vol. 7, pp. 37–55, Mar. 1967.

[8] Internal communication with Ken Wood, QMC company.

[9] ANTD-44.00.00.00-004-A-REP, ALMA Front End Optics Design Report for ACA 7-m Antenna. 2011.

A FULL VERSION OF THIS PROCEEDINGS ARTICLE HAS BEEN SUBMITTED TO THE IEEE TRANSACTIONS OF TERAHERTZ SCIENCE AND TECHNOLOGY (TST) FOR PEER-REVIEW, PLEASE CHECK THE WEBSITE: HTTPS://WWW.MTT.ORG/TERAHERTZ

As grown ultra-thin MgB_2 films for superconducting detectors

Evgenii Novoselov* and Sergey Cherednichenko

Chalmers University of Technology, SE-41296, Gothenburg, Sweden
*Contact: evgenii@chalmers.se

Abstract—For both hot-electron bolometers (HEBs) and superconducting single-photon detectors (SNSPDs) high quality thin superconducting films are of crucial importance. Using MgB_2 with a critical temperature (T_c) of 39K (vs 15-16K for NbN) much higher operation temperatures (15-30K) could be achieved. In this case compact cryocoolers could be used, hence extending missions life time significantly. Furthermore, utilization of MgB_2 for HEB mixer improves gain and noise bandwidths due to shorter electron-phonon interaction time and better acoustic matching to the substrate. Unfortunately, reduction of film thickness is usually followed by reduction of T_c (in 3-5nm NbN films T_c is 9-11K), while for HEBs and SNSPDs a combination of both a small thickness and a high T_c is desirable. Low film roughness and high homogeneity are other importance merits. A hybrid physical chemical vapour deposition (HPCVD) method has been reported to be much more efficient for high quality thin MgB_2 film depositions compared to e.g. molecular beam epitaxy (MBE), co-evaporation etc. It has been shown before than a gain bandwidth (GBW) of 6GHz can be achieved for HEBs made from 15-20nm MgB_2 films. Our goal is to develop a deposition method providing MgB_2 films thinner than 10nm and with a T_c >30K in order to extend GBW to >10GHz.

Here we present our recent results on ultra-thin MgB_2 film deposition using our (in-house built) HPCVD system. To study film properties on submicron level films were patterned in bridges with dimensions varying from $0.3\times0.3\mu m^2$ to $1\times1\mu m^2$. 20nm thick films had a room temperature resistivity ρ_{295K} of $50\mu\Omega\cdot cm$ ($13\mu\Omega\cdot cm$ for un-patterned films) with a T_c of 39K and a critical current density J_c (4.2K) up to $1.2\times10^8 A/cm^2$. A deposition rate of 0.8Å/s is much lower compared to previously reported values (vs 3Å/s). We obtained MgB_2 films as thin as 5-7nm with a T_c of 31-34K, a ρ_{295K} (in sub-micron scale bridges) of $\sim100\mu\Omega\cdot cm$, and a J_c of $(1\text{-}3)\times10^7 A/cm^2$. Using such films antenna integrated HEB mixers ($1\times1\mu m^2$) have been fabricated which showed both a low noise and a noise bandwidth >10GHz (see also "MgB_2 HEB mixer with an 11GHz bandwidth" on this conference).

InGaAs Schottky technology for THz mixers

D. Pardo[1*], H. Wang[1], H. Sanghera[2], B. Alderman[1&2], B. Ellison[1], P. Huggard[1], and J. Grajal[3]
[1]*STFC-Rutherford Appleton Laboratory, OX11 0QX, Oxfordshire, UK*
[2]*Teratech Components Ltd, Oxfordshire, UK*
[3]*Universidad Politécnica de Madrid, E.T.S.I. Telecomunicación, 28040, Madrid, Spain*
*Contact: diego.pardo@stfc.ac.uk

Abstract— This paper presents a theoretical analysis of the capabilities of InGaAs Schottky diodes as mixers in the THz band. InGaAs diodes are interesting because of their low barrier height compared with GaAs diodes, which yield a reduction in required LO power. In order to provide a reliable and accurate description of the electrical and noise performance of InGaAs mixers, a Monte Carlo model of the diode coupled to a multi-tone harmonic balance technique has been used in this work. Progress towards the development of THz InGaAs Schottky diode mixers at STFC-RAL Space is also presented.

I. INTRODUCTION

GaAs Schottky barrier diode frequency mixers are used in Earth observation and planetary science heterodyne receivers and wider applications such as security imaging and non-destructive testing of materials. Schottky mixers offer good sensitivity in the THz band with the important capability of operation at either room or cryogenic temperatures. An important drawback of these devices is the higher level of required local oscillator (LO) power when compared with superconducting mixer equivalents, an attribute that is exacerbated as the frequency of operation increases [1, 2]. As a means of LO power mitigation, Schottky mixers can be configured as sub-harmonically pumped (SHP) devices and for which the LO frequency is one half that of the signal to be detected. Additionally, DC biasing of the diode can be introduced, though this challenging to implement at frequencies above ~1THz. Despite these development strategies, mixer LO power generation remains a significant issue, particularly for spaceborne applications, with requirements typically in excess of a few mW above 1 THz.

Addressing the above limitation requires an examination of the properties of the Schottky barrier formed at the semiconductor-mental junction interface. By use of an alternative semiconductor material to GaAs, for example InGaAs, a reduction in the barrier height can be achieved and this, in turn, lowers the point at which the mixing action occurs, i.e. the LO power required to effectively pump the diode is reduced [3, 4]. This allows a simplification of the device embedding circuity since the InGaAs does not need to be biased and thereby assisting with higher frequency implementation. InGaAs SHP mixers at 183 GHz have been previously demonstrated and have achieved a conversion loss of 6.6 dB and noise temperature of 700 K with only 0.34 mW applied LO power. When compared with the equivalent state-of-the-art GaAs mixer performance of 5.7 dB and 450 K respectively [5], the InGaAs sensitivity is inferior. But, the low power required for the GaAs is almost an order of magnitude higher and places considerable demands on the LO generation scheme. The performance of InGaAs Schottky mixers at higher frequencies has not, however, been demonstrated and only numerical results based on simple lumped equivalent circuit models of the diode and the conversion matrix formalism are available [3, 4]. It is known that these simulation tools are limited in their capability to model the electrical and noise performance of the InGaAs diode at high frequencies.

This paper reports on progress towards the development of THz InGaAs Schottky diode mixers at the STFC-RAL Space Millimetre Wave Technology group (MMTG). Monte Carlo (MC) modelling of the diode together with a multi-tone harmonic balance (HB) technique have been used to provide an accurate and reliable description of the electrical and noise performances of InGaAs mixers above 1 THz. The DC characteristics of InGaAs diodes fabricated at the MMTG will be presented.

II. MODELLING OF INGAAS DEVICES

In order to develop THz InGaAs circuits, it is fundamental to model accurately the electrical and noise performances of the InGaAs diode at high frequencies. The design of THz Schottky circuits is usually carried out using simple lumped equivalent circuit models of the diode where the diode parameters are set based on experimental results. This modeling strategy is not available for InGaAs based circuits because of the lack of experimental knowledge of InGaAs diodes at the THz band [5, 6].

In this work, we have use a Monte Carlo model of the diode developed by the Tor Vergata University to describe the performance of the diode. The MC model provides a unified and self-consistent description of the electrical and the noise performances of the device without the necessity of any additional analytical or empirical model. This model provides a solution for the Boltzmann transport equation by simulating the trajectories of individual carriers as they move through a device under the influence of electric fields and random scattering forces [7]. Therefore, this technique provides an accurate description of physical phenomena in the device up to THz frequencies.

The MC model has been successfully used to model the performance of devices under high frequency conditions for several semiconductors, GaAs and InGaAs among them [7, 8, 9]. In order to simulate mixer circuits, the MC diode model has been coupled to a multi-tone harmonic balance (MCHB). This simulation tools allows evaluating the conversion loss and noise temperature of mixers as described in [9].

A. Electrical performance of InGaAs diodes

The selection of the *In* fraction in the InGaAs semiconductor has an important impact on the electrical properties of the semiconductor, and therefore, on the performance of InGaAs circuits. According to the experimental results in [10, 11], the low field electron mobility of InGaAs increases as the *In* fraction increases which contributes to reduce the series resistance of the InGaAs devices. On the other hand, the higher the *In* fraction, the lower the ideal barrier height ϕ_b of a Schottky contact on InGaAs. A reduction of ϕ_b will reduce the LO power required to pump an InGaAs Schottky mixer [4], but it will also contribute to increase the nonlinear capacitance of the diode. This increase of the capacitance will lead to higher shunting effect at THz frequencies. Also, the higher the ϕ_b, the higher the tunnel current through the barrier, which will increase the ideality factor of the diode [3, 4].

Despite the possibility of optimizing the *In* fraction in a InGaAs semiconductor for a particular application [4], we have selected an *In* fraction of 0.53 since the resulting lattice constant matches InP subtrate, simplifying the fabrication technology of $In_{0.53}Ga_{0.47}As$ devices [3]. From now on, every time we speak of InGaAs we will be referring to $In_{0.53}Ga_{0.47}As$.

Figs. 1 and 2 show a comparison of measurements [10, 11] and Monte Carlo simulations of the low field electron mobility and the velocity-field curves of bulk InGaAs (doping concentration $1x10^{14}$ cm^{-3}). The mobility of GaAs is included for comparison. The higher mobility of InGaAs than GaAs - Fig. 1- will lead to Schottky diodes with lower dc series resistance. In addition, InGaAs shows larger velocity saturation than GaAs, see Fig. 2, improving the performance of the diode under high field conditions.

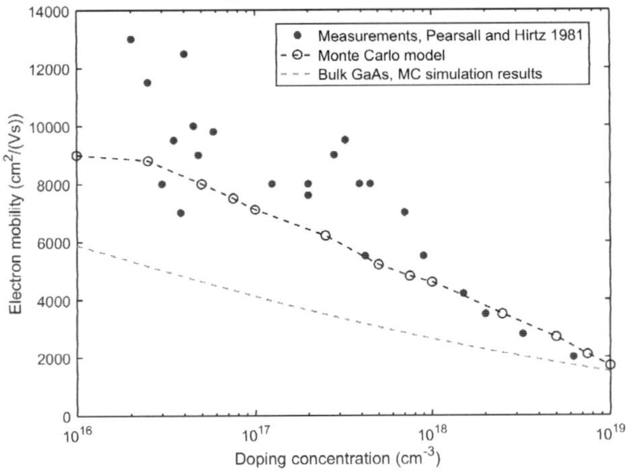

Fig. 1 Measurements [10] and MC simulations of the low field electron mobility of InGaAs as a function of the doping density.

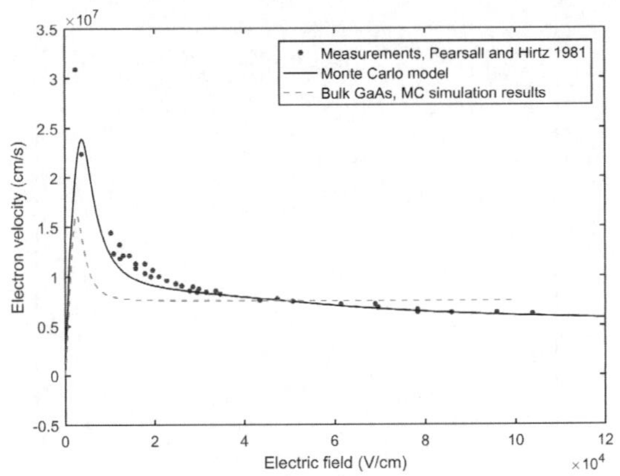

Fig. 2 Measurements [11] and MC simulations of velocity versus field curves for bulk InGaAs with doping concentration $1x10^{14}$ cm^{-3}.

III. ANALYSIS OF THZ INGAAS MIXERS

This section presents an analysis of the capabilities of InGaAs SHP mixers at THz frequencies in comparison with GaAs SHP mixers. The analysis is based on MCHB circuit simulations. The circuit impedances at the LO, RF and intermediate (IF) frequencies used in the simulations of the InGaAs mixers have been chosen to minimize the conversion loss at the LO power of minimum loss while for the GaAs mixers they have been set to the values provided in the literature. The evaluation of the equivalent input noise temperature of the mixers includes the contribution of the losses on the feed horn, filters and mixer waveguides, L_{rf}, as well as the resistive and mismatch losses in the IF matching circuit, L_{if}. Values of L_{rf} and L_{if} provided for GaAs Schottky mixers in the literature have been assumed in the simulations. In order to compare InGaAs and GaAs mixers, the diode parameters presented in the literature for GaAs mixer diodes have been also used for the InGaAs diodes. Table I shows the main circuit and diode parameters of the analyzed mixers.

TABLE I
DIODE AND CIRCUIT PARAMETERS OF THE SIMULATED INGAAS AND GAAS SHP MIXERS

	168 GHz SHP Mixer [12]		**360 GHz SHP mixer [13]**		**1.2 THz SHP mixer [1]**	
	GaAs	InGaAs	GaAs	InGaAs	GaAs	InGaAs
f_{RF} (GHz)	168		360		1230	
f_{LO} (GHz)	72		168		600	
V_{bias} (V)	0		0		0.5	0
Epi thick (nm)	75		85[a]		100[a]	
Epi doping (cm^{-3})	$2x10^{17}$		$2x10^{17}$		$5x10^{17}$ [a]	
Barrier height (eV)	0.89	0.21	0.85	0.21	0.8	0.21
Anode area (μm^2)	0.78		0.8		0.5	
$C_j(0)$ (*f*F)	1.28	2.57	1.3	2.6	1.21[a]	2.4
R_s (Ω)	21.4	13.6	18.0	12.4	20[a]	12.7
Z_{RF} (Ω)	60+j73	42+j36	83+j53	21+j21	27+j14[a]	4.5+j11.5
Z_{LO} (Ω)	95+j240	100+j75	147+j207	44+j37	38+j33[a]	7.5+j24
Z_{IF} (Ω)	100		100		50[a]	
L_{rf}, L_{if} (dB)	0.7, 1[a]		0.7, 1		0.7, 1.2[a]	

[a]Some data for the GaAs diodes are not available in the literature. In those cases, they have been optimized with MCHB or extrapolated from available data for similar mixers, see [9].

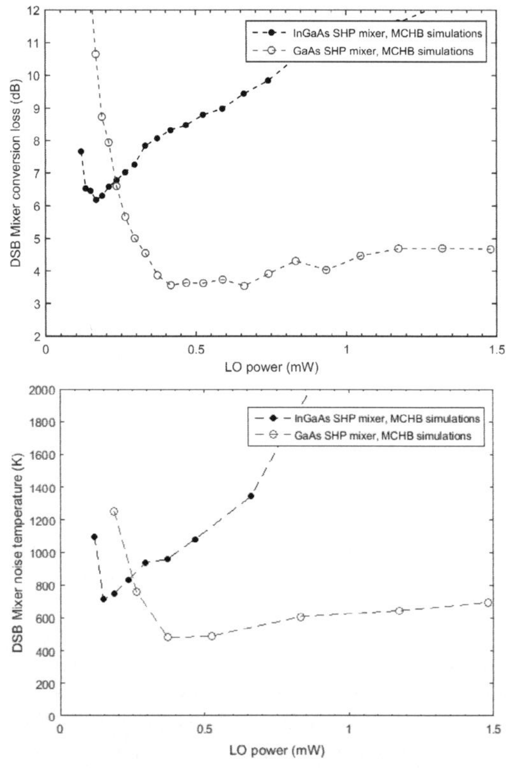

Fig. 3 DSB conversion loss and mixer noise temperature of the 168 GHz SHP mixer based on InGaAs and GaAs [12] diodes. Results obtained with MCHB.

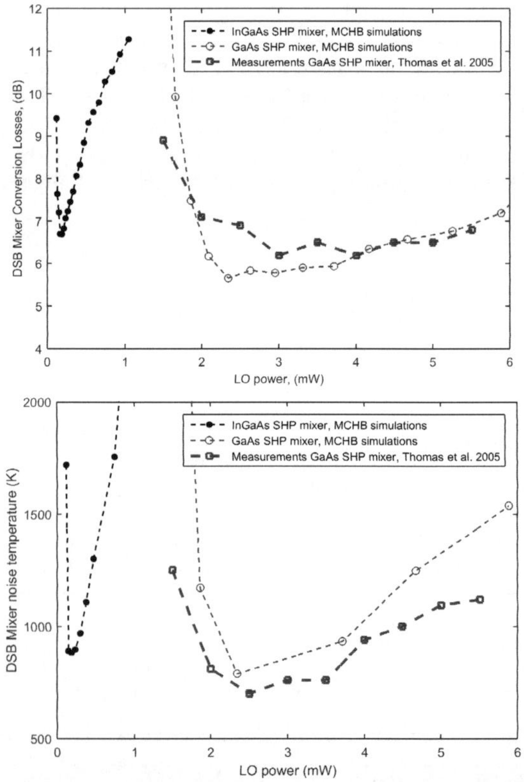

Fig. 4 DSB conversion loss and mixer noise temperature of the 360 GHz SHP mixers based on InGaAs and GaAs diodes, obtained with the MCHB tool. Measurements from [12] are also included.

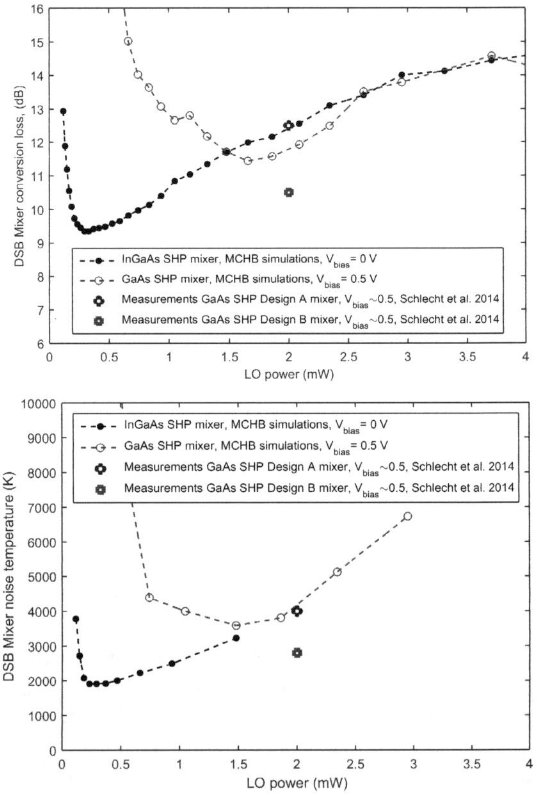

Fig. 5 DSB conversion loss and mixer noise temperature of the 1.2 THz SHP mixer based on InGaAs and GaAs SHP mixers obtained with MCHB. Measured results from [1] are included.

Figs. 3 to 5 show the mixer conversion loss and noise temperature for the InGaAs and GaAs SHP mixers described in Table I calculated with the MCHB tool. The most important conclusions from these figures are:

- The LO power to reach the minimum conversion loss and noise temperature of the InGaAs SHP mixers is around 1/4 of the corresponding GaAs SHP mixers because of the lower barrier height of the former.
- For the simulated mixers at frequencies below 1 THz, InGaAs mixers show higher conversion loss and noise temperature than the corresponding GaAs mixers because of the higher junction capacitance of the InGaAs diodes, which increases shunting effect (in this initial analysis, tunnel effect was not taken into account in the MC diode model).
- The GaAs 1.2 THz SHP mixer in [1] is biased in order to reduce the LO power requirements. According to simulations results in Fig. 5, an unbiased InGaAs 1.2 THz SHP mixer will require lower LO power than the biased GaAs mixer and it will provide better performance.

IV. FABRICATION AND TEST OF INGAAS DIODES

The MMTG is developing Schottky direct detectors and mixers based on InGaAs semiconductor at THz frequencies. Fig. 6 shows current versus voltage measurements of our InGaAs diodes with anode diameter 0.9 μ m. Simulation results using physics based diode models including tunnel transport and barrier lowering due to image force are also included in the figures, showing a good agreement with the measured results.

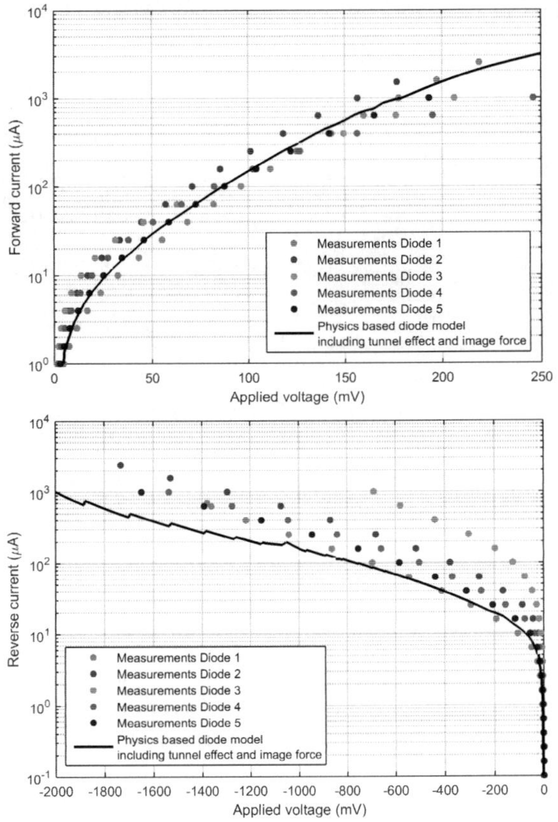

Fig. 6 Current versus voltage measurements under reverse and forwards conditions for InGaAs diodes of anode diameter 0.9 μm fabricated at the MMTG. Simulations results from physics based diode models are also presented.

V. Conclusion

A MCHB circuit simulator has been used to analyse the electrical and noise performances of InGaAs SHP Schottky mixers at the THz band. At frequencies below 1 THz, InGaAs mixers require LO powers around 1/4 of that required by GaAs mixers, but they show higher conversion loss and noise temperature. The use of InGaAs technology is expected to be very competitive at frequencies above 1 THz: while GaAs SHP mixers need to be biased, an InGaAs SHP mixer can operate without bias, requiring 1/4 of the LO power of the GaAs mixer and showing better performance.

Some initial fabrication results of InGaAs diodes at the MMTG have been presented, showing an excellent dc performance when compared with the expected performance provided by physics based diode models.

Acknowledgment

This project has received funding from the European Union's Horizon 2020 research and innovation programme under the Marie Skłodowska-Curie grant agreement No 665593 awarded to the Science and Technology Facilities Council, the Spanish National Research and Development Program under project TEC2014-53815-R, and the Madrid Regional Government under project S2013/ICE-3000 (SPADERADAR-CM).

References

[1] E. Schlecht et al., "Schottky diode based 1.2 THz receivers operating at room-temperature and below for planetary atmospheric sounding," *IEEE Trans. THz Sci. Technol.*, vol 4, no. 6, 2014.

[2] J. Treuttel et al., "A 520 -620 GHz Schottky receiver front-end for planetary science and remote sensing with 1070 K -1500 K DSB noise temperature at room temperature," *IEEE Trans. THz Sci. Technol.*, vol 6,no 1, 2016.

[3] U. V. Bhapkar, T. A. Brennan, and R. J. Mattauch. InGaAs, "Schottky barrier mixer diodes for minimum conversion loss and low LO power requirements at terahertz frequencies," *in Proc. 2nd Int. Symp. Space Terahertz Tech.*, 1991.

[4] E. Schlecht and R. Lin, "Schottky diode mixers on gallium arsenide antimonide or indium gallium arsenide?," *in Proc. 19st Int. Symp. Space Terahertz Tech.*, 2008.

[5] S. Khanal et al., "Characterisation of low-barrier Schottky diodes for millimeter wave mixer applications," *in Global Symposium on Millimeter Waves*, 2016.

[6] I. Oprea, A. Walber, O. Cojocari, H. Gibson, R. Zimmermann, and H. Hartnagel, "183 GHz mixer on InGaAs Schottky diodes," *in Proc. 21st Int. Symp. Space Terahertz Tech.*, Oxford, UK, Mar 2010.

[7] C. Jacoboni and P. Lugli, The Monte Carlo Method for Semiconductor Device Simulation. New York, NY, USA: Springer-Verlag, 1989.

[8] J. Mateos, T. Gonzalez, D. Pardo, V. Hoel, H. Happy and A. Cappy, "Improved Monte Carlo algorithm for the simulation of δ -doped AlInAs/GaInAs HEMTs," *IEEE Transactions on Electron Devices,* vol. 47, no. 1, pp. 250-253, Jan 2000.

[9] D. Pardo, J. Grajal, and S. Pérez, "Electrical and noise modeling of GaAs Schottky diode mixers in the THz band," *IEEE Trans. THz Sci. Technol.*, vol. 6, no. 1, pp. 69–82, Jan. 2016.

[10] T. P. Pearsall and J. P. Hirtz, "The carrier mobilities in $Ga_{0.47}$ $In_{0.53}As$ grown by organo-mettalic CVD and liquid-phase epitaxy," *Journal of Crystal Growth*, vol. 54, pp. 127–131, Jul. 1981.

[11] T. Windhorn, L. Cook, and G. Stillman, "The electron velocity-field characteristic for n-$In_{0.53}Ga_{0.47}As$ at 300 K," *IEEE Electron Device Lett.*, vol. 3, no. 1, pp. 18–20, Jan. 1982.

[12] B. Alderman et al., "Integrated Schottky Structures for Applications Above 100 GHz," *European Microwave Integrated Circuit Conference,* pp. 202-205, Amsterdam, 2008.

[13] B. Thomas, A. Maestrini, and G. Beaudin, "A low-noise fixed-tuned 300–360 GHz sub-harmonic mixer using planar Schottky diodes," *IEEE Microw. Wireless Compon. Lett.*, vol. 15, no. 12, pp. 865–867, Dec. 2005.

Frequency Triplers at 94 GHz and 300 GHz for Millimeter-Wave Radars

Carlos G. Pérez-Moreno* and Jesús Grajal
Signals, Systems, and Radiocommunications Department, Universidad Politécnica de Madrid,
E.T.S.I. Telecomunicación, Av Complutense 30, 28040 Madrid, Spain
*Contact: carlos@gmr.ssr.upm.es

Abstract— In the framework of a ground-based cloud profiling radar at 94 GHz and a 3-D high-resolution imaging radar at 300 GHz developed at the Universidad Politécnica de Madrid, signal sources based on generation by direct digital synthesis, power amplification and frequency multiplication are designed. This work focuses on the last frequency multiplication stage of the source chain of both systems. The design of frequency triplers at 94 GHz and 300 GHz based on Schottky diode technology is reported. For the cloud profiling radar, the simulated room-temperature performance of the designed 94 GHz frequency tripler shows a conversion efficiency of around 11% in the 90-98 GHz frequency band for an input power of 100 mW, with a 3-dB bandwidth from 75 to 102 GHz. For the imaging radar, the simulated room-temperature performance of the designed 300 GHz frequency tripler predicts over 2% of conversion efficiency between 260 and 300 GHz for an input power of 100 mW.

I. Introduction

Two millimeter-wave radars have been recently developed at the Universidad Politécnica de Madrid for weather monitoring and security applications. On the one hand, a ground-based cloud profiling radar at 94 GHz [1] and, on the other hand, a 3-D high-resolution imaging radar at 300 GHz [2]. In both systems, the transmitted signal is generated by direct digital synthesis and several stages of frequency multiplication and power amplification. This work focuses on the design of two frequency triplers at 94 GHz and 300 GHz based on Schottky diodes for the last frequency multiplication stage of the transmitter chains of both radars.

The effects of current saturation and self-heating are the main performance limiting factors for millimeter-wave Schottky diode frequency multipliers. A physics-based electro-thermal model developed at the Universidad Politécnica de Madrid [3] which takes into account these effects is used in the design of the frequency triplers reported in this work. The design methodology details are presented in Section II, whereas the predicted performance of the designed frequency triplers at 94 GHz and 300 GHz are displayed in Section III and Section IV respectively.

II. Design Methodology

The design methodology combines in-house simulation tools with conventional design software like Ansys HFSS and Keysight ADS. A physics-based numerical electro-thermal model for Schottky diodes coupled with a harmonic-balance code is used to optimize the diode electrical and geometrical parameters together with output power, conversion efficiency, and the impedances at the specific harmonics. This tool takes into account self-heating and the temperature dependency of the material parameters. In addition, it is able to calculate internal temperature distributions allowing the evaluation of the thermal effects on the circuit performance. Once the diode structure is selected, the synthesis of the circuit topology is performed with HFSS and ADS. The results obtained from the HFSS electromagnetic simulations of the actual 3-D circuit structure are included into ADS together with an appropriate diode model in order to determine the overall circuit performance. The design process is iterative (Fig. 1) and can be divided into three steps.

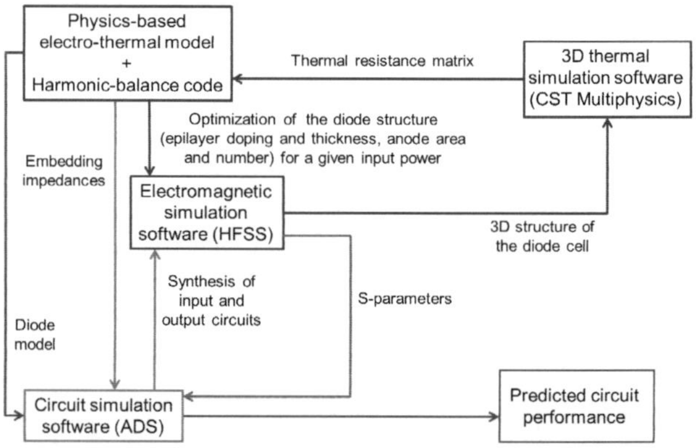

Fig. 1. Iterative design process utilized to optimize circuit performance.

A. Step 1

In the first step, the diode structure is optimized by using the physics-based numerical electro-thermal model integrated into a circuit simulator based on the harmonic balance technique. Parameters like epitaxial layer doping and thickness, and anode area and number are optimized for a given input power. Then, the diode cell is configured.

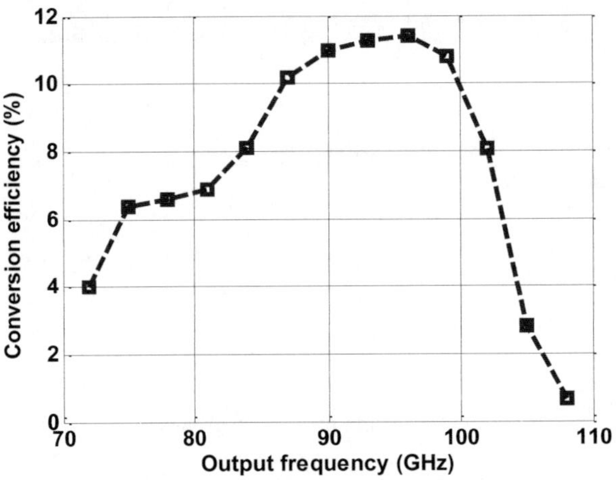

Fig. 2. 94 GHz frequency tripler: conversion efficiency as a function of output frequency for an input power of 100 mW.

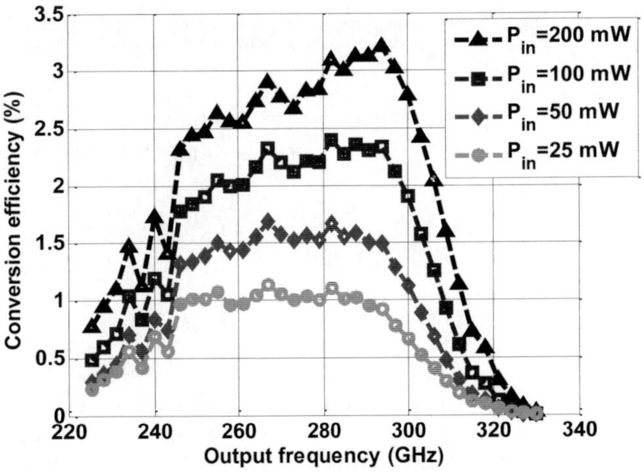

Fig. 3. 300 GHz frequency tripler: conversion efficiency as a function of output frequency for the 25-200 mW input power range.

A thermal resistance matrix [3] is extracted by simulating the 3-D structure of the diode cell with CST Multiphysics and applying the dissipated power obtained with the physics-based electro-thermal model.

Finally, the embedding impedances and diode model to be included into ADS are determined by using the physics-based numerical electro-thermal model in combination with the harmonic balance code. The diode parameters are reoptimized if necessary.

B. Step 2

In the second step, the matching networks of the input and output circuits are linearly simulated with ADS by using the embedding impedances extracted in Step 1 and the S-parameters of the diode cell.

C. Step 3

In the third step, the circuit performance is predicted by carrying out a nonlinear circuit simulation with ADS and using the diode model extracted in Step 1 and the S-parameters extracted from the HFSS simulations of the diode cell, and the input and output circuits.

III. 94 GHz Frequency Tripler

For the source of the cloud profiling radar, a frequency tripler providing around 4 mW of output power in the 93-95 GHz frequency band for an input power of 100 mW in the 31-31.66 GHz frequency band is required [1].

The predicted room-temperature performance of the designed 94 GHz frequency tripler shows a conversion efficiency of around 11% in the 90-98 GHz frequency band for an input power of 100 mW, with a 3-dB bandwidth from 75 to 102 GHz (Fig. 2), and a maximum anode temperature of around 400 K. This frequency tripler uses a discrete GaAs diode chip from Teratech Components Ltd. (UK) with six planar Schottky varactors in a series configuration.

IV. 300 GHz Frequency Tripler

The design of a 300 GHz frequency tripler for the 3-D high-resolution imaging radar is also presented. The design goal is to maximize the bandwidth for a wide range of input powers. This tripler features four anodes integrated into a GaAs membrane. Fig. 3 shows the predicted conversion efficiency for input powers between 25 and 200 mW. The 3-dB bandwidth is higher than 50 GHz for the whole simulated input power range.

V. Conclusions

The design of two frequency triplers at 94 GHz and 300 GHz by following a design methodology that includes an integrated thermal management approach has been presented. The predicted performance of both triplers fulfills the design requirements and will be compared against measurements once the circuits are fabricated.

Acknowledgment

The authors would like to thank Dr. Jeff Powell and Dr. Byron Alderman, Teratech Components Ltd., for very helpful discussions on the designed frequency triplers. This work was supported by the Spanish National Research and Development Program under Project TEC2014-53815-R and by the Madrid Regional Government under Project S2013/ICE-3000 (SPADERADAR-CM).

References

[1] G. Rubio-Cidre, J. Grajal, A. García-Pino, and O. Rubiños, "Design and preliminary results of a ground-based cloud profiling radar at 94 GHz," in *IEEE Radar Conf.*, 2016.

[2] J. Grajal, A. Badolato, G. Rubio-Cidre, L. Úbeda-Medina, B. Mencia-Oliva, A. García-Pino, B. González-Valdés, and O. Rubiños, "3-D high-resolution imaging radar at 300 GHz with enhanced FoV," *IEEE Trans. Microw. Theory Techn.*, vol. 63, no. 3, pp. 1097–1107, Mar. 2015.

[3] C. G. Pérez-Moreno and J. Grajal, "Physical electro-thermal model for the design of Schottky diode-based circuits," *IEEE Trans. THz Sci. Technol.*, vol. 4, no. 5, pp. 597–604, Sep. 2014.

A cryogenic solid state LO source at 1.9THz

Nicolas Reyes[1], Rolf Güsten[2], Urs Graf [3], Christophe Risacher [2], Oliver Ricken[2]
[1]*Universidad de Chile, Santiago, Chile*
[2]*Max Planck Institut für Radioastronomie, 53121, Bonn, Germany*
[3]*Universität zu Köln, 50923, Köln, Germany*
*Contact: nireyes@uchile.cl

Abstract— The Low Frequency Array instrument at SOFIA is a heterodyne array designed to observe in the complete range between 1.9 and 2.5 THz. It is a 14 pixel instrument, working in a dual polarization configuration. To provide enough LO power for the 14 HEB mixers a cryogenic solid state LO source is used. It consists of two identical set of millimeter wave amplification and multiplication chains, one per polarization. The frequency coverage is limited to 1.830 to 2.070 GHz, centered on one of the main scientific target of the instrument, the CII at 158mm. We present an overview of the design of this LO source and present the experimental results obtained.

INTRODUCTION

The Low Frequency Array instrument at SOFIA is a heterodyne array designed to observe in the complete range between 1.9 and 2.5 THz. It is a 14 pixel instrument, working in a dual polarization configuration. The instrument is operative since 2015[1] and have produced valuable scientific data [2]. To provide enough Local Oscillator (LO) power for the 14 HEB mixers a cryogenic solid state source is used. It consists of two identical set of millimeter wave amplification and multiplication chains, one per polarization. The solid state sources were produced at Virginia Diodes inc. [3] and are constantly improved for better performance. Their current frequency coverage is 1.830 to 2.070 GHz, centered on one of the main scientific target of the instrument, the CII emission line at 158mm. Current operational modes allows for separated tuning of the two sources, to have each polarization working at a different frequency. In this work we present an overview of the LO system as installed during the commissioning flights of the instrument.

LO MULTIPLICATION CHAINS

The LO solid state chains from VDI use novel technologies, such as diamond substrates for thermal management, in-phase combining networks, and power amplification at 30 GHz. They achieve a peak output power of 30 mW at ambient temperature. To further increase the output power, the last two passive triplers are cooled to 80 K, effectively doubling the output power to about 60 mW.

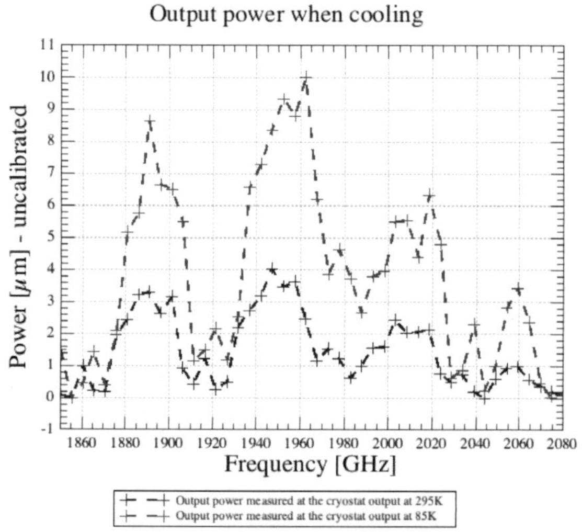

Fig. 1 Measured power at ambient temperature (blue) and at 85 K (red). Power scale is not calibrated and includes the coupling efficiency of the detector horn, vaccum window losses and free space propagation.

LO SYSTEM

The two solid-state chains are integrated with the control electronics, power supplies, and cooling machine in to a compact module. Triplers are cooled to 80 K by a Stirling cooler based on a moving magnet. It provides 2 W of cooling power at 80 K. The cold tip from the cooler is thermally connected to the triplers via flexible copper straps. To thermally isolate the triplers from the room temperature components, two 1” stainless steel WR-5 waveguides are used. The waveguides were copper plated (thanks to NRAO) to reduce the losses to only 0.5dB. The small LO cryostat optical window is a 1-mm-thick high resistivity Silicon window coated on both sides with Parylene. The window is optimized for 1.9 THz, reaching a transmission above 88%. The solid state multiplier chains need a reference signal in the 13 GHz range. The characteristics of the reference signal are critical, as the phase and AM noises of the signals contribute to the total system performance. Several synthesizer modules were tested. When comparing the overall receiver

A FULL VERSION OF THIS PROCEEDINGS ARTICLE HAS BEEN SUBMITTED TO THE IEEE TRANSACTIONS OF TERAHERTZ SCIENCE AND TECHNOLOGY (TST) FOR PEER-REVIEW, PLEASE CHECK THE WEBSITE: HTTPS://WWW.MTT.ORG/TERAHERTZ

performance, the best results are achieved with a YIG synthetizer from VDI. In that configuration the receiver sensitivity compares to lab-bench synthesizers. Using other models, such as VCO-based synthesizers, the receiver noise temperature can degrade by up to 20%.

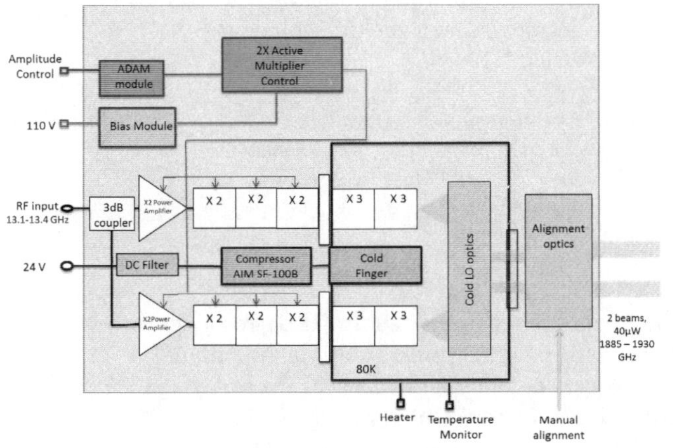

Fig. 2 Electrical diagram of the system. Multiplication factor of the chains is 144. The two LO sources use different optical paths to avoid crosstalk and spurious generation.

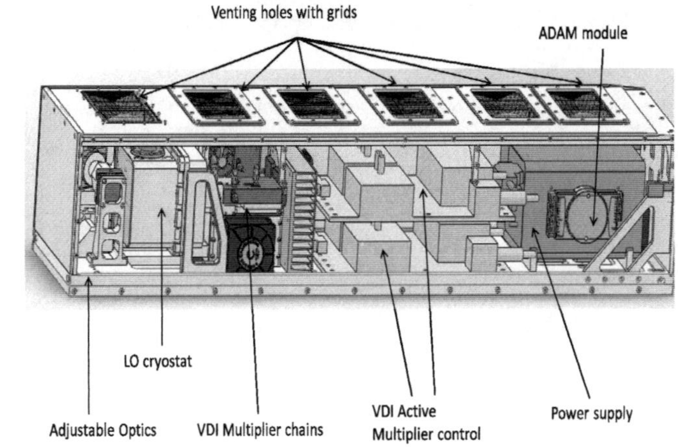

Fig. 3 CAD schematic of the produced system. Thermal management of the power amplifiers and Stirling cooler is critical. Current design optimize the air-flow on the limited space available.

OPTICAL PATH

In the first optical design, the outputs for both chains had orthogonal polarizations and were combined through a coupling grid for a common propagation, and then they were separated again by another grid to be coupled to each HEB array. However this scheme generated unacceptable levels of spurious. A revised optical design was implemented providing separate optical paths for both LO beams. An optic adjustment stage allows directing the beams to the phase gratings where the 7 beams per polarization are generated. The Fourier phase grating design details can be found in []. It is designed to operate at a center frequency of 1.9 THz with about 10% usable bandwidth. Every output beam contains about 12.8% of the incident power, i.e. 90% efficiency. The phase grating was fabricated by direct milling at U.Köln. Figure 4 shows laboratory measurements of the output beams, It was noticed during measurements that alignment of the structure is critical to achieve the required performance.

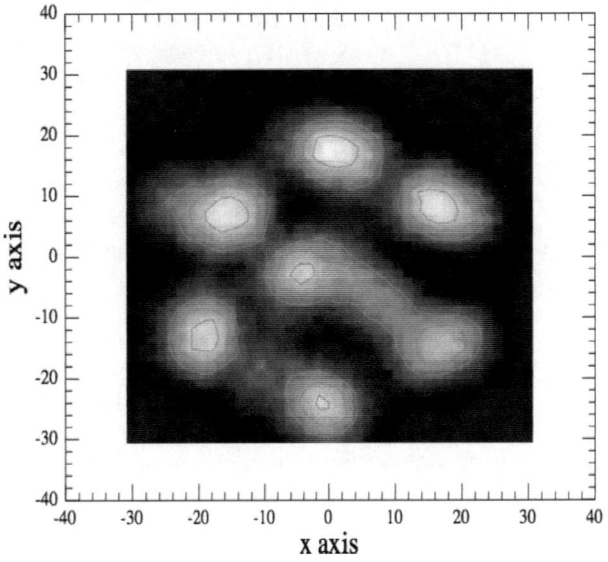

Fig. 3 Measured beam pattern of the 7 LO signal beams. Alignment of the system is critical to obtain equal power beams.

CONCLUSIONS

The LO unit is currently on use at the LFA instrument on board of SOFIA. The LO chains have been replaced to take advantage of the latest improvements on output power achieved by this technology. This article presents an overview of the design of this 1.9THz source.

REFERENCES

[1] C. Risacher, et al. "The upGREAT 1.9THz multi-pixel high resolution spectrometer for the SOFIA observatory", Astronomy and Astrophysics, Volume 595, November 2016

[2] C. Risacher, et al. "First Supra-THz Heterodyne Array Receivers for Astronomy With the SOFIA Observatory", IEEE Transaction on Terahertz Science and Technology. Vol. 6, N°2, March 2016.

[3] https://www.vadiodes.com, accessed: June 2017.

[4] U. U. Graf and S. Heyminck, "Fourier gratings as submillimeter beam splitters," *IEEE Trans. Antennas Propag.*, vol. 49, no. 4, pp. 542–546, Apr. 2001.

A 211-275 GHz receiver prototype

Kirill Rudakov[1,2,3], Pavel Dmitriev[3], Andrey Baryshev[2,4],Andrey Khudchenko[2], Ronald Hesper[2], Oleg Kiselev[3], Valery Koshelets[3]

[1]*Moscow Institute of Physics and Technology, Dolgoprudny, Russia*
[2] *Kapteyn Astronomical Institute, University of Groningen, Groningen, the Netherlands*
[3]*Kotel'nikov Institute of Radio Engineering and Electronics RAS , Moscow, Russia*
[4]*SRON Netherlands Institute for Space Research, Groningen, the Netherlands*

*Contact: mailto:kirill.i.rudakov@gmail.com

Abstract— We have made a preliminary design of a sideband separating (2SB) receiver for frequency range 211-275 GHz, which can be implemented for LLAMA observatory in Argentina [1], and for the Millimetron space mission [2]. The receiver is conceptually similar to the ALMA Band 9 2SB receiver [3][4], it will be based on single ended superconductor-insulator-superconductor (SIS) mixers connected to a waveguide RF hybrid block. The design of the single ended SIS mixer was developed using electromagnetic modeling in CST microwave studio. We use a waveguide orthogonal probe and have found optimal waveguide sizes and quarts substrate width. The mixer is based on Nb/AlN/NbN SIS junctions embedded in a Nb/SiO2/Nb microstrip line. In order to improve design stability with respect to manufacturing tolerances we will investigate extension of the 64GHz bandwidth utilizing high current density AlN barrier SIS junctions. The detailed receiver description will be presented at the conference.

[1] Large Latin American Millimeter Array, http://www.iar.unlp.edu.ar/llama-web/
[2] Make a link to some latest paper about Millimentron
[3] Hesper, R., Gerlofsma, G., Mena, F. P., Spaans, M. C., and Baryshev, A. M., "A sideband-separating mixer upgrade for alma band 9," in [Twentieth International Symposium on Space Terahertz Technology, Charlottesville], 257–260 (apr 2009).
[4] A. Khudchenko, R. Hesper, A.M. Baryshev, J. Barkhof and F.P. Mena, "Modular 2SB SIS Receiver for 600-720 GHz: Performance and Characterization Methods", IEEE Tr. on TST, vol. 7, No. 1, 2017.

AC-Biased Superconducting NbN Hot-Electron Bolometer for Frequency-Domain Multiplexing

Sergey V. Seliverstov[1,*], Anastasia A. Rusova[1], Natalya S. Kaurova[1], Boris M. Voronov[1] and Gregory N. Goltsman[1]
[1]*Physics Department, Moscow State Pedagogical University, 29 Malaya Pirogovskaya St., Moscow, 119992, Russia*
*Contact: sv.seliverstov@mpgu.edu

Abstract— We present the results of characterization of fast and sensitive superconducting antenna-coupled THz direct detector based on NbN hot-electron bolometer (HEB) with AC-bias. We discuss the possibility of implementation of the AC-bias for design the readout system from the multi-element arrays of HEBs using standard technique of frequency-domain multiplexing. We demonstrate experimentally that this approach does not lead to significant deterioration of the HEB sensitivity compared with the value obtained for the same detector with DC-bias. Results of a numerical calculations of the HEB responsivity at AC-bias are in a good agreement with the experiment.

INTRODUCTION

Multi-element arrays of superconducting bolometers are being developed for a number of astronomical observations in far-IR and mm range. The currently used detectors in such a systems have an ultimate sensitivity limited by photon noise that is achieved because of the millikelvin operating temperatures. Our NbN hot-electron bolometer (HEB) is one of the well-known devices operated at more convenient temperatures (about 10 K) [1]. It can be used as an element of such arrays for the far-IR (THz) range. However, integration of a large number of HEBs into array on a single chip presents a difficult problem in terms of design the readout system. The obvious solution of this problem is using of AC-bias for the detectors. This method allows an easy extension to multi-element arrays with the standard frequency-domain multiplexing (FDM) technique.

At the same time, THz frequency range is very attractive for a number of important practical applications, such as security systems and medicine. As one can see from real practice, developers all over the world prefer using of pulsed THz sources, which are reliable, more powerful, stable in operation and user-friendly with respect to the CW sources of THz radiation. For that reason, the practical THz systems should be equipped with sensitive and fast detectors. Both features are combined in a direct detector based on HEB.

AC CURRENT-BIAS

To establish proper operating conditions for the HEB direct detector, the bath temperature is raised by a heater from the liquid-helium temperature up to the value close to the critical temperature T_c. The HEB is then DC-biased to the operating point with the highest sensitivity. The same regime is used for TES. In a common SQUID readout system for TES, the negative electrothermal feedback is provided by using the voltage-bias, so any small change in the absorbed optical power is compensated by a change in the Joule power.

Alternatively, one can apply AC-bias to the HEB, in which case the Joule power oscillates between minimum and maximum at twice the bias frequency. This technique was previously discussed in [2] for devices with inverse time constant lower than the bias frequency. It was shown that the load curves taken with AC- and DC-bias were nearly identical, indicating that there was no degradation in the performance of the bolometer due to the AC-bias. This approach was also discussed without direct experimental demonstration in [4].

In this paper, we demonstrate operation of a superconducting THz detector based on the NbN HEB with an AC-bias at a frequency much lower than the detector inverse time constant. For the fast bolometer, this regime differs from the described above. In this case, the electron temperature of the bolometer changes between maximum and minimum value just like the bias current without any delay. Thus, not only the bias power oscillates between minimum and maximum value, but the resistance of the bolometer as well. We also discuss the possibility of implementation of the AC-bias for organizing the readout from the multi-element arrays of HEBs using FDM.

DEVICES AND SETUP

The main aim of our experimental study was characterization of responsivity and noise-equivalent power (NEP) of NbN HEB with AC-bias. We used an elliptical lens made of high-resistivity Si, without an antireflection coating. The HEB chip was glued to the flat surface of the lens made of the same material as the chip substrate. This structure was installed into the lens holder that was mounted onto the cold plate of the optical liquid-helium cryostat with a high-density polyethylene window and a cold Zitex-104 filters cutting off room-temperature background radiation.

In out measurements, we used a cryogenic low-noise amplifier with a gain of 24 dB across a bandwidth of 0.01-200 MHz. An AC-bias at 1 MHz was applied to the bolometer through a 1-kΩ resistor with the function generator. The input impedance of the amplifier was 50 Ω at the biasing frequency. For that reason, we did not use the bias-T in our

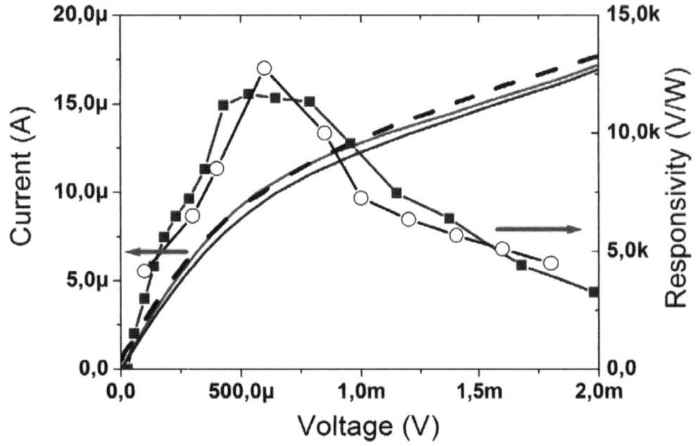

Fig. 1 Results of the experiment and the calculations. Blue and red curves show the calculated IV curves with the RF power and without it, correspondingly. Dash black curve shows the experimental IV curve. Wine and black curves show the calculated and measured HEB responsivity, correspondingly.

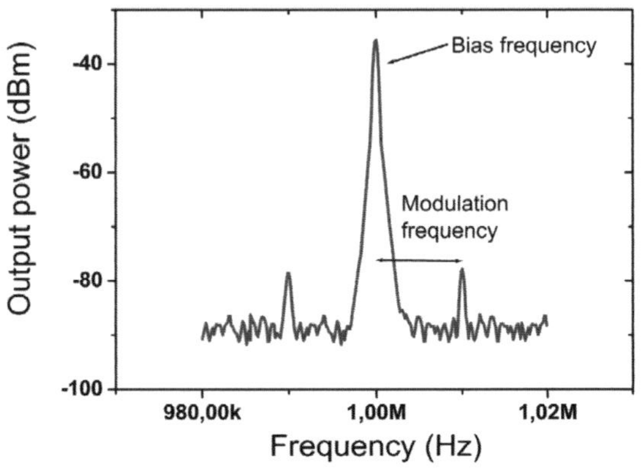

Fig. 2 Spectrum of the detector output power obtained from the output port of the low-noise amplifier. The bias frequency was 1 MHz. The input power was amplitude modulated with frequency of 10 kHz.

measurements. A gas discharge laser operating on a 2.5 THz H_2O line was used as a radiation source. The amplified signal from the bolometer was measured with the lock-in amplifier.

RESULTS AND DISCUSSION

A basic principle of signal formation in the detector operated with the AC current bias differs from that in the case of DC-bias. The signal V_{out} that we measured was a difference between corresponding amplitudes of voltage oscillations from the bolometer with the applied RF power and without it. The power of the laser output was measured with a Golay cell. We calculated the input signal power P_{inc} using the measured transmission coefficient of the filters and the window. The input RF power was so adjusted that the detector was within its dynamic range. With these data we were able to calculate the optical responsivity of the entire receiver as $S_{Rx} = V_{out} / P_{inc}$. We measured S_{Rx} at different values of the amplitudes of the bias current along the optimal current-voltage curve at a bath temperature close to T_c. Then the input-signal power was attenuated until we were no longer able to detect any signal. We assumed that the remaining part of the lock-in amplifier readings was the noise level of the receiver.

Theoretical analysis of the described bias technique was based on the numerical solution of the well-known steady-state one-dimensional heat-balance equation applied to the HEB. The obtained results of our calculations, together with the measured values, are also shown in Fig. 1. Experimental results are shown with correction on the optical losses. The correction factor is about 5. This value is determined by the optical losses of the components of the experimental setup.

The noise level of the bolometer at the optimal operating point amplified with the cryogenic amplifier was measured to be about 0.4 $\mu V \cdot Hz^{-0.5}$, which is just about 2.3 higher than that for the same bolometer operated at DC-biased mode. We associate small degradation of the sensitivity with the instability of the employed function generator. So, as one can see, our mode of biasing the detector does not lead to significant degradation of its sensitivity.

The obtained results open the possibility of using an AC current-bias in conjunction with FDM [3] for constructing multi-element arrays of HEBs. Each element should be integrated with LC-filter with a specific resonance frequency. The inductance L_i of each LC-filter must be at least 24 μH for an operating resistance of about 60 Ω to obtain the bandwidth $\Delta f_i = R/(2\pi L_i)$ of 400 kHz. The resonance frequency of each LC-filter is selected by capacitors with values C_i chosen to keep the bias frequencies in the necessary range, using the relation $C_i = 1 / (4\pi^2 L_i f_i^2)$. For example, the necessary value of capacitance is equal to 10 pF for the bias frequency of 10 MHz. These inductors and capacitors can be easily manufactured using standard deposition technique.

The experimental demonstration of this idea at bias frequency of 1 MHz was performed using the amplitude modulation of the input power with the frequency of 10 kHz. The measured spectrum of the detector output power obtained from the output port of the LNA is shown in Fig. 2. As one can see, two sidebands located at the offset of the modulation frequency are clearly visible.

In summary, we have demonstrated successful application of the AC-bias to the HEB as a direct detector. This technique can be implemented for FDM for the multi-element arrays of HEBs. This work was supported in part by the Russian Foundation for Basic Research under Grant No. 16-32-00622 and the Russian Ministry of Education and Science under State Contract Nos. 14.B25.31.0007 and 11.2423.2017/PP.

REFERENCES

[1] Seliverstov, S. et al (2015). Fast and sensitive terahertz direct detector based on superconducting antenna-coupled hot electron bolometer. *IEEE Transactions on Applied Superconductivity*, 25(3), 1-4.

[2] Yoon, J. et al (2001). AC voltage-biased superconducting bolometer for a frequency-domain SQUID multiplexer. *IEEE transactions on applied superconductivity*, 11(1), 562-565.

[3] Lanting, T. M. et al (2005). Frequency-domain multiplexed readout of transition-edge sensor arrays with a superconducting quantum interference device. *Applied* Physics *Letters*, 86(11), 112511.

[4] Kuzmin, A. A., Shitov, S. V., Scheuring, *et al.* (2013). TES Bolometers With High-Frequency Readout Circuit. *IEEE Transactions on Terahertz Science and Technology*, **3**(1), 25-31.

Design and Fabrication of a Dual-Polarization, Balanced SIS Mixer Integrated Circuit

Wenlei Shan[1*,2], Shohei Ezaki[1], Keiko Kaneko[1], Matthias Kroug[1], Alvaro Gonzalez [1], Takafumi Kojima[1], Yasunori Fujii[1], Shinichiro Asayama[1], Takashi Noguchi[1]

[1] *National Astronomical Observatory of Japan*, 181-8588, Mitaka, Tokyo, Japan,*
[2] *Purple Mountain Observatory, 210008, Nanjing, JiangSu, China*

[*]Contact: wenlei.shan@nao.ac.jp

Abstract—We have been carrying out a study on waveguide SIS mixers for large format multibeam receivers. The concept design was presented in ISSTT last year (Wenlei Shan, etc., "Concept Design of a Dual-Polarization Sideband-Separating Multi-Pixel SIS receiver," ISSTT 2016). The main feature of this concept lies in the highly compact configuration with dual polarization receiving capability. The central part of the receiver is an integrated SIS mixer chip, on which planar orthomode transducers (OMTs) are formed on dielectric membranes. The LO introduced through metal waveguides is also coupled by membrane based probes. In this way, the metal waveguide network for LO distribution embedded in the mixer holder becomes much simpler and thus can be manufactured with conventional machining.

We have designed a single-pixel verification model operating in the 125 - 163 GHz band. This verification model incorporates all the necessary features that allow extending the design to an array. It is a dual-polarization and balanced mixer, and a sideband separation scheme will be fulfilled when a thin-film resistor process is established. We have fabricated membranes from SOI wafers with a deep reactive-ion etching process. Flat silicon membranes with 6-micrometer thickness have been fabricated in 3 mm x 3 mm windows. In this symposium, we are going to present the details of the circuit design and the fabrication process of the mixer chips.

Design of Large-Band Room-Temperature On-Chip Diplexed Schottky Receivers for Planetary Science

Jose V. Siles, Choonsup Lee and Imran Mehdi

[1]*Jet Propulsion Laboratory, California Institute of Technology, Pasadena, CA 91109*
*Contact: Jose.V.Siles@jpl.nasa.gov

Abstract— Planetary Water, chemistry and energy/heat are the three key aspects to consider when it comes to address habitability in planetary bodies such as Europa and Enceladus. Most of these lines, salts (**NaCl, KCl, MgCl, NaOH, KOH, MgO**), carbon molecules (**CO, CN, HCN, H_2C, CH_3CN, CH_3OH**), water (**H_20, $H_2{}^{18}O$, $H_2{}^{17}O$, HDO**), and sulfur molecules (**H_2S, SO_2**), span across a large frequency band: 216-580 GHz, which imposes a large required RF bandwidth for a receiver. Such a receiver needs to be room-tcapable of high-resolution (<100 kHz) molecular spectroscopy to measure line shapes and Doppler shifts of molecular emissions from gases in environments as harsh as the Jupiter system. This can be accomplished using Schottky diode heterodyne receivers, which offer high resolution based on their temporal stability and dynamic range. The GaAs diodes are inherently radiation immune. Fly-bys mission require fast integration times, which also implies high-receiver sensitivities. By using a single large-band receiver with no need to continuously switch the LO, data can be retrieved faster. To be able to detect simultaneously most of the key molecules without switching the LO, a 3-50 GHz IF is required. In addition, the use of a single-receiver avoids the need of polarizing grids to separate the telescope beam into multiple receivers, which reduces RF losses and increases the sensitivity.

In this work we will present the progress towards developing a large-band receiver in the 216-581 GHz featuring Schottky mixers based on a novel frequency mixer design topology, on-chip frequency diplexer-mixer that will be able to extend the bandwidth of current submillimeter-wave receivers from the current 15% (state-of-the-art) up to around 90%. The mixers will be pumped by frequency multiplied local oscillator sources based on the same on-chip duplexing concept. This novel concept is an evolution of the JPL's on-chip power-combined frequency multiplier concept [1] and it consists of a single-substrate multiplier or mixer chip with half of the diodes tuned to the lower half of the target frequency band, and the other half tuned to the upper half of the band. All diodes couple the multiplied signal to the same output waveguide using two on-chip E-probes, each one tuned to one half of the total band.

Millimetron Space Observatory

A. Smirnov[1*], Th. de Graauw[1], A. Baryshev[2], P. de Bernardis[3], S. Likhachev[1], E. Golubev[1], S. Pilipenko[1], T. Kosmovich[1] and N. Kardashev [1] on behalf of Millimetron team

[1] *Astro Space Center of P.N. Lebedev Physical Institute, 117997, Moscow, Russian Federation*
[2] *Kapteyn Astronomical Institute, University of Groningen, 9747 AD, Groningen, The Netherlands*
[3] *Sapienza University of Rome, 00185, Roma, Italy*
[*] Contact: asmirn@asc.rssi.ru

Abstract— The Millimetron Space Observatory (MSO) is to be the next space observatory for the FIR and Submm wavelength range. The observatory is based on a 10-m cryogenically cooled telescope, to be deployed in orbit after having arrived at a L2 orbit of the Earth-Sun system. MSO will be launched at ambient temperatures and cooled down in orbit through a combination of effective passive and active cooling using onboard mechanical coolers. This combination will cool the telescope to temperatures less than 10K as a goal. The latter and aperture size provides an unparalleled achievement in terms the sensitivity of the astronomical instrument. FIR camera, imaging spectrometers and heterodyne instrument on-board of MSO will provide high-resolution imaging and spectroscopy and allow investigating the coldest objects in the Universe – star forming regions, molecular clouds, dust in our and distant galaxies, outer parts of protoplanetary disks, etc. Moreover the observatory will operate not only as a single aperture telescope enabling high-resolution imaging and spectroscopy but also as an element of Space-Earth Very Large Baseline Interferometer (SVLBI). MSO as an element of Space-Earth VLBI will provide an unprecedented sub-microarcsecond angular resolution which is necessary to study the most compact objects in the Universe - supermassive black holes, jets, etc. The MSO is a new scientific instrument with breakthrough astronomical capabilities. We will present a status and progress in the development of the payload module.

Cryogenic IF Balanced LNAs Based on Superconducting Hybrids for Wideband 2SB THz receivers

Erik Sundin[1,*], Andrey Ermakov[1], Vincent Desmaris[1], Hawal Rashid[1], Denis Meledin[1], Maria Bylund[1] and Victor Belitsky[1]
[1]*Group for Advanced Receiver Development, Department of Space, Earth and Environment, Chalmers University of Technology, Sweden*
*Contact: erik.sundin@chalmers.se

Abstract—Future heterodyne 2SB THz receivers for Radio astronomy shall require better sensitivity and higher IF bandwidth. This paper investigate, in a modular approach, the use of balanced amplifiers based on superconducting quadrature hybrids, as an option to realize wide band low noise amplifiers with improved input matching without compromising the noise performance. Measurements of the proposed wideband balanced amplifier shows very good agreement with the standalone amplifier in terms of gain and noise temperature. The input matching of the balanced amplifiers are limited by the input matching of the hybrids, still showing return loss better than 15 dB.

INTRODUCTION

Future heterodyne receivers operating at mm-wave or THz frequencies would require higher sensitivity and higher IF bandwidth [1]. A larger IF bandwidth of the receiver is however not possible without the availability of wideband IF low noise amplifiers (LNA). Unfortunately, no suitable amplifiers, even from specialized suppliers, are currently available with acceptable noise performance and adequate power consumption over a large bandwidth.

LNAs aimed for operation at cryogenic temperatures and used for radio astronomy are usually optimized for best noise performance and low power consumption at the cost of poor input matching. This forces the use of isolators between the mixers and amplifiers, adding potentially extra noise but certainly limiting the bandwidth over which amplification can be achieved because of the inherent small bandwidth of the circulators.

One approach to solve the bandwidth limitation without compromising the noise performances is to use a balanced scheme where two quadrature hybrids are used, one in front, and one after the LNAs [2].

In this paper we report on the noise performance and input reflection performances of balanced amplifiers fabricated using a modular approach and using available superconducting hybrids developed for ALMA Band 5 [3] and LNAs from the ALMA Band 5 pre-production [4]. The performance of the balanced amplifiers could then be compared with the performance of the standalone amplifiers.

BALANCED AMPLIFIER

The balanced amplifier were built by cascading a superconducting quadrature hybrid at the input followed by two amplifiers, and a superconducting quadrature hybrid at the output [2] as in Fig. 1.

In this experiment we tested two different sets of superconducting hybrids: a 4-8 GHz hybrid with built-in bias-T, currently used in ALMA band 5 [5], and a prototype 4-12 GHz hybrid. Both types of hybrids are fabricated in-house using Silicon substrate and superconducting transmission lines and air-bridging technology [6].

The 4-12 GHz units contains just the hybrid chip, while the 4-8 GHz units also have built in bias-T and have additional Si-chips to match the longer center-to-center pitch of the coaxial connectors needed in the mixer assembly where they are used. Clearly for this application, the built in bias-T are not needed, and are instead a parasitic, however, this are the units available for test with no modification required.

The cryogenic LNAs used in this study consists of 3-stage amplifiers, where the first stage is made of InP HEMT whereas the two others are realized using GaAs HEMTs.

MEASUREMENTS

The noise performance of the standalone LNAs and balanced amplifiers were measured in a cryostat at about 4 K using the cold attenuator method with an Agilent N4000A ENR noise diode and an Agilent MXA N9020A spectrum analyzer with Noise Figure Measurement option. In practice, as the main purpose of these initial measurements is to find the relative differences between the different configurations and not necessarily the correct absolute noise temperature, we carried out a simplified characterization of the cryostat losses and the temperature sensor to be used for the cold attenuator.

The optimum noise performances of the different LNAs (balanced and standalone) were determined by independently tuning the bias conditions of the different transistor stages.

The LNAs were later characterized by means of S-parameters measurements at their optimum noise bias conditions using an Agilent PNA E8364B vector network analyzer. Those measurements were performed without the cold attenuator in front of the amplifier, thus requiring additional cooling cycle.

The two balanced amplifier chain configurations used is the 4-8 GHz version shown in Fig. 1, and the 4-12 GHz shown in Fig. 2.

Fig. 1 Balanced amplifier assembly based on 4-8 GHz hybrids

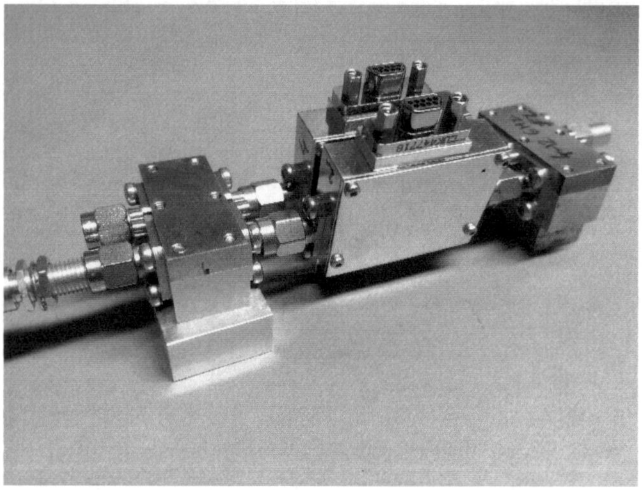

Fig. 2 Balanced amplifier based on the 4-12 GHz hybrids.

RESULTS AND DISCUSSION

The measured noise performances of the different LNAs are presented on Fig. 3. It clearly shows that using superconducting hybrids and a balanced mode does not increase significantly the noise performance compared to the standalone LNAs. In fact, the 4-12 GHz version result in an increase of about 0.2 K (from 5.49 K to 5.69 K) when used in the balanced configuration described. For the 4-8 GHz version, a noticeable increase is seen around the frequencies where the imbalance between the through and coupled signals of the hybrids was maximal (0.7 dB), which could not be compensated by a difference in biasing conditions of the two LNAs.

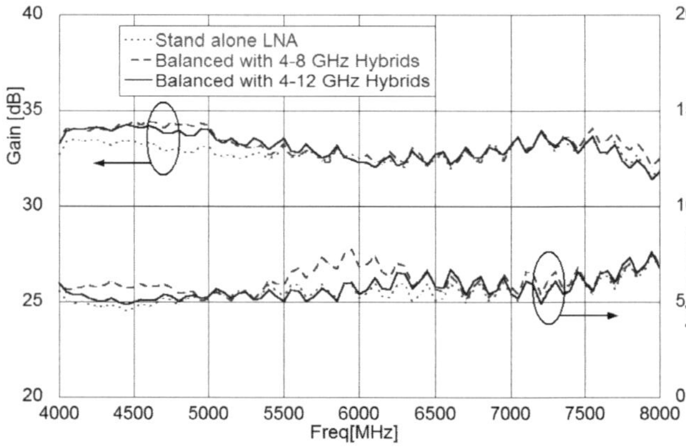

Fig. 3 Noise performance of the Balanced and standalone amplifiers

The Return loss of the different LNAs is presented in Fig. 4. The improvement of the input reflection using balanced amplifiers is clearly demonstrated and the difference between the two balanced configuration can be ascribed to the difference in input reflection of the hybrid versions. In addition, measurement results showed that the return loss of the wideband hybrid is the limiting factor for the return loss of the balanced amplifier. The obtained results shows that very good input matching can be achieved with such balanced amplifier at cryogenic temperatures. The receiver design can potentially be simplified by removing the need of the typically used isolator [7].

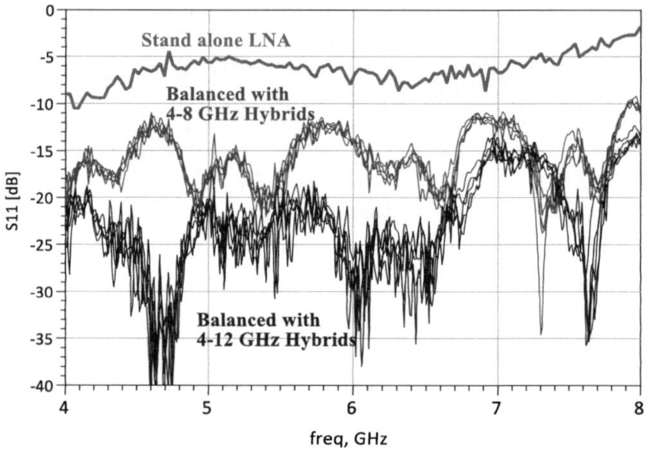

Fig. 4 Return loss performance of the Balanced and standalone amplifiers

CONCLUSION

The use of superconducting hybrids for the demonstration of cryogenic balanced LNAs in a modular approach is successfully demonstrated. The balanced LNAs showed negligible degradation of the noise performance of balanced amplifiers compared to the standalone versions. In addition, our results show that the improvements in terms of return loss are considerable (better than 15 dB over the whole frequency band) and that return loss of the hybrids sets the limits for the input matching of the balanced LNAs.

REFERENCES

U.U. Graf, C.E. Honingh, K Jacobs, and J. Stutzki, "Terahertz Heterodyne Array Receivers for Astronomy," *Journal of Infrared, Millimeter, and Terahertz Waves*, vol. 36, pp. 896–921, 2015.

K. Eisele, R. Engelbrecht and K. Kurokawa, "Balanced transistor amplifiers for precise wideband microwave applications," *1965 IEEE International Solid-State Circuits Conference. Digest of Technical Papers*, pp. 18–19, 1965.

B. Billade et al., "Performance of the First ALMA Band 5 Production Cartridge," *IEEE Transactions on Terahertz Science and Technology*, vol. 2, pp. 208–214, March. 2012.

V. Belitsky, I. Lapkin, B. Billade, E. Sundin, A. Pavolotsky, D. Meledin, V. Desmaris, M. Strandberg, R. Finger, O. Nyström, M. Fredrixon, S. Ferm, H. Rashid och D. Henke, "Prototype ALMA Band 5 Cartridge:Design and Performance," *Proceedings of the 20TH INTERNATIONAL SYMPOSIUM ON SPACE TERAHERTZ TECHNOLOGY*, pp. 2–5, April 2009.

H. Rashid, D. Meledin, V. Desmaris, A. Pavolotsky, and V. Belitsky, "Superconducting 4–8-GHz Hybrid Assembly for 2SB Cryogenic THz Receivers," *IEEE Transactions on Terahertz Science and Technology*, vol. 4, pp. 193–200, March 2014.

G. Matthaei, L. Young and E.M.T. Jones, *Microwave Filters, Impedance-Matching Networks, and Coupling Structures*, Norwood, USA: Artech House, 1980.

V. Belitsky, I. Lapkin, V. Vassilev, R. Monje, A. Pavolotsky, D. Meledin, D. Henke, O. Nyström, V. Desmaris, C. Risacher, M. Svensson, M. Olberg, E. Sundin, M. Fredrixon, D. Dochev, S-E. Ferm and H. Olofsson, "Facility Heterodyne Receiver for the Atacama Pathfinder Experiment Telescope," *2007 Joint 32nd International Conference on Infrared and Millimeter Waves and the 15th International Conference on Terahertz Electronics*, pp. 326–328, 2007.

An 8-Pixel Compact Focal Plane Array with Integrated LO Distribution Network

Boon-Kok Tan[1,*] and Ghassan Yassin[1]

[1]*Department of Physics (Astrophysics), University of Oxford, Keble Road, Oxford OX1 3RH, UK*

*Contact: boonkok.tan@physics.ox.ac.uk

Abstract— We present the design of an 8-pixel Superconductor-Insulator-Superconductor (SIS) array centred at 650 GHz, which comprises two nearly identical 1×4 planar array chips, stacked together to form a 2×4 focal plane array. The array is fed by a single local oscillator (LO) source, and the array size is extendable by either increasing the number of mixing elements in the array chip or the number of stacking. The LO and RF signals for each mixer in the array are combined on-chip via a microstrip-coplanar waveguide (CPW) crossover which allows control of the RF/LO coupling level for each mixing element. The use of this planar beam splitter enables us to simplify greatly the design of the array mixer chip, as well as the design of the mixer block, which is important for future large pixel arrays. In this paper, we describe the design of the various components of the array chip, and the design of the mixer array block including the simplified LO distribution network.

INTRODUCTION

The time available for observing at short sub-millimetre wavelengths in the heterodyning mode, above 600 GHz, is limited even at high dry sites such as the Atacama or South Pole. It is therefore vital to fully maximise the scientific returns within the available observing time by constructing a large focal plane array. However, the key technologies required to build heterodyne receivers with large number of pixels at high frequencies is still very challenging [1], which requires innovative ideas such as compact LO injection scheme [2], modular array design that is easily extendable [3], and highly integrated IF amplifiers [4]. In this paper, we aim to address these challenges by using advanced superconducting planar circuit technology, to build an 8-pixel demonstrator at 650 GHz that could be extended to form a large format focal plane array. These ideas were chosen such that we can target on developing the novel solutions that would have large impact in realising compact and scalable receiver architectures, which will enable the construction of kilo-pixel heterodyne array at THz frequencies.

PLANAR COMPACT ARRAY CHIP

A major challenge for constructing a compact arrays is that there are only two interfaces for each mixer element which requires three input/output ports i.e., the RF, LO and IF ports. Traditional arrays solve this by using either multiple split blocks with a large waveguide coupler network [5], series of free-space beam splitter [6] or LO beam multiplexing [7], to inject the RF and LO signal to the mixer's input port. These solutions are often bulky and complicated. More importantly, the required LO power to feed all pixel may not be available by a single LO source. This is particularly vital for high-frequency applications where the LO power is limited. In our design, we replace these methods with a simple planar beam splitter that can be fabricated on-chip, directly integrated with

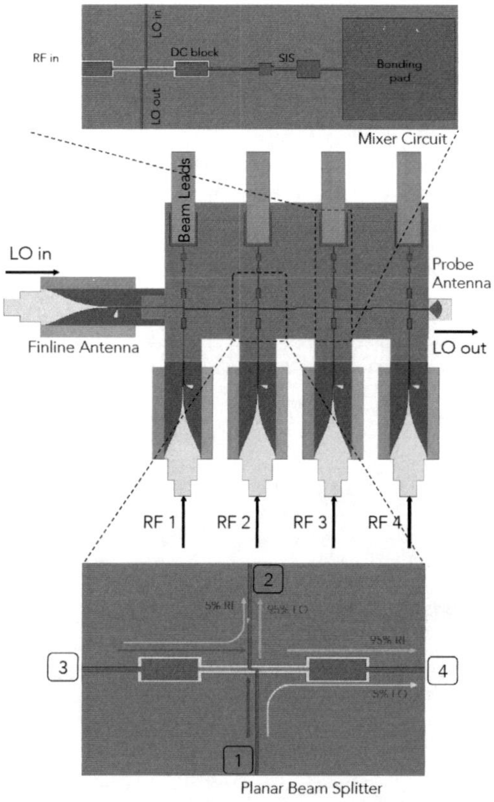

Fig. 1 Planar on-chip 4-pixel array with integrated planar beam splitters and mixer circuits deposited on a 15 μm thick silicon-on-insulator (SoI) substrate. The planar circuits are formed with 400 nm top and bottom Niobium layer sandwiching a 475 μm thick silicon monoxide dielectric. The bottom panel shows the transmission properties of the planar beam splitters, which divides power in a similar way as the traditional dielectric beam splitter.

A FULL VERSION OF THIS PROCEEDINGS ARTICLE HAS BEEN SUBMITTED TO THE IEEE TRANSACTIONS OF TERAHERTZ SCIENCE AND TECHNOLOGY (TST) FOR PEER-REVIEW, PLEASE CHECK THE WEBSITE: HTTPS://WWW.MTT.ORG/TERAHERTZ

the mixer detector circuit, and therefore eliminate the need for large array block or bulky optical arrangement.

Fig. 1 shows the design layout of a 1×4 array chip. The LO power is coupled from the finline antenna on the left and delivered to each mixer through a planar beam splitter, and exit the array chip via a probe antenna. Each mixing pixel comprises an input finline antenna, a planar beam splitter, and the mixer itself. The finline antenna and the mixer circuit design are similar to our conventional SIS mixer which operate at wide RF/IF bandwidths [8]. The planar beam splitter, on the other hand, is the key component that enables us to integrate multiple individual mixer elements onto a single substrate.

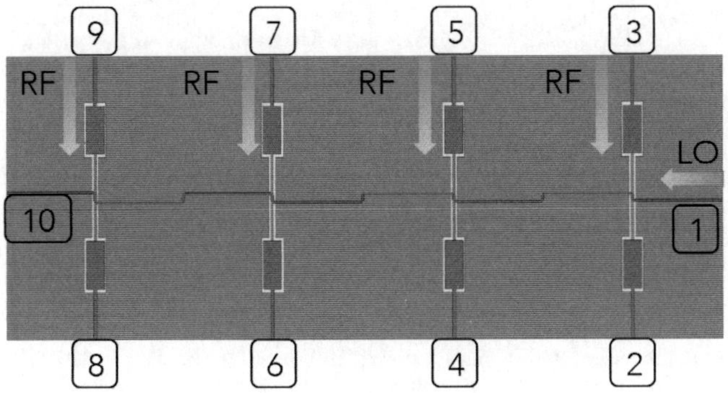

(a)

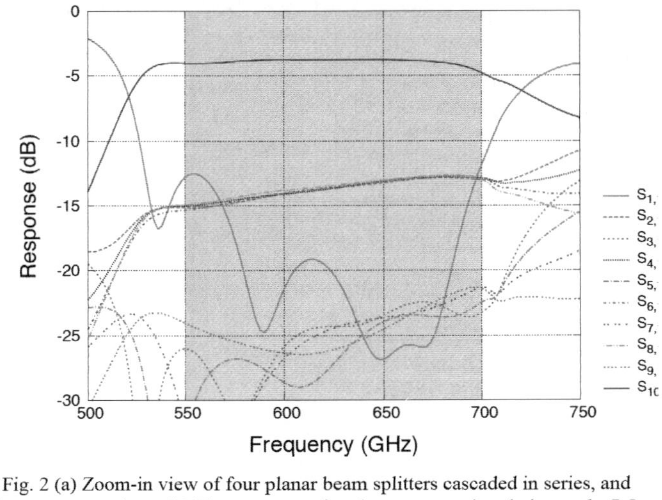

Fig. 2 (a) Zoom-in view of four planar beam splitters cascaded in series, and the port annotations. (b) The response of each output port in relation to the LO port (refer the port notation to panel above). Note that the LO power coupled to all the mixer ports ($S_{2,1}$, $S_{4,1}$, $S_{6,1}$ and $S_{8,1}$) are identical from 550–700 GHz.

The bottom panel of Fig. 1 shows the construction of the planar beam splitter, which is in fact similar to a coplanar waveguide (CPW) crossover reported in [9], except that in this case, instead of minimising the power coupling between the two crossing transmission lines, we aim to couple a small amount of power between them in a controllable fashion. This is achieved by bending a small section of the orthogonally oriented microstrip to align it with the CPW at the bottom layer (hereinafter Z-bend section). The detailed electromagnetic design of the planar beam splitter can be found in [10], so we concentrate here on how it can be used to form a complete on-chip array.

One important feature of the planar beam splitter is that the coupling coefficient between the RF and LO paths can be controlled simply by altering the length of the microstrip Z-bend section. This is especially useful for constructing the LO power distribution network that needs to split the input power evenly between mixer branches of the array. In particular, the drop of LO power after each planar beam splitter can now be compensated for, by gradually increasing the Z-bend length, hence ensuring the power delivered to all mixers remains constant. As shown in Fig. 2, the performance of the planar beam splitter is broadband, with leakage well below –20 dB from 500–700 GHz, and the return loss is as low as –15 dB from 580–700 GHz. The coupling coefficient stays constant at –13 dB within the same range for all the mixers. The power coupling also gradually increases with frequency to compensate for the gradual loss of LO power at the higher frequency end.

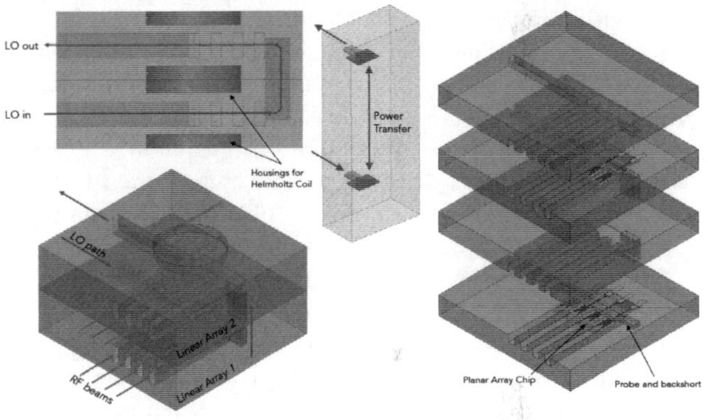

Fig. 3 Illustration of the array block design. Each 1×4 array can be stacked and the LO distribution network could be connected via a pair of radial probe antennas at the end of each array chip.

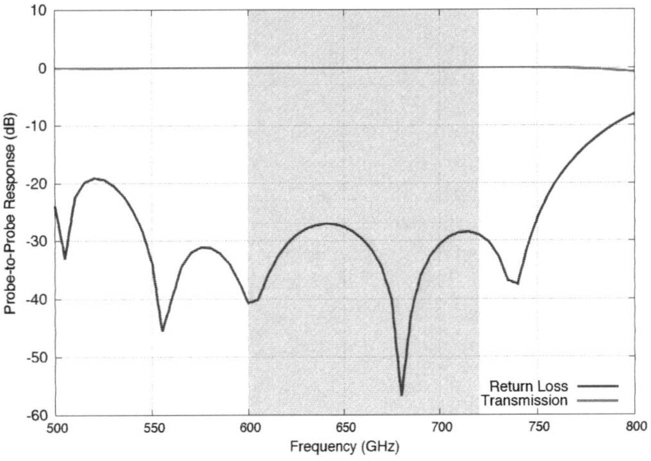

Fig. 4 The return loss and transmission characteristics of the probe-to-probe coupler, showing a broadband response.

Note that the remaining LO power after passing through the four SIS pixels arrays is still strong at $S_{10,1} = -4$ dB, and more importantly, remains flat across the operating frequency range. Hence, by cascading more mixers of the array chip in series,

A FULL VERSION OF THIS PROCEEDINGS ARTICLE HAS BEEN SUBMITTED TO THE IEEE TRANSACTIONS OF TERAHERTZ SCIENCE AND TECHNOLOGY (TST) FOR PEER-REVIEW, PLEASE CHECK THE WEBSITE: HTTPS://WWW.MTT.ORG/TERAHERTZ

the remaining LO power that pass through the planar beam splitter with small loss can be recycled for coupling to more mixer elements. This allows us to harness all of the available LO power and enable the use of a single LO source to feed multiple SIS pixels in a large focal plane array.

2×4 Focal Plane Array

The simple nature of the planar array chip design enables us to fabricate the array using the conventional split block technology, with the chip positioned at the E-plane of the input/output rectangular waveguides as shown in Fig 3. Two of these array blocks could be stacked together to form a 2×4 array. The array chip can also be populated with more pixels by simply duplicating more individual units of planar beam splitters and mixer circuits in series, and many more of these arrays can be stacked on top of each other to form an ever larger focal planar array. The spacing between each mixer unit could easily be adjusted to suite the Nyquist sampling of the required beams for a particular telescope. In the example shown in Fig. 3, the entire 1×4 array chip is only a few mm long, and compared to the traditional mixer array, this translates to almost an order of magnitude smaller array block. In order to feed the two arrays with a single LO source, the LO power could be coupled between each array via the radial probes at the end of the array chips. This is shown in Fig. 3 where two probes, one from each array chip, form a back-to-back coupling antenna with the aid of a pair of rectangular waveguide backshorts. The remaining LO power from the bottom array chip is coupled through the probe and the rectangular waveguide, to the probe of the top array chip. As shown in Fig. 4, this design is broadband with well below –25 dB return loss from near 500–750 GHz. Finally, a pair of Helmholtz superconducting coils can be inserted between the array to suppress the unwanted Josephson effects for the entire array of SIS junctions.

Conclusions

We have presented the design of an array chip comprising four SIS mixers fabricated on an SoI substrate. The LO distribution network for the array was integrated on-chip with the input antennas and the mixer circuits, through a series of cascaded planar beam splitters. We have shown that by gradually increasing the length of the microstrip Z-bend section of the planar beam splitter, we could compensate for the LO power loss after each mixer unit and ensure that the LO power coupled to each individual mixer is equal. Two of these array chips along with their housing blocks can be stacked together to form a 2×4 compact array. In order to fully utilise the LO power, we coupled the remaining LO power after the first array chip to the second chip via a set of back-to-back probe antennas. The preliminary prediction from the electromagnetic simulation shows promising performance with broad bandwidth. Due to the simple design of the array chip and the block, the compact array size can be extended easily either by adding more mixer units along the array chip or increasing the number of stacked array blocks. This enables the construction of hundred or kilo-pixel array in the near future, in particular near the THz frequency region.

References

[1] C. H. Wheeler, C. E. Groppi, H. Mani, P. McGarey, L. Kuenzi, S. Weinreb, D. S. Russell, J. W. Kooi, A. W. Lichtenberger, C. K. Walker, *et al.*, "The Kilopixel array pathfinder project (KAPPa), a 16-pixel integrated heterodyne focal plane array: characterization of the single pixel prototype," in *SPIE Astronomical Telescopes+ Instrumentation*, pp. 91530K–91530K, International Society for Optics and Photonics, 2014.

[2] J. Garrett, B.-K. Tan, F. Boussaha, C. Chaumont, and G. Yassin, "A 220 GHz finline mixer with ultra-wide instantaneous bandwidth," in *Twenty Sixth International Symposium on Space Terahertz Technology*, 2015.

[3] J. Leech, B. K. Tan, G. Yassin, P. Kittara, and S. Wangsuya, "Experimental investigation of a low-cost, high performance focal-plane horn array," *IEEE Transactions on Terahertz Science and Technology*, vol. 2, no. 1, pp. 61–70, 2012.

[4] S. Montazeri, P. K. Grimes, C.-Y. E. Tong, and J. C. Bardin, "A 220- GHz SIS mixer tightly integrated with a sub-hundred-microwatt SiGe IF amplifier," *IEEE Transactions on Terahertz Science and Technology*, vol. 6, no. 1, pp. 133–140, 2016.

[5] D.Maier, "230 GHz sideband-separating mixer array," in *Proc. 20th Int. Symp. on Space Terahertz Technology*, pp. 14–18, 2009.

[6] H. Smith, R. E. Hills, S. Withington, J. Richer, J. Leech, R. Williamson, H. Gibson, R. Dace, P. Ananthasubramanian, R. W. Barker, *et al.*, "HARP-B: a 350-GHz 16-element focal plane array for the James Clerk Maxwell telescope," in *Astronomical Telescopes and Instrumentation*, pp. 338–348, International Society for Optics and Photonics, 2003.

[7] C.Groppi,C.Walker,C.Kulesa,D.Golish,J.Kloosterman,S.Weinreb, G. Jones, J. Bardin, H. Mani, T. Kuiper, *et al.*, "Test and integration results from SuperCam: a 64-pixel array receiver for the 350 GHz atmospheric window," in *SPIE Astronomical Telescopes+ Instrumentation*, pp. 77410X–77410X, International Society for Optics and Photonics, 2010.

[8] B.-K. Tan, G. Yassin, P. Grimes, and K. Jacobs, "650 GHz SIS mixer fabricated on silicon-on-insulator substrate," *Electronics Letters*, vol. 49, no. 20, pp. 1273–1275, 2013.

[9] A. Abbosh, S. Ibrahim, and M. Karim, "Ultra-wideband crossover using microstrip-to-coplanar waveguide transitions," *IEEE Microwave and Wireless Components Letters*, vol. 22, no. 10, pp. 500–502, 2012.

[10] B.-K. Tan and Yassin, "A Planar Beam Splitter for Millimetre and Sub-Millimetre Heterodyne Mixer Array," *IEEE Transaction on Terahertz Science and Technology*, 2017. Submitted.

A FULL VERSION OF THIS PROCEEDINGS ARTICLE HAS BEEN SUBMITTED TO THE IEEE TRANSACTIONS OF TERAHERTZ SCIENCE AND TECHNOLOGY (TST) FOR PEER-REVIEW, PLEASE CHECK THE WEBSITE: HTTPS://WWW.MTT.ORG/TERAHERTZ

An All Solid-State Receiver at 2 THz for Atmospheric Sounding

J. Treuttel[1&2], E. Schlecht[1], C. Lee[1], J-V. Siles[1], D. Hayton [1], R.Lin[1], D. Wu [3] and I. Mehdi[1]

[1]Jet Propulsion Laboratory, USA
[2]Observatory of Paris, LERMA, France
[3]NASA Goddard Space Flight Center

Abstract— Schottky diode front-end receivers have been demonstrated up to 2.5 THz with a CO2-pumped methanol gas laser local oscillator sources [1, 2]. In this work we report on the progress towards demonstrating a fully solid-state heterodyne receiver in the 2 THz range. A subharmonically pumped mixer chip pumped by a cascaded multiplier chain as the LO has been designed, developed and tested. The mixer chip [3] is packaged in a split waveguide block with an integrated diagonal horn for RF coupling. To reduce losses, current implementation is based on last stage tripler and mixer housed together. Recent results obtained at room temperature will be discussed. The LO scheme can provide 1-2 mW of power around 1 THz which is enough to make preliminary receiver sensitivity measurements indicating a DSB mixer noise temperature of better than 15,000 K. On-going work is focused on improving the mixer sensitivity and making cryogenically cooled measurements.

[1] P. Siegel, R. Smith, M. Gaidis, and S. Martin, "2.5-THz GaAs monolithic membrane-diode mixer," *IEEE Trans. Microw. Theory Techn.* , vol. 47, no. 5, pp. 596–604, May 1999.

[2] H.-W. Hübers, T. W. Crowe , G. Lundershausen , W. C. B. Peatman , H. P. Röserl "Noise Temperatures and Conversion Losses of Submicron GaAs Schottky—Barrier Diodes," 4th *International Symposium on Space Terahertz Technology*, 1993.

[3] B. Thomas, A. Maestrini, J. Ward, E. Schlecht, G. Chattopadhyay, J. Gill, C. Lee, R. Lin, and I. Mehdi, "Terahertz cooled sub-harmonic Schottky mixers for planetary atmospheres," *5th ESA Workshop Millimeter Wave Tech. Apps*, 3/30/2015.

Theoretical consideration of SIS up-converters for frequency division multiplexing

Yoshinori Uzawa[1,2*], Wenlei Shan[2], Takafumi Kojima[2], Alvaro Gonzalez[2], and Matthias Kroug[2]

[1]*National Institute of Information and Communications Technology, 184-8795, Tokyo, Japan*
[2]*National Astronomical Observatory of Japan, 181-8588, Tokyo, Japan*
[*]Contact: uzawa@nict.go.jp

Abstract— In radio astronomy there is a strong interest in large-scale multi-pixel heterodyne receivers, which enable wide field-of-view observations with high spectral resolutions. So far, multi-pixel heterodyne receivers have been developed by several institutes. The maximum pixel number is 64 achieved by "SuperCam" to explore astrophysically important emission and absorption lines within the 850 micron atmospheric window. Further increase of the pixel number to as many as 1000 is now being considered. In this case one of the issues would be the total power consumption of a large number of cryogenic low noise amplifiers (LNAs). Given that a typical LNA consumes several milliwatts, for example, refrigerators with several watts of cooling capacity at the 4-K stage are necessary, which would be impractical. To overcome this issue, we propose to use a frequency division multiplexing (FDM) technique for reducing the number of amplifiers at 4 K. The concept is the following; Each SIS mixer corresponding to each pixel outputs an intermediate frequency (IF) signal at microwave frequencies that is up-converted with different local oscillator (LO) frequencies (e.g. frequency comb signals) where the frequency differences of LOs (the comb frequency interval) should be larger than the IF signal bandwidth. These up-converted signals of multiple pixels could be combined by an FDM, then amplified by an amplifier which has a wideband at a much higher frequency than the IF. One of the key elements to make this scheme possible is the up-converter which should have a gain larger than unity to prevent degradation of the receiver performance due to extra losses before the LNA. We consider SIS mixers as a good candidate for the up-converter because positive conversion gains could be expected by quantum effects. For the feasibility study, we calculated up-conversion characteristics of SIS mixers based on the Tucker theory. Analytical results will be shown to find operating conditions of SIS mixers as up-converters for a workable FDM scheme.

Pre-prototype ALMA Band 2+3 Down-Converter & Local Oscillator System

H. Wang[1*], B. Ellison[1], E-M Butroid[1], GA Fuller[2], D. George[3], D Cuadrado-Calle[2,3], P. Yagoubov[4]

[1]*STFC-Rutherford Appleton Laboratory, Oxfordshire, OX11 0QX, UK*
[2]*School of Physics & Astronomy, The University of Manchester, Manchester, M13 9PL, UK*
[3]*School of Electrical and Electronic Engineering, The University of Manchester, Manchester, M13 9PL, UK*
[4]*European Southern Observatory, Karl-Schwarzschild-Str. 285748, Garching, Germany*
*Contact: hui.wang@stfc.ac.uk

Abstract— A merger of the individual ALMA front-end receiver Bands 2 and 3 into a single receiver system (Band 2+3) has the advantage of allowing simultaneous observations of spectral lines across a wide frequency range and thereby enhancing ALMA science. It also provides operational advantage by increasing access to the ALMA front-end cryostat system and thus allowing the possibility of expansion of the array front-end capability in the future.

The wide operational bandwidth of a Band2+3 system requires demonstration, however, and key technologies need to be proven. In support of this, the UK Science and Technology Facilities Council (STFC) has funded the development of a pre-prototype down-converter chain suitable for use with Band 2+3 front-end LNAs and in support of the European Southern Observatory (ESO) objective of exploring the feasibility of developing a combined Band 2+3 receiver front-end.

The ALMA Band 2+3 system is designed to operate over an input signal frequency range of 67-116GHz. The receiver system uses a cryogenic low noise amplifier (LNA) at its input, the output of which must be frequency translated (down converted) to the ALMA intermediate frequency (IF) range of 4-12GHz. The down-converter comprises a subharmonic sideband separating mixer (one for each polarisation), local oscillator (LO) and IF amplifier chain. The LO must be frequency tunable and provide sufficient output power to pump the mixer. The LO chain includes a voltage controlled oscillator (VCO) which is amplified and harmonically up converted and encompasses the frequency range 39-52.5GHz, with typical +8dBm output power. A digital interface connected to a control computer via a standard universal serial bus (USB) is used to set the VCO, and hence LO output frequency.

The sideband separating subharmonic mixer uses RAL fabricated Schottky barrier diodes. Signal input and LO division is accomplished within single mixer block housing. Design simulations indicate the expected mixer noise and conversion loss performance to be approximately 1000K (SSB) and -8dB respectively for a typical LO power level of +8dBm. The mixer block and internal circuitry have been manufactured at RAL. Details of the measurement results including mixer noise temperature and conversion loss will be presented.

Noise Temperature of a Wideband Superconducting HEB mixer

Kangmin Zhou[1,2], Daowei Wang[1,2,3], Wei Miao[1,2], Wen Zhang[1,2], Yan Delorme[4] and Shengcai Shi[1,2,*]

[1]*Purple Mountain Observatory, Chinese Academy of Sciences, 210008 Nanjing, China*

[2] *Key Laboratory of Radio Astronomy, 210008 Nanjing, China*

[3] *Shanghai Normal University, 200234 Shanghai, China*

[4] *Observatoire de Paris, 75014 Paris, France*

*Contact: scshi@pmo.ac.cn

***Abstract*—In this paper we report on the development of a log-spiral antenna coupled superconducting HEB mixer which has a RF bandwidth of 0.2-2 THz. The DSB receiver noise temperature (T_{rec}) has been measured from 0.2 THz to 1.4 THz with same measurement setup at 4 K bath temperature, the uncorrected T_{rec} shows good performance in the whole operating frequency and is 700 K at 0.2 THz, 700 K at 0.5 THz, 750 K at 0.6 THz, 750 K at 0.85 THz and 1000 K at 1.34 THz. The calibrated intrinsic noise temperature of HEB device shows a frequency independent performance across the gap frequency.**

***Index Terms*—Superconducting HEB mixer, receiver noise temperature, wide band, frequency dependence**

I. Introduction

Superconducting hot-electron bolometer mixers [1], with the advantages of high sensitivity and low LO power requirement, have been already used in ground-based [2] and space-based observatories [3]. Unlike superconducting SIS mixers, superconducting HEB mixers don't suffer from a cut-off frequency set by the superconductor's energy gap (gap frequency) [4], therefore they can be used in whole THz frequency range if integrating with different antenna feeds. A non-uniform absorption model of superconducting HEB mixer has been proposed recently [5], the frequency dependence of intrinsic noise temperature across the frequency gap has been theoretical investigated but hasn't been experimentally proved.

A planar log-spiral antenna [6] coupled superconducting HEB mixer can achieve very wide input bandwidth, making it good choice to study the noise temperature performance far below and beyond the gap frequency. In this paper, we report on the design and characterization of a spiral antenna coupled superconducting NbN HEB mixer whose frequency range is from 0.2 THz to 2 THz. The receiver noise temperatures across the gap frequency are studied in particular.

II. HEB Mixer Design

The 0.2-2 THz superconducting HEB mixer consists of a 2 μm wide, 0.2 μm long and 5.5 nm thick NbN microbridge based on a highly resistive and natively oxidized Si substrate, and a self-complementary log-spiral planar feed which couples the input RF and LO signal to the NbN microbridge.

A scanning electron microscope (SEM) micrograph of the measured device is shown in Fig. 1. The parameter $a = 0.32$ was chosen to build the log-spiral arms of the feed according to $r = ke^{a\phi}$ with ϕ being the azimuth angle and a being the distance from the geometric center of the spiral. The size of a spiral portion of the feed is defined by the outer and inner diameter. The outer diameter (D = 300 μm) is the diameter of the smallest circle that encompasses the spiral structure. The inner diameter (d = 8.4 μm) is the diameter of the smallest circle at which the arms still obey the spiral equation.

The superconducting HEB device was fabricated at LERMA based on an in-situ process. Unlike conventional processes [7] with a superconducting layer adopted to improve the contacting layer quality, the Au contact layer is directly deposited on the NbN film after an in-situ cleaning by the argon plasma. A 5 nm titanium layer is used as an adhesion layer. Then a lift-off process is performed on the Au layer to form the HEB's electrodes, the antenna, the transmission lines and the contact pads. The width of the microbridge is determined by reactive ion etching through a mask made of nickel. Finally, a dielectric SiN layer was deposited on the whole substrate for protection [8].

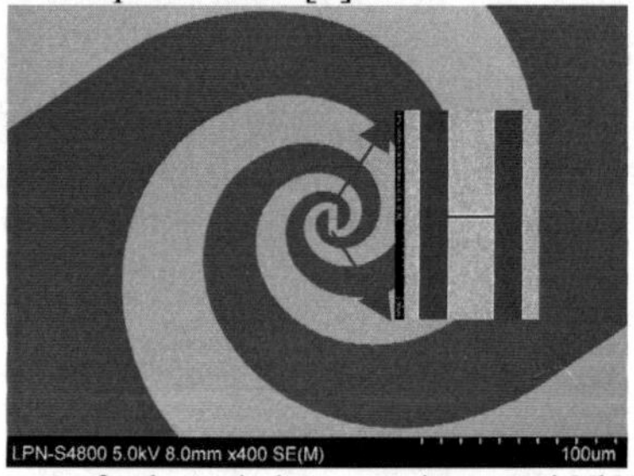

Fig. 1. SEM image of a log-spiral antenna integrated NbN HEB mixer on silicon. The light gray area is the gold antenna structure, while the dark area is the Si substrate. The close-up picture shows the structure of micro-bridge.

III. Receiver Noise Temperature

The noise temperatures of this HEB mixer are characterized at 0.2 THz, 0.5 THz, 0.6 THz, 0.85 THz and 1.34 THz using the Y-factor method. The measurement setup is illustrated in Fig. 2. The HEB device is glued to the backside of an elliptical Si lens which has a diameter of 10 mm and an extension length of 1.149 mm, and is mounted into a copper mixer block anchored on the 4 K cold plate of a close-cycled cryostat. The LO signals are provided by different frequency multiplier chains. The beam from the LO source is collimated by an objective Teflon lens, and then coupled into the cryostat through a 25 μm Mylar beam splitter. The cryostat has a 1.5 mm thick high-density polyethylene (HDPE) window and a Zitex G104 infrared filter. The mixer intermediate frequency (IF) output signal goes through a bias-tee and an isolator, and is then amplified by a cryogenically cooled low-noise amplifier (LNA) of 35 dB gain and a room temperature amplifier of 40 dB gain. The IF output signal is filtered at 1.5 GHz within a bandwidth of 80 MHz and recorded by a square-law detector.

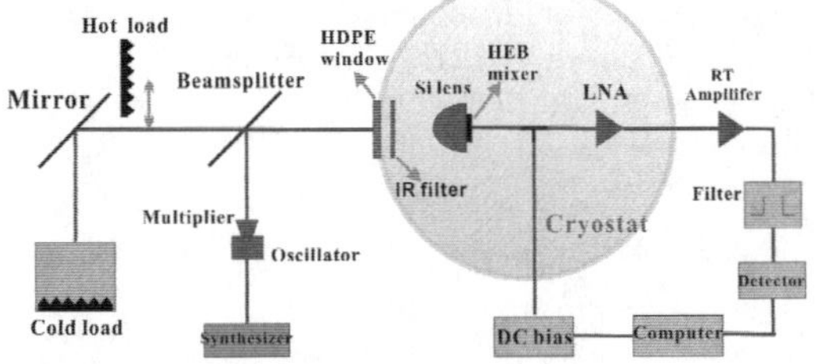

Fig. 2. Diagram of the experimental setup.

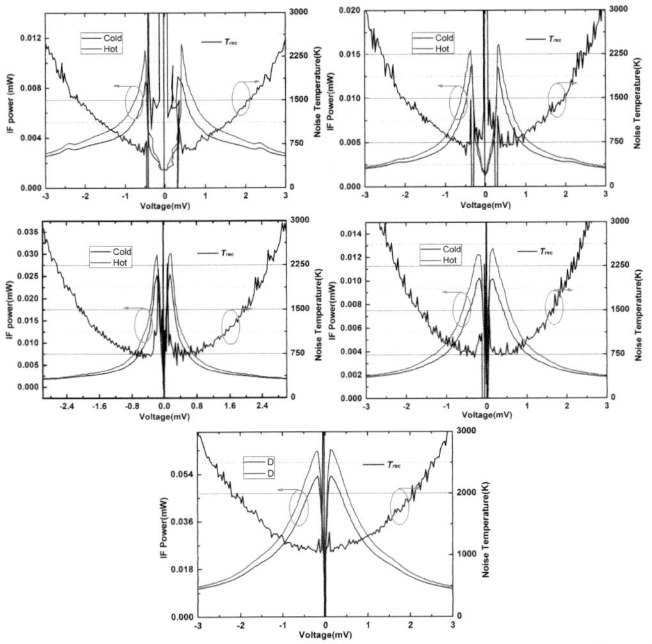

Fig. 3. Measured receiver IF output power (left axis) for the hot-and cold-load at an optimal LO power and the corresponding DSB receiver noise temperature (right axis) at 0.2 THz, 0.5 THz, 0.6 THz, 0.85 THz and 1.34 THz, shown as a function of the mixer dc-bias voltage.

Fig. 3 shows the measured receiver IF output powers at 295 K and 77 K and the corresponding receiver noise temperature as a function of the HEB dc-bias voltage. The temperatures of the hot and cold blackbody (295 K and 77 K) were calibrated to the effective radiation temperatures in terms of the Callen-Welton definition [9]. In addition, the direct detection effect was compensated by adjusting the LO power [10] to make the IV curves unchanged between the hot-load and cold-load measurement. The lowest uncorrected DSB receiver noise temperatures at these frequencies are listed in Table I. The result indicates that this HEB mixer has high performance in a broad frequency range. The calibrated intrinsic noise temperature of HEB device shows a frequency independency.

TABLE I
Lowest DSB receiver noise temperatures

Frequency (THz)	0.2	0.5	0.6	0.85	1.34
Uncorrected $T_{rec}(K)$	700	700	750	750	1000
Corrected $T_{heb}(K)$	290	295	290	295	300

IV. Conclusion

We have successfully developed a spiral antenna coupled superconducting NbN HEB mixer which has a high performance over a frequency range of 0.2-2 THz. The corrected intrinsic noise temperature shows a frequency independency over the whole frequency range across the frequency gap.

Acknowledgment

This work was supported in part by the NSFC under Grants11190012, 11673073, 11603081, 11473075 and by Chinese Academy of Sciences under Grant XDB04010300 and XDB23020200. The authors are grateful to K. Zhang, Q. J. Yao and W. Y. Duan of PMO for their assistance.

References

[1] E.M. Gershenzon, G.N. Gol'tsman, I.G. Gogidze, A.I. Elant'ev, B.S. Karasik, and A.D. Semenov, "Millimeter and submillimeter range mixer based on electronic heating of superconducting films in the resistive state," *Soviet Physics Superconductivity*, Vol. 3, pp. 1582-1597, 1990.

[2] D. Meledin, A. Pavolotsky, V. Desmaris, I. Lapkin, C. Risacher, V. Perez, D. Henke, O. Nystrom, E. Sundin, D. Dochev, M. Pantaleev, M. Fredrixon, M. Strandberg, B. Voronov, G.N. Gol'tsman, and V. Belitsky, "A 1.3-THz balanced waveguide HEB mixer for the APEX telescope," *IEEE Transactions on Microwave Theory and Techniques,* Vol. 57, No. 1, pp. 89-98, 2009.

[3] S. Cherednichenko, V. Drakinskiy, T. Berg, P. Khosropanah, and E. Kollberg, "Hot-electron bolometer terahertz mixers for the Herschel Space Observatory," *Review of scientific instruments,* Vol. 79, p. 034501, 2008.

[4] J.R. Tucker and M.J. Feldman, "Quantum detection at millimeter wavelengths," *Rev. Mod. Phys.*, 1985 (57), pp. 1055–1113.

[5] W. Miao, W. Zhang, J. Q. Zhong, S. C. Shi, Y. Delorme, R. Lefevre, A. Feret, and T. Vacelet, "Non-uniform absorption of terahertz radiation on superconducting hot electron bolometer microbridges," *Applied Physics Letters,* 104.5 (2014): 052605.

[6] T. H. Büttgenbach, R. E. Miller, M. J. Wengler, D. M. Watson, and T. G. Phillips, "A broadband low-noise SIS receiver for submillimeter astronomy," *IEEE Trans. Microw. Theory Tech.*, vol. 36, no. 12, pp.1720–1726, Dec. 1988.

[7] J. J. A. Baselmans, M. Hajenius, J. R. Gao, T. M. Klapwijk, P. A. J. de Korte, B. Voronov and G. Gol'tsman, "Doubling of sensitivity and bandwidth in phonon cooled hot electron bolometer mixers," *Appl. Phys. Lett.* 84(11), 2004.

[8] R. Lefevre, Y. Delorme, A. Feret, F. Dauplay, W. Miao, L. Pelay, B. Lecomte, B. Guillet, G. Beaudin, and J.M. Krieg, "Development of membrane based NbN-HEBs for submillimeter astrophysical applications," *Proc. of 19th International Symposium on Space Terahertz Technology, Groningen, Netherlands*. pp. 413-415, 2008.

[9] A.R. Kerr "Suggestions for revised definitions of noise quantities, including quantum effects," *IEEE Transactions on Microwave Theory and Techniques*, Vol. 47, No. 3, pp. 325-329, 1999.

[10] Baselmans, J. J. A., et al. "Direct detection effect in small volume hot electron bolometer mixers." *Applied Physics Letters,* 86.16 (2005): 163503-163503.

GREAT's Internal Beam Scanner

Urs U. Graf*

[1]*University of Cologne, 50937, Cologne, Germany*
*Contact: graf@ph1.uni-koeln.de

Abstract—We present the design of a compact measurement device to determine the position of a beam in a radio optical setup. The unit is used to align the Terahertz optics of the GREAT/upGREAT instrument on the airborne astronomical observatory SOFIA.

Resonant Modes in Parallel Josephson Junction Arrays for Submm Oscillator Applications

F. Boussaha, J-G. Caputo, C. Chaumont, A. Féret, and T. Vacelet

***Abstract*—We report on the development of submm oscillators using parallel Josephson junctions unevenly distributed within a superconducting stripline. To optimize their RF operation within a desired bandwidth, a straightforward method has been used. The radiation emitted from the arrays is detected using integrated SIS twin-junctions. Despite the small junction number (N=21), the *I-V* characteristics of the detector exhibit clearly photon-assisted quasiparticule steps when the array is biased upon Josephson resonances ranging from 370 to 500GHz. The array has a moderate current density of ~ 6kA/cm2 and provides an output power around 0.28μW which is sufficient to achieve sensitive integrated heterodyne receivers. Furthermore, the parallel array approach can also be extended to higher frequencies using high critical temperature superconducting materials.**

I. Introduction

Parallel SIS array concept has already been theoretically and experimentally investigated with the aim of implementing tunable oscillators [1-4]. However, most studies were done by considering only arrays without taking into account some important engineering aspects related, for example, to the power coupling towards an external circuit (antenna, RF coupler), expected frequencies of operation, bandwidth of interest, etc. Indeed, the typical studied arrays feature an identical distance between two adjacent junctions which is likely randomly set. To achieve efficient and usable oscillators, we designed, fabricated and characterized parallel arrays comprising a few small Nb/AlO_x/Nb-based SIS junctions (N=21) [5] which are parallel-connected by a superconducting Nb/SiO/Nb microstrip line. The array is capacitively coupled, through an RF/DC block, to a sub-mm SIS-based detector to detect the emitted radiations. Using the electrical model of fig.1, the array is optimized as a *RLC* circuit where $R=R_j$ is the quasiparticle current, $L=L_j$ is the Josephson inductance and $C=C_j$ the intrinsic capacitance junction. This lead to an irregular distribution of junctions within a superconducting Nb/SiO/Nb microstrip line. As a first order, we linearized the Josephson nonlinear admittance produced by the Josephson current $I=I_c sin(\varphi)$ where φ is the phase difference of the two superconductor electrodes by a constant inductance $L_j=\Phi_0/2\pi I$ where $I_c=J_c S_{j\_array}$ is the critical current of the SIS junction. In this paper, we experimentally study the different Josephson modes of the array that optimized as an oscillator in 320-600GHz band.

II. Circuit and Experimental Results

Figure 2 presents the 21-junction array coupled to a SIS-based detector through a slotline/microstrip transition which acts as a RF/DC block. Further details can be found in [5]. Fig. 3-a shows the *I-V* curve of the array which displays resonances right before jumping to the voltage gap (not shown in the figure), for different values of the applied magnetic field generated by a current I_{MG_array} passing through a control line. I_{MG_array} ranges from 85 to 174mA. For clarity, the measured resonance voltages are converted in frequencies using $f_{res}=V_{res}/\Phi_0$.

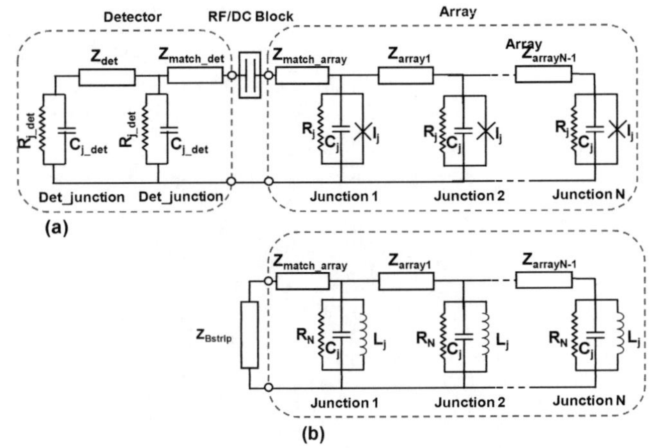

Fig. 1. (a) Equivalent model of the circuit consisting of a SIS junction array capacitively coupled to a SIS-based detector through a RF/DC block. Each SIS junction is modeled by its intrinsic capacitance C_j, quasiparticles conductance $G_j=1/R_j$ and Josephson current $I_j=I_c sin(\varphi)$ in parallel. (b) The Josephson nonlinear admittance is linearized by a constant inductance $L_j=\Phi_0/2\pi I_c$ and quasiparticle resistance R_j is taken close to the normal resistance R_N.

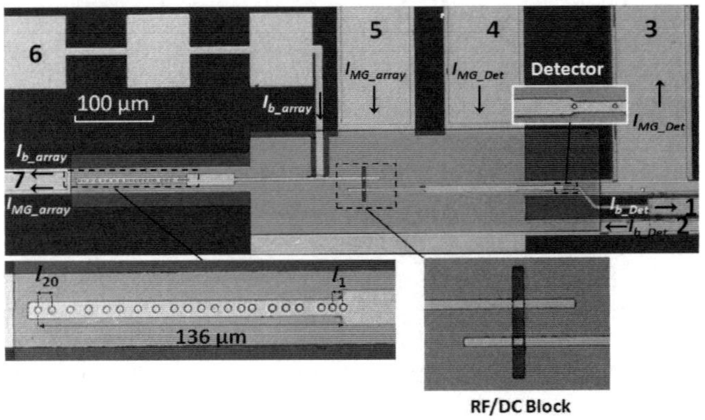

Fig. 2. Photograph of the typical optimized parallel array-based circuit which consists of: twin Nb/AlOx/Nb SIS junction detector, dc-block made and parallel arrays of 21 Nb/AlOx/Nb junctions. $l_{1..20}$=5, 5, 10, 6, 6, 9, 5, 6, 6, 6, 6, 6, 8, 8, 8, 6, 8, 8, 8, 6μm.

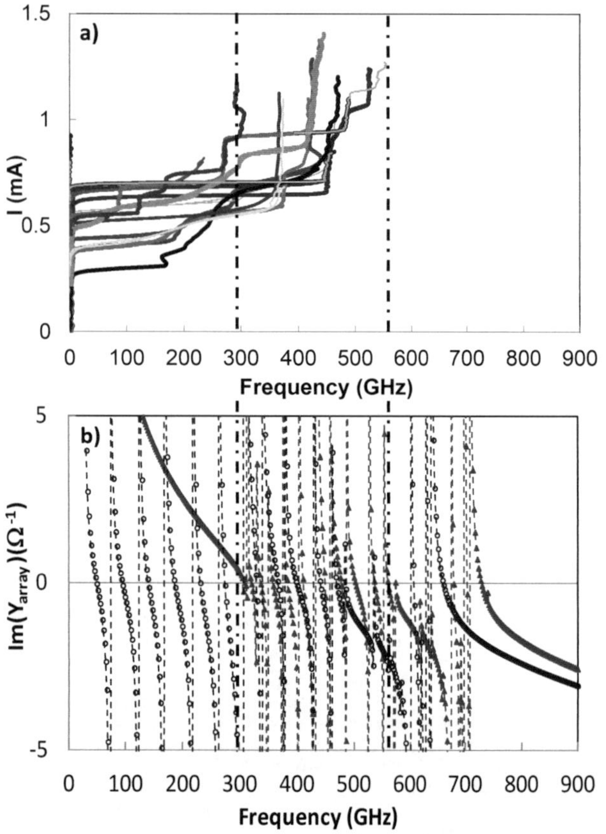

Fig. 3. Array's *I-V* curve showing resonances excited by magnetic fields generated by currents I_{MG_array} ranging from 85 to 174mA. The array admittance $Y_{array}(\omega)$ includes the linearized Josephson admittances $(jL_J\omega)^{-1}$. To compare with the simulated imaginary $Im[Y_{array}(\omega)]$ (curve below), the voltages of resonances are converted in frequencies using $V_{res}\times483.59$GHz.

Depending on the current magnitude, we can distinguish 2 resonant modes. The first mode occurs up to ~266 GHz and shows up six steps featuring low current magnitude, between V_{res}=0.1 and 0.55 ± 0.01mV. In frequency, this leads to $f_{res}=V_{res}/\Phi_0$=48 GHz and 266 ± 4.83GHz, respectively. The resonances appear with nearly regular voltage step of ΔV_{res}=0.08 ± 0.01mV (Δf_{res}=38±4.83GHz) at the same voltage locations whatever the magnetic field strength. The second resonant mode displays sharp resonances with significantly higher current magnitude occurring from V_{res}=0.55 to 1.14 ± 0.01mV. This yields $f_{res}=V_{res}/\Phi_0\approx$266 to 551GHz which corresponds to the optimized bandwidth limits [5]. Unlike the first mode, there are not equidistant in voltage and their respective voltage position depends strongly upon I_{MG_array} values (i.e., magnetic field). In order to access the resonant frequencies, we plot the imaginary part of the array admittance $Y_{array}(\omega)$ as function of the frequency. Thus, the resonances occurred when $Im[Y_{array}(\omega)]$ goes to 0 [6]. The blue curve (circle) shows the $Im[Y_{array}(\omega)]$ in the absence of the Josephson inductive element while the red curve (triangle) shows the $Im[Y_{array}(\omega)]$ when it is taken into account. Further detailed analysis of both modes will be reported elsewhere. To access whether output powers come out of the array, we exited resonances whose frequencies are expected within the detector bandwidth. The radiation emitted from the arrays is detected using the integrated SIS twin-junctions. The I-V characteristics of the detector clearly exhibit photon-assisted quasiparticule steps when the array is biased upon its Josephson resonances ranging from 370 to 500GHz. Figure 4 displays the detector's *I-V* curve when the array is biased at V_{res}=0.79, 0.85 and 0.91 ± 0.01mV . These correspond to emitted frequencies $f_{res}=V_{res}/\Phi_0$=381, 411, 440 ± 4.83GHz, respectively. The resonances are excited with magnetic fields generated by currents $I_{\_MG}$=91, 97 and 105mV, respectively. The unpumped curves are obtained when arrays are biased out of resonances either at V_{bias}=0 or at $V_{bias}>V_{gap}$. From the *I-V* pumped curve, the calculation of frequencies using the photon-assisted steps $f_{ph}=e(V_{gap}-V_{ph})/h$ gives same frequencies than the emitted ones. The maximum pumping is reached at 440 ± 4.83GHz with V_{res}=0.91±0.01mV and I_{MG}=105mA. The array has a moderate current density (Jc≈6kA/cm2) and provided an output power around 0.28μW which would be sufficient to achieve sensitive heterodyne reception [5].

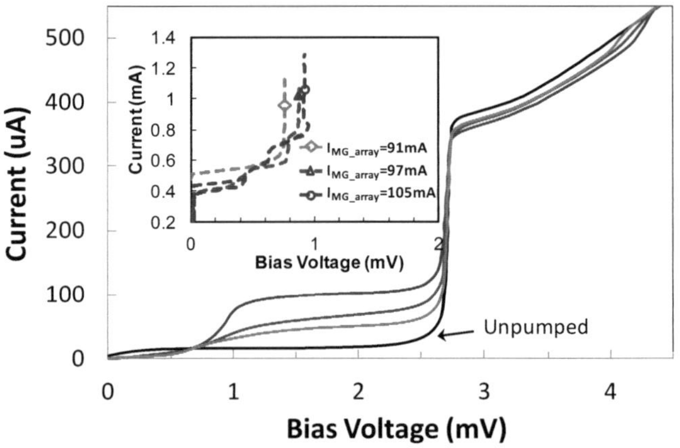

Fig. 4. Measured detector's *I-V* characteristics pumped at 381, 411, 440GHz when the 21-junction array is tuned at V_{res}=0.79, 0.85, 0.91mV respectively. The inset shows the resonances obtained for I_{MG_array}=91, 97 and 105mA, respectively. The measured unpumped curve is obtained when the array is biased out of the resonance at V_{bias}=0 or $V_{bias}>V_{res}$.

REFERENCES

[1] A. V. Ustinov, M. Cirillo, B. H. Larsen, V. A. Oboznov, P. Carelli, and G. Rotoli, "Experimental and numerical study of dynamic regimes in a discrete sine-Gordon lattice", Phys. Rev. B 51, 308, 1995.

[2] H.S.J. Van der Zant, E.H. Visscher, D. R. Curd, T.P. Orlando, "Vortex dynamics in one-dimensional parallel arrays of underdamped Josephson junctions," IEEE Trans. On Appl. Superconductivity, Vol. 3, No. 1, 1993.

[3] Y. M. Zhang and P. H. Wu, "Numerical calculation of the height of velocity-matching step of flux-flow type Josephson oscillator," J. Appl. Physics 68, 4703, 1990.

[4] S. Han, B. Bi, W. Zhang, and J. E. Lukens, "Demonstration of Josephson effect submillimeter wave sources with increased power," Appl. Phys. Lett. 64, 1424, 1994.

[5] F. Boussaha, C. Chaumont, A. Féret, J-G. Caputo and T. Vacelet "Optimized Non-Uniform Nb SIS parallel Josephson junction array oscillators", IEEE Transactions on Applied Superconductivity, Vol. 26, Issue 3, 2016.

[6] M. Salez and F. Boussaha, "Fluxon modes and phase-locking at 600 GHz in superconducting tunnel junction nonuniform arrays," J. Appl. Phys. 107 (013908), (2010).

THz Heterodyne Sensors Using Superconducting MgB_2

Boris S. Karasik[1,*], Daniel Cunnane[1], Narendra Acharya[2], Wenura K. Withanage[2], Xiaoxing Xi[2]

[1]*Jet Propulsion Laboratory, California Institute of Technology, Pasadena, CA 91109, USA*
[2]*Department of Physics, Temple University, Philadelphia, PA 19122, USA*
[*]Contact: boris.s.karasik@jpl.nasa.gov

Abstract—Since its discovery in 2001, the expectations have been high for magnesium diboride (MgB_2) to find multiple applications in superconducting electronics, including radiation detectors. Operation at 20+ K rather than at liquid helium temperatures is the key here, especially for sensors intended for space use. However, until recently not much has been done mostly due to the challenge of obtaining thin films with high critical temperature (T_C) close to the bulk value of 39 K. The Hybrid Physical-Chemical Vapor Deposition (HPCVD) technique has been a breakthrough in the thin-film technology yielding superior c-axis oriented MgB_2 films as thin as 5 nm and with a critical temperature of 36-38 K. A combination of a high T_C, an unusually high sound velocity in MgB_2 ($\approx$ 8 km/s), and a good acoustic impedance match to Al_2O_3, MgO, and SiC substrates greatly accelerates the thermal relaxation of non-equilibrium electrons thus enabling promising applications in THz Hot-Electron Bolometer (HEB) mixers. Indeed, an 7-8 GHz intermediate frequency (IF) in HEB mixers has been already demonstrated. Our more recent achievement is a demonstration of an operation of the spiral antenna coupled MgB_2 HEB working between 4 K and 15 K without any sensitivity degradation. The respectable receiver's (no correction) DSB noise temperature T_M = 2,000 K @ 600 GHz and T_M = 3,600 K @ 1.89 THz has been measured. The ultimate goal here is to achieve a broadband mixer for [OI] @ 4.7 THz line detection where an IF bandwidth $\approx$ 8 GHz is required for capturing complete line spectra across the galaxy. An ongoing material development work addresses manipulation of the lattice disorder in the film in order to reduce the granular structure in the film thus increasing the uniformity of the resistive state and overall HEB mixer performance.

Besides the HEB, we have achieved high-quality planar Josephson junctions (JJ) with an I_CR_N product approaching 5.3 mV have been already studied as radiation mixers. Even without optimization, they demonstrated a noise temperature $\approx$ 600 K @ 600 GHz. The mixing was observed all the way up to 2 THz where $T_M \approx$ 2,000 K. The noise temperature degrades slightly (a 20% increase) between 10 K and 20 K. The other advantages of the JJ mixer are the unlimited IF bandwidth and smaller than in HEB required LO power. It still remains to be seen whether the quasiparticle tunneling branch can be utilized for mixing but the possibility to use MgB_2 for the device ground plane, antenna, and wiring is quite real. This may help to improve the state-of-the-art of Superconductor-Insulator-Superconductor (SIS) mixers. Another application area for the MgB_2 JJ's may be in the on-chip integrated THz sources (e.g., frequency multipliers, flux-flow osscillators).

In this talk, we will overview the results obtained so far and for the above detectors and discuss the ways to improve their performance.

Shot Noise in NbN/AlN/NbN Superconducting Tunneling Junctions

Dong Liu[1,2,3,*], Jie Hu[1,2,3], Ming Yao[1,2,3], Jing Li[1,2], and Shengcai Shi[1,2]

[1]*Purple Mountain Observatory, Nanjing, China*
[2]*Key Laboratory of Radio Astronomy, CAS, Nanjing, China*
[3]*Graduate School of Chinese Academy of Sciences, Beijing, China*
[*]Contact: dliu@pmo.ac.cn

Abstract—With sensitivity approaching the quantum limit, superconductor-insulator-superconductor (SIS) mixers play an important role in radio astronomy and atmospheric science at millimeter and submillimeter wavelengths. As one of the intrinsic noise sources in superconducting tunneling junctions, shot noise is still not well understood, particularly for those of relatively high energy gap (e.g., NbN/AlN/NbN).

In this paper, we mainly study the Multiple Andreev reflection (MAR) enhanced shot noise of two different NbN SIS junctions (i.e., junction array and long junction) as well as their temperature dependence. Barrier strength Z is taken into consideration for theoretical analysis of effective charge of the MAR effect based on the Blonder-Tinkham-Klapwijk (BTK) theory with the Andreev clusters method. It has been found that the effective charge of the MAR effect is inversely proportional to temperature, the barrier strength Z and transparency T are proportional to temperature. Detailed measurement results and analysis will be presented.

The specific capacitance of Nb/Al-AlO$_x$/Nb SIS junctions with extremely low R$_n$A product

P. Yadranjee Aghdam*, Hawal Rashid, Alexey Pavolotsky, Vincent Desmaris, Victor Belitsky
Department of Earth and Space Sciences, Chalmers University of Technology, Gothenburg 41296, Sweden
*Contact: parisa.aghdam@chalmers.se

Abstract— This paper provides new insight regarding the specific capacitance of Nb/Al-AlO$_x$/Nb SIS junctions with low R$_n$A product. Employing the direct junction capacitance measurement method, the specific capacitance (C$_s$) and R$_n$A of several junctions with various R$_n$A values ranging from 8.8 to 68 Ω.µm^2 was studied. We noticed non-negligible scatter in the measured R$_n$C (normal resistance times junction capacitance) product for the junctions with the same R$_n$A value. We demonstrated that the local variations in the thickness distribution of the tunnel barrier could have resulted in the scatter of the R$_n$C data. We also show that, even at such low microwave frequencies as in our direct measurement method, the previously neglected nonlinear susceptance should be accounted, especially for junctions with low R$_n$A values. We present the measured C$_s$ vs R$_n$A data for Nb/Al-AlOx/Nb junctions.

Introduction

The ever growing need for having wider RF and IF bandwidth in radio astronomical receiver sets stringent requirements on the SIS junction properties such as having extremely low R$_n$A (< 20 Ω.µm^2) values and submicron area of the device. The junction capacitance affects both the RF and IF bandwidth of SIS mixers and plays a crucial role in designing tuning circuitry [1]. At low R$_n$A values, the accurate value of the junction capacitance cannot be ascertained mainly due to the disagreement among the previous indirect measurements of the specific capacitance (C$_s$) reported in the literature [2]–[4]. In order to achieve improved accuracy compared with the previous methods, we presented a direct microwave measurement method employing a dedicated cryogenic calibration technique [5]. Our method with uncertainties down to ±2%, has an advantage over the previously used approaches, which for instance, involved extraction of the SIS junction capacitance from model of a complex superconducting resonant structure.

In this paper, great care was taken to extract the true geometrical junction capacitance. It was found that even at such low frequencies (f = 4 GHz), the susceptance [6], [7] obtained by the Kramers-Kronig transformation of the imaginary part of the response function (i.e. quasiparticle dc IV characteristics), should be calculated and subtracted from the measured capacitance. Interestingly, we found that this susceptance, which hereafter will be referred to as the nonlinear susceptance, is significant for junctions with low R$_n$A values. Additionally, the scatter in C$_s$ resulted from the effect of local non-uniformities in the tunnel barrier [8] should also be considered while determining the junction capacitance. Employing the direct method in this study and extracting the true geometrical specific capacitance, the C$_s$ vs R$_n$A data is obtained.

Experiment and Results

We fabricated 34 Nb/Al-AlO$_x$/Nb SIS junctions with R$_n$A and junctions' nominal areas range of 8.8–68 Ω.µm^2 and 3.6–20 µm^2, respectively. The details of the trilayer deposition parameters and the junction fabrication process are available in [5], [9]. The range of AlO$_x$ oxygen exposure parameter for these junctions was 1530–13000 Pa.s, which is presented in details for each batch of junctions in our recent publication [8]. In order to obtain the R$_n$A value, the true junction size (A) (which accounted for the dimension variation due to the fabrication process) was estimated and the junction normal resistance R$_n$ was extracted as explained in [5], [8]. The complex impedance of these junctions were directly measured at 4 GHz center frequency and at 4 K temperature. The calibration at 4 K was performed using the time-domain measurement techniques and the gap voltage biased junction was used as the short-circuit reference [5], [10]. Then an Agilent ADS equivalent circuit model [5] was employed in which the model parameters were adjusted based on the calibration. Later the SIS junction capacitance was extracted once the best fit was achieved between the model and the experimental data. More details on this procedure can be found in [5].

The measured junction capacitances (C$_m$) for the batch with R$_n$A=9.4 Ω.µm^2 are presented in Figure 1 and Figure 2. In this study, we calculated the contribution of the reactive component of the tunneling current, which results from the extreme nonlinearity of this dc I-V relation at the gap voltage [6], [7]. The resulted nonlinear susceptance calculated through the Kramers-Kronig transform of the dc I-V curve [6], [7] was capacitive under our measurement conditions [11] and was non-negligible even at such low operating frequencies (e.g. C$_n$ in Figure 1). As can be seen in Figure 1, at low R$_n$A values, C$_n$ is comparable to the C$_m$. More details regarding the nonlinear capacitance calculation and its important contribution can be found in [6], [7], [11]–[13]. It should be noted that the

A FULL VERSION OF THIS PROCEEDINGS ARTICLE HAS BEEN SUBMITTED TO THE IEEE TRANSACTIONS OF TERAHERTZ SCIENCE AND TECHNOLOGY (TST) FOR PEER-REVIEW, PLEASE CHECK THE WEBSITE: HTTPS://WWW.MTT.ORG/TERAHERTZ

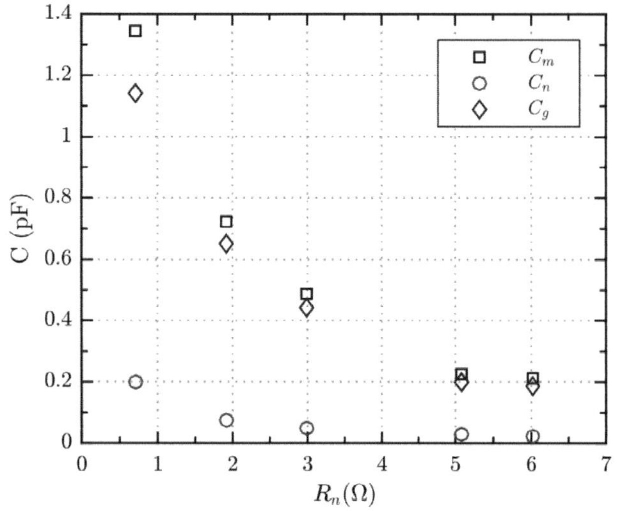

Fig. 1. The calculated nonlinear capacitance C_n (red circles) and the measured capacitance C_m (black squares) for the batch with R_nA of 9.4 $\Omega.\mu m^2$. The blue diamond data points represent the true geometrical capacitance C_g, which is obtained by the difference of C_m and C_n for each R_n. Graph is reproduced from [11].

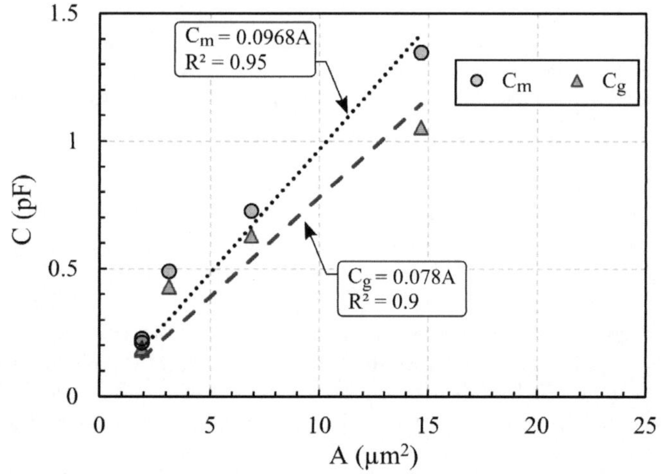

Fig. 2. The measured (C_m) and the true geometrical capacitance (C_g) as a function of the estimated area (A). From the slope of the linear fit to the C_m and C_g, the resulting specific capacitance values are obtained. Graph is reproduced from [11] .

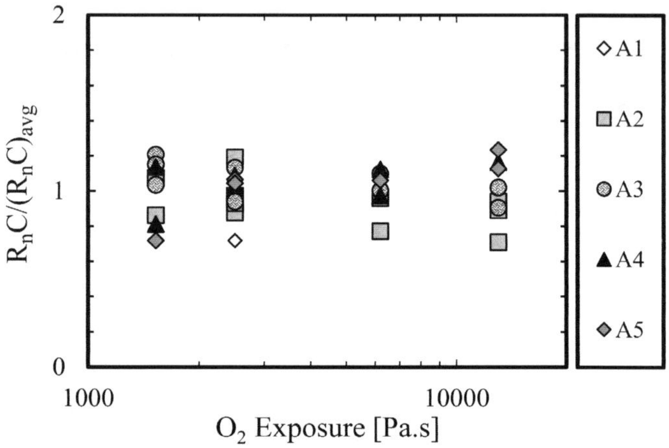

Fig. 3. $R_nC_g/(R_nC_g)_{avg}$ as a function of O_2 exposure for all the characterized junctions with areas ranging from A1=3.6 μm^2 to A5=20 μm^2, after [8].

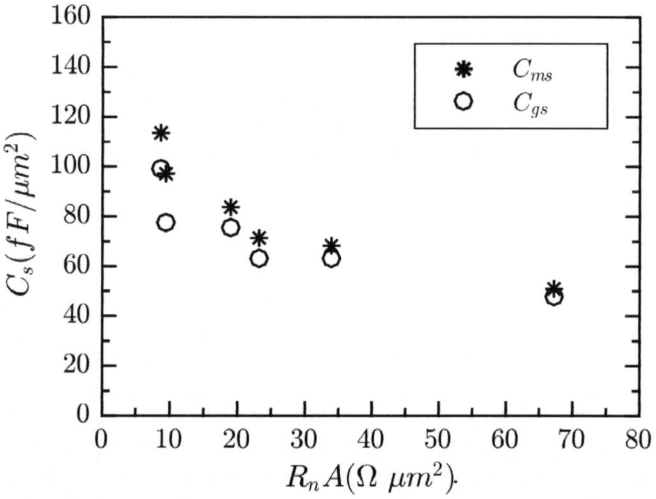

Fig. 4. The measured (C_{ms}) (asterisk) and the corrected (C_{gs}) (circle) specific capacitance as a function of R_nA. Graph is reproduced from [11].

nonlinear capacitance was found to be more significant for junctions with low R_nA values [11]. Also, for junctions with the same R_nA, C_n was higher for those with lower R_n (see Figure 1). The true geometrical capacitance (C_g in Figure 1) was then obtained by subtracting the C_n from C_m. The measured and the true geometrical junction capacitances as a function of the estimated junction area (A) are presented in Figure 2. The specific junction capacitance is the slope of the fitted lines to the C (A) data points. As can be seen, the true geometrical junction capacitance results in a lower junction specific capacitance.

In [8], we investigated the origin of the scatter of the area-independent R_nC_g product (see Figure 3). We found that the scatter of as much as 40% could not only be attributed to the area estimation and the measurement uncertainty of the R_nC_g, which varies depending on the junction area and is between just ±2% to ±11.2% for the largest and smallest junctions among all the batches, respectively. Employing an illustrative model [8] and the obtained local thickness distribution of the AlO_x tunnel barrier in Nb/Al-AlO_x/Nb trilayer (using high resolution transmission electron microscopy) we demonstrated that these variations in the thickness distribution of the tunnel barrier result in the scatter of the R_nC_g data, which is consistent with our measurements. It should be noted that the scatter of the R_nC translates into the scatter in both specific capacitance and R_nA. The "local" nature of such variations over the wafer area states that averaging out the effect of these non-uniformities in junction parameters, R_nC, specific capacitance, and R_nA is less probable at the junction size scaled down. Also, such variations could be different depending on the trilayer deposition system, and hence resulting in different reports of the specific capacitance for the same R_nA [11].

The data of the measured (C_{ms}) and the true geometrical specific capacitance (C_{gs}) as a function of the R_nA value for each batch is illustrated in Figure 4. The obtained C_{gs} (R_nA) data is compared with the previously reported experimentally obtained relations in [11]. Reference [11], also proposes an improved and more accurate model for the C_s (R_nA) relation, which can greatly improve the performance of SIS mixers.

Conclusions

In this paper, the results of the directly measured junction capacitance of 34 Nb/Al-AlO_x/Nb SIS junctions with various oxygen exposure parameters were considered. It was shown that the nonlinear capacitance obtained from the Kramers-Kronig transform of the dc I-V curve is significant for junctions with low R_nA even at frequencies of a few GHz. The resulting C_{gs} (R_nA) relation was presented. Our findings show that for junctions with low R_nA values and submicron area, the scatter of C_s as a result of local thickness non-uniformities becomes more significant and should be considered.

References

[1] A. R. Kerr, "Some fundamental and practical limits on broadband matching to capacitive devices, and the implications for SIS mixer design," *IEEE Trans. Microw. Theory Tech.*, vol. 43, no. 1, pp. 2–13, 1995.

[2] H. S. J. van der Zant, R. a. M. Receveur, T. P. Orlando, and a. W. Kleinsasser, "One-dimensional parallel Josephson-junction arrays as a tool for diagnostics," *Appl. Phys. Lett.*, vol. 65, no. 16, p. 2102, 1994.

[3] M. Maezawa, M. Aoyagi, H. Nakagawa, I. Kurosawa, and S. Takada, "Specific capacitance of Nb/AlOx/Nb Josephson junctions with critical current densities in the range of 0.1–18 kA/cm2," *Appl. Phys. Lett.*, vol. 66, no. 16, p. 2134, 1995.

[4] D. M. J. Lea, "Integrated test structures for characterization of thin films for superconductor-insulator-superconductor devices and circuits," Univ. Virginia, 1996.

[5] P. Y. Aghdam, H. Rashid, A. Pavolotsky, V. Desmaris, D. Meledin, and V. Belitsky, "Direct Measurement of Superconducting Tunnel Junction Capacitance," *IEEE Trans. Terahertz Sci. Technol.*, vol. 5, no. 3, pp. 464–469, May 2015.

[6] J. Tucker and M. Feldman, "Quantum detection at millimeter wavelengths," *Rev. Mod. Phys.*, vol. 57, no. 4, pp. 1055–1113, Oct. 1985.

[7] H. Rashid, V. Desmaris, A. Pavolotsky, and V. Belitsky, "Harmonic and reactive behavior of the quasiparticle tunnel current in SIS junctions," *AIP Adv.*, vol. 6, no. 4, p. 45109, Apr. 2016.

[8] P. Yadranjee Aghdam, H. Marouf Rashid, A. Pavolotsky, V. Desmaris, and V. Belitsky, "Dependence of the scatter of the electrical properties on local non-uniformities of the tunnel barrier in Nb/Al-AlOx/Nb junctions," *J. Appl. Phys.*, vol. 119, no. 5, p. 54502, Feb. 2016.

[9] A. B. Pavolotsky, D. Dochev, and V. Belitsky, "Aging- and annealing-induced variations in Nb/Al–AlOx/Nb tunnel junction properties," *J. Appl. Phys.*, vol. 109, no. 2, p. 24502, 2011.

[10] H. Rashid, D. Meledin, V. Desmaris, A. Pavolotsky, and V. Belitsky, "Superconducting 4–8-GHz Hybrid Assembly for 2SB Cryogenic THz Receivers," *IEEE Trans. Terahertz Sci. Technol.*, vol. 4, no. 2, pp. 193–200, Mar. 2014.

[11] P. Yadranjee Aghdam, H. Rashid, A. Pavolotsky, V. Desmaris, D. Meledin, and V. Belitsky, "Specific Capacitance Dependence on the Specific Resistance in Nb/Al-AlOx/Nb Tunnel Junctions," *IEEE Trans. Terahertz Sci. Technol.*, vol. 7, no. 5, 2017. DOI: 10.1109/TTHZ.2017.2723803

[12] H. Rashid, S. Krause, D. Meledin, V. Desmaris, A. Pavolotsky, and V. Belitsky, "Frequency Multiplier Based on Distributed Superconducting Tunnel Junctions: Theory, Design, and Characterization," *IEEE Trans. Terahertz Sci. Technol.*, vol. 6, no. 5, pp. 1–13, 2016.

[13] H. Rashid, V. Desmaris, A. Pavolotsky, and V. Belitsky, "THz Frequency Up-Conversion using Superconducting Tunnel Junction," *IEEE Microw. Wirel. Components Lett.*, vol. 26, no. 10, pp. 831–833, Oct. 2016.

Titanium nitride for kinetic-inductance detectors: a problematic material or an engineering opportunity?

Eduard F. C. Driessen[*1]
[1]*Institut de Radioastronomie Millimétrique (IRAM), 38400 St Martin d'Hères, France*
[*]Contact: driessen@iram.fr

Abstract—Titanium nitride and related materials with a large normal-state resistivity have been proposed and used for kinetic inductance detectors at various wavelengths from near-IR to millimeter. The expected advantages of these materials are manifold, e.g. a tunable critical temperature, a large kinetic inductance fraction, and better matching to the incident radiation due to the large resistivity. However, whereas even large (2" diameter) telescope-ready detector arrays have come into sight, the detailed behavior of TiN detectors still shows many puzzling features, at odds with the perfectly understood behavior of "conventional" aluminum KIDs.

In this contribution I will give an overview of the different studies that have been performed on TiN resonators. In this overview, I will emphasize the unconventional behavior of the material, and the differences with aluminum. Among these differences are a smooth detection gap edge, an increasing sensitivity with optical power, and a quality factor that does not change with loading. I will argue that these differences are unavoidably linked to the large normal-state resistivity of the material and its accompanying intrinsic electronic inhomogeneity, and that they should be fully taken into account when considering this material for detectors.

Finally, I will discuss the status of TiN for use in (sub)-mm instruments and I will argue that some of the observed unconventional behavior might in fact prove an engineering opportunity for ground-based observation.

MgB_2 THz HEB mixer with an 11GHz bandwidth

Evgenii Novoselov, Sergey Cherednichenko*

Terahertz and Millimetre Wave Laboratory, Department of Microtechnology and Nanoscience, Chalmers University of Technology, SE-412 96 Gothenburg, Sweden

*Contact: serguei@chalmers.se

Abstract—We have developed a hot-electron bolometer (HEB) mixer with a noise bandwidth (noise temperature rises by a factor of 2) of 11GHz. Mixers were tested with both a 0.69THz and a 1.63THz local oscillators (LO) where noise temperatures, Tr of 800-900K and 900-1000K were demonstrated at 5K operation. Mixers were working with a low noise performance up to 20K. Devices were fabricated from ultrathin magnesium diboride (MgB_2) superconducting films as thin as 5-7nm with a superconducting transition (in the device) at 30-32K. Films were deposited with a custom made Hybrid Physical Chemical Vapor Deposition (HPCVD) system, which will be discussed in a separate presentation at this conference. Comparison to state of the art NbN HEB mixers shows that with MgB_2 HEB mixers noise bandwidth increases by a factor of 2-3. We have managed to reproduce results in at least two separate batches at the moment. MgB_2 HEB mixers show to be very robust against ESD. Current mixers are fabricated in the quasioptical scheme, and we are working towards waveguide based device technology adaptation as well.

Gain bandwidth of NbN HEB mixers on GaN buffer layer operating at 2 THz local oscillator frequency

S. Antipov[1]*, A. Trifonov[1], S. Krause[2], D. Meledin[2], V. Desmaris[2], V. Belitsky[2], and G. Gol'tsman[1]
[1] *Moscow State Pedagogical University, Moscow, 119992, Russia*
[2] *Group for Advanced Receiver Development, Chalmers University of Technology, SE-41296, Gothenburg, Sweden*
*Contact: sv_antipov@mail.ru

Abstract— In this paper, we present IF bandwidth measurement results of NbN HEB mixers, which are employing NbN thin films grown on a GaN buffer-layer. The HEB mixers were operated in the heterodyne regime at a bath temperature of approximately 4.5 K and with a local oscillator operating at a frequency of 2 THz. A quantum cascade laser served as the local oscillator and a reference synthesizer based on a BWO generator (130-160 GHz) and a semiconductor superlattice (SSL) frequency multiplier was used as a signal source. By changing the LO frequency it was possible to record the IF response or gain bandwidth of the HEB with a spectrum analyzer at the operation point, which yielded lowest noise temperature.

The gain bandwidth that was recorded in the heterodyne regime at 2 THz amounts to approximately 5 GHz and coincides well with a measurement that has been performed at elevated bath temperatures and lower LO frequency of 140 GHz. These findings strongly support that by using a GaN buffer-layer the phonon escape time of NbN HEBs can be significantly lower as compared to e.g. Si substrate, thus, providing higher gain bandwidth.

INTRODUCTION

The superconducting NbN hot electron bolometer (HEB) mixer continues to be a key technology for the heterodyne instruments operating at frequencies above 1 THz, offering high spectral resolution combined with record sensitivity at Terahertz frequencies [1, 2]. However, the intermediate frequency (IF) bandwidth of NbN HEB mixer still remains limited to typically 3-4 GHz for recent operational receivers [1, 3, 4]. The employment of on-substrate buffer-layers to promote the growth of high quality NbN films and better acoustic matching to facilitate the phonon escape are considered essential to overcome the IF limitation [5-6]. The latest progress in this direction has been achieved by fabrication of high-quality NbN ultra-thin films with the use of GaN buffer layer [7-8]. The study of IF bandwidth of NbN/GaN HEB mixers was performed under quasi-equilibrium conditions at $T \approx T_c$ using a direct measurement method when HEB mixer operates at elevated bath temperature, at low sub-THz LO frequency, and signal beating oscillations [9]. The impact of the improved NbN films on the RF and especially IF performance of HEB mixers at THz frequencies at low temperature and optimal LO power has not been demonstrated. In this paper, we present the gain IF bandwidth measurement of NbN/GaN HEB mixer operating at 2 THz LO frequency, and study the influence of the GaN buffer layer on IF bandwidth.

EXPERIMENT

A. NbN/GaN HEB mixer

The fabrication of submicrometer devices such as HEB mixers with bridge length of about 100-300 nm is a crucial and technologically challenging process. Thus, we produced independently two batches of HEB mixers with identical design at Chalmers and MSPU in order to exclude effects from fabrication related issues. In both cases, the mixers were fabricated starting from NbN thin films deposited at Chalmers by means of reactive DC magnetron sputtering on a GaN (0001) buffer-layer on sapphire and featured a single crystal structure due to the small lattice mismatch with high Tc of 12.5 K for 4.5 nm of thickness [9].The bolometer bridges with dimensions of 0.18 μm x 2.3 μm and the log-spiral antenna were defined by e-beam lithography and subsequent dry etching. The normal state resistance of the bolometer is close to 100 Ω.

B. Experimental setup

To record the IF output power of the HEB mixer, the heterodyne technique at T≈4.5 K was applied by mixing the THz radiation of LO and signal sources. The measurements were performed at the optimal level of the LO power at a fixed bias corresponding to the smallest noise temperature.

A 2 THz quantum cascade laser as a local oscillator and a signal source consisting of a reference synthesizer based on BWO generator (130-160 GHz) and a semiconductor superlattice (SSL) frequency multiplier operating at the 13th SSL harmonic were used as LO and signal source, respectively.

The THz radiation was coupled to the HEB mixer by an extended hemispherical Si lens. The mixer with the lens was cooled to 4.5 K in a cryostat with optical access through a 2 mm-thick high-density polyethylene (HDPE) window and a 200 μm thick Zitex G108 infrared filter mounted on the 77 K screen. The LO beam was diplexed into the signal path using a 12 μm thick Mylar beam splitter. The HEB mixer was IF-coupled to a 1-inch coplanar line with a subsequent transition to a coaxial cable which led the IF signal to a cryogenically cooled bias-T and HEMT low-noise amplifier with a

A FULL VERSION OF THIS PROCEEDINGS ARTICLE HAS BEEN SUBMITTED TO THE IEEE TRANSACTIONS OF TERAHERTZ SCIENCE AND TECHNOLOGY (TST) FOR PEER-REVIEW, PLEASE CHECK THE WEBSITE: HTTPS://WWW.MTT.ORG/TERAHERTZ

bandwidth 0.4-12.0 GHz and a gain of about 30 dB. This was followed by a similar broadband room-temperature amplifier. The calibration of the IF chain was determined by measuring the portion of the RF white noise that went through the amplifier chain. The white Johnson noise in excess to the intrinsic noise of the first amplifier was produced by the very same HEB mixer driven in the normal state by additional heating to elevated bath temperature close to $T \approx Tc$. The IF power versus intermediate frequency for HEB mixer was measured for constant optimal level of the LO power at a fixed bias conditions. While sweeping the signal source frequency, the IF output power level was measured with a spectrum analyzer.

Results and Discussion

In Fig. 1 we present the relative gain bandwidth of NbN/GaN HEB mixer (red star) that was measured for the 2 THz LO frequency for the optimal operation conditions (colored point in IVC) providing the highest IF signal to noise ratio. In addition, the gain bandwidth of NbN/Si HEB mixer (green circle) is depicted. Standing waves due to reflections between the HEB mixer and cryogenic LNA resulted in the resonance peaks in the IF dependence of the relative conversion efficiency. The peaks almost disappeared when we used the calibration procedure.

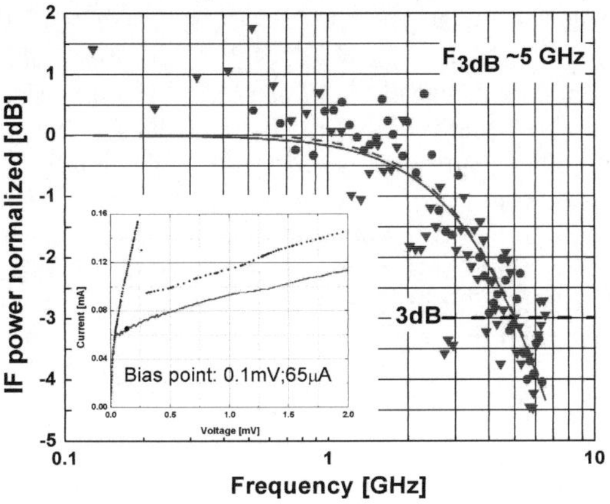

Fig. 1 Gain bandwidth in heterodyne regime at 2 THz at optimal pumping and bias conditions for NbN/GaN HEB mixer. The normalized IF power versus frequency shows a roll-off at approximately 5 GHz. Set of unpumped (blue) and optimally pumped (red) IVCs along with the operation point (0.1 mV; 65 μA) in dc voltage bias regime for NbN/GaN HEB mixer. Comparison of gain bandwidths of NbN/GaN HEB mixers in heterodyne regime at 2 THz (circles) vs elevated bath temperature regime at 140 GHz (triangles).

The normalization considerably enhanced the roll-off frequency determination accuracy of the IF bandwidth. We obtained the 3 dB roll-off frequency f_{3dB} by fitting our experimental IF dependence of the relative gain with single-pole Lorentzian. The gain bandwidth amounts to approximately 5 GHz for the NbN/GaN HEB at optimal pumping and bias conditions.

NbN/GaN HEB mixer shows a significant enhancement of IF roll-off frequency compared to commonly used Si substrate due to the reduction of phonon escape time (τ_{es}) for the NbN/GaN material combination. The typical NbN/Si HEB mixer with slightly large dimensions (5 nm × 0.3 μm × 3 μm) demonstrates of 3 dB roll-off frequency $f_{3dB} \approx 3.4$ GHz in heterodyne regime at 2 THz at optimal pumping and bias conditions.

When NbN/GaN HEB mixer operates at elevated bath temperatures close to the critical temperature of the NbN ultra-thin film, the contributions from electron-phonon processes and self-heating effects are relatively small, therefore IF roll-off will be governed by the phonon-escape. As the roll-off is similar for both the heterodyne and direct detection regime (Fig. 1), it may be concluded that in NbN/GaN HEBs the electron-phonon contribution is negligible compared to the phonon escape one. Thus, the IF bandwidth is determined mostly by the phonon escape, which seems to be significantly improved as compared to commonly used Si substrates. In [6], it was presented, that the improved phonon transparency is due to the good acoustic matching between NbN and GaN, as well as the improved low defect interface and enhanced superconducting properties of the NbN film.

Conclusions

The THz heterodyne measurements, which allowed the study the gain bandwidths of NbN/GaN HEB mixer at optimal bias condition, are presented. Utilizing a GaN buffer-layer has resulted in significant enhancement of gain bandwidth frequency of phonon-cooled HEB mixers based on NbN material compared to commonly used Si substrate. Matching of the gain bandwidth results at elevated bath temperature at 140 GHz and in heterodyne regime at 2 THz indicates the reduction of phonon escape time (τ_{es}) due to the GaN buffer layer.

References

[1] D. Meledin, et al., "A 1.3-THz Balanced Waveguide HEB Mixer for the APEX Telescope," IEEE Transactions on Microwave Theory and Techniques, vol. 57, no. 1, 2009.

[2] A. V. Smirnov, et al., "The current stage of development of the receiving complex of the millimetron space observatory", Radiophysics and Quantum Electronics, Vol. 54, Issue 8–9, pp 557–568, January, 2012

[3] P. Pütz, et al., "Waveguide hot electron bolometer mixer development for upGREAT," in 39th International Conference on Infrared, Millimeter, and Terahertz waves (IRMMW-THz), Tucson, AZ, 2014.

[4] C. Risacher, et al., "The upGREAT 1.9 THz multi-pixel high resolution spectrometer for the SOFIA Observatory", Astronomy and Astrophysics, 595 A34 (2016)

[5] D. Meledin, et al., "Study of the IF bandwidth of NbN HEB mixers," IEEE Trans. Appl. Supercond., vol. 13, no. 2, pp. 164–167, Jun. 2003.

[6] D. Dochev, et al., "Growth and characterization of epitaxial ultra-thin NbN films on 3C-SiC/Si substrates for terahertz applications," Supercond. Sci. Technol., vol. 24, no. 3, 2011.

[7] S. Krause, et al., "Epitaxial growth of ulta-thin NbN films on AlxGa1-xN buffer-layers," Supercond. Sci. Technol., pp. vol. 27, no. 6, Apr 2014.

[8] S. Krause, et al., "Ambient temperature growth of mono- and polycrystalline NbN nanofilms and their surface and composition analysis," IEEE Trans. Appl. Supercond., vol. 26, no. 3, Apr. 2016.

[9] S. Krause, et al., "Reduction of Phonon Escape Time for NbN Hot Electron Bolometers by Using GaN Buffer Layers", IEEE Transactions on Terahertz Science and Technology, vol. 7, issue 1, pp. 53 – 59, Jan. 2017.

A FULL VERSION OF THIS PROCEEDINGS ARTICLE HAS BEEN SUBMITTED TO THE IEEE TRANSACTIONS OF TERAHERTZ SCIENCE AND TECHNOLOGY (TST) FOR PEER-REVIEW, PLEASE CHECK THE WEBSITE: HTTPS://WWW.MTT.ORG/TERAHERTZ

Design of a wideband balanced waveguide HEB mixer employing a GaN buffer-layer for the 1-1.5 THz band

S. Krause*, D. Meledin, V. Desmaris, V. Belitsky

Group for Advanced Receiver Development, Chalmers University of Technology, SE-41296, Gothenburg, Sweden

*Contact: sascha.krause@chalmers.se

Abstract— We present the design and implementation of a wideband balanced waveguide NbN HEB mixer employing a GaN substrate to be operated in the frequency range of 1 - 1.5 THz.

The balanced receiver scheme consisting of a 90° RF hybrid, a pair of NbN phonon-cooled HEB mixers and a 180° IF hybrid has major advantages over the single-ended configuration. Furthermore, the usually small IF bandwidth of phonon-cooled NbN HEB mixers has been addressed by employing a GaN substrate instead of a conventional Si or quartz substrate. It has recently been shown that using GaN substrate reduces the escape time of phonons from NbN bridge to the substrate and thus, prospectively enhances the overall cooling rate of hot electrons and yielding larger IF bandwidth. The mixer housing is implemented in a back-end configuration and has been fabricated by means of a micro-machining method, providing excellent control of the dimensions and smoothness of the all-metal waveguide components. The expected RF performance of the proposed HEB design as well as its fabrication and DC characterization are presented.

Introduction

Heterodyne instruments for high-resolution spectroscopy at the terahertz frequency range primarily employ Hot Electron Bolometer (HEB) mixers (Leisawitz, 2000), (Meledin, o.a., 2009) due to their superior sensitivity and low local oscillator power requirement when compared to the competing Schottky and SIS mixers. Despite the strong absorption by water vapor of the incident THz radiation, it is possible to observe astronomical objects from the ground in three frequency windows in the band of 1-1.5 THz containing several important CO transition lines (Pardo, Serabyn, & Cernicharo, 2001). The demand for a wide IF bandwidth is paramount for the efficient use of valuable observation time as well as for the study of distant objects with strong spectral line broadening.

The working principle of hot electron bolometer relies on the thermal energy exchange between "hot" electrons in the NbN film and the underlying substrate (Gershenzon, Gol'tsman, Gogidze, Elant'ev, Karasik, & Semenov, 1990). Thus, the IF roll off frequency and corresponding bandwidth of receivers in operation is limited to typically 3-4 GHz for NbN based HEB mixers using a conventional substrate such as Si (Il'in, Milostnaya, Verevkin, Gol'tsman, Gershenzon, & Sobolewski, 2000), (Pütz, Büchel, Jacobs, Schultz, Honingh, & Stutzki, 2014), SiN (Cherednichenko, o.a., 2007), quartz (Meledin, o.a., 2003) or sapphire (Kooi, o.a., 2007). Hereby, it is considered that the phonon escape from the film to the substrate plays a major role in limiting the IF bandwidth (Kooi, o.a., 2007). The employment of buffer-layers such as MgO (Miki, Uzawa, Kawakami, & Wang, 2001), (Meledin, o.a., 2003) and SiC (Dochev D. , o.a., 2011) were found to promote the epitaxial growth of NbN. Moreover, it has recently been shown that also the use of the hexagonal GaN (Krause, o.a., 2014), (Krause, o.a., 2016), (Krause, o.a., 2017) allows for high-quality single-crystal NbN films with improved superconducting properties and enhanced phonon escape time. This study investigates the possibility of using this promising GaN buffer-layer to be used in waveguide based NbN HEBs with prospectively increased IF bandwidth and low noise performance.

Mixer Design

The balanced receiver scheme consists of a 90° RF hybrid, a pair of HEB mixers with desirably identical IV characteristics and a 180° IF hybrid, whereas one port is used to combine the resulting IF output of the mixers and the other to terminate the difference signal, in fact suppressing the LO AM noise and yielding higher receiver stability (Meledin, o.a., 2009). Despite the higher level of complexity, the major advantage is that all available LO power can be used as in contrast to, e.g., a beam splitter for LO and RF combination.

The full-metal RF hybrid, waveguides and mixer block were manufactured using a micromachining technique, where thick photoresist is used as a sacrificial mold (Desmaris, Meledin, Pavolotsky, Monje, & Belitsky, 2008). This technique provides excellent surface accuracy of below 15 nm and fulfils the high demands for THz waveguide components (Desmaris, Meledin, Dochev, Pavolotsky, & Belitsky, 2011).

A. Fullwave 3D simulation

The mixer design was optimized in the 3D fullwave FEM simulator HFSS in the frequency band between 900 GHz to 1600 GHz. Particular emphasis was put on a wideband design that would be able to cover 3 important frequency windows with a single receiver. The HEB impedance was assumed to be 90 Ohm and the E-probe was optimized accordingly to provide proper matching in the band of interest as illustrated in Fig. 1.

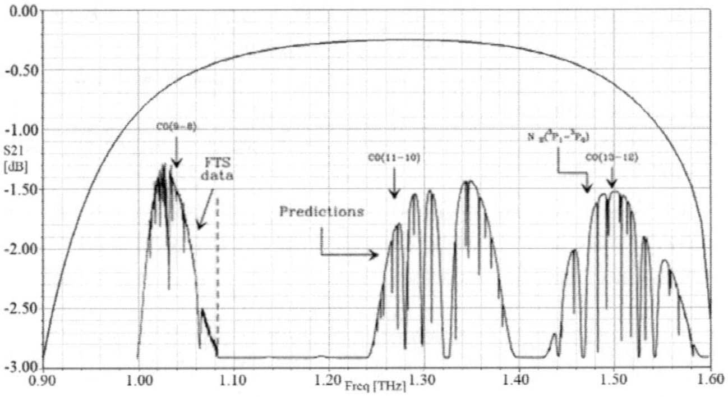

Fig. 1: Important absorption lines in three windows between 1 and 1.5 THz (Pardo, Serabyn, & Cernicharo, 2001). The red curve presents the expected coupling (S21) to the HEB bridge. The E-probe was optimized for a real impedance of 90 Ohm.

In order to facilitate the crucial alignment of the mixer chip, so called alignment notches were added to each side of the probe and should be used as an optical guide as well as limiting the maximum misalignment of the mixer chip inside the waveguide. Simulation indicates, as presented in Fig. 2, that the frequency band of 1.05-1.5 THz still can be covered even in the worst alignment scenario.

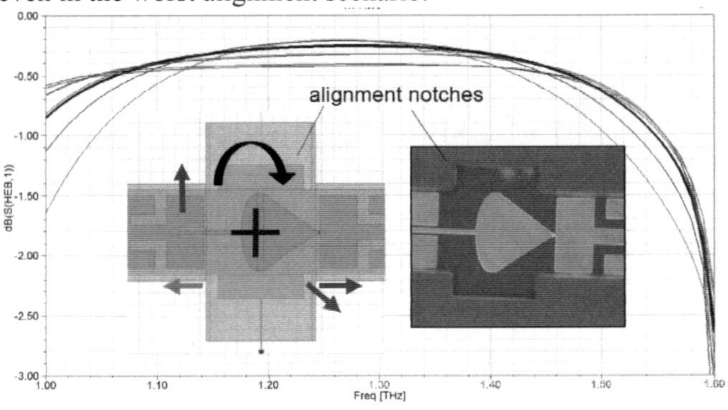

Fig. 2: Misalignment scenarios (worst case) and their effect on coupling efficiency. The frequency band from 1.05 to 1.5 THz can still be covered even in the case of the most unfavourable misalignment.

B. Fabrication

The fabrication of HEBs used for waveguide based applications is in general more challenging than for its quasi-optical counterpart, as the mixer chip eventually needs to be thinned down to be placed inside the waveguide without electrically loading the waveguide too much. This also implies that the NbN film with thickness of a few nanometer is at a higher risk of degradation during the thinning process. Thus, it is important to use high quality films exhibiting high critical temperature. We have grown epitaxial films with 4.5 nm thickness onto GaN buffer-layer with Tc of 12.5 K using reactive DC magnetron sputtering in a nitrogen/argon atmosphere at elevated temperatures (Krause, o.a., 2014).

E-beam lithography has been employed to define the RF structure as well as the bolometer and contact pads with great accuracy. All HEB chips have been characterized in an RT measurement prior to shaping the GaN membrane and crucial thinning down steps in order to track eventual degradation. A photoresist was used as an etch mask to pattern the 5.5 μm thin GaN buffer-layer into long membranes with the described alignment notches. In order to separate the devices from the remaining wafer, the bulk Si was removed from the backside by dry etching in SF6 chemistry. Subsequently, the mixer chips were placed in the mixer housing, carefully aligned and electrically contacted with a conductive adhesive (Dochev D. , Desmaris, Pavolotsky, Meledin, & Belitsky, 2011), (Meledin, o.a., 2009).

Results and Discussion

The electrical characterization of the HEB bridges revealed that the critical temperature of bolometers with bridge length as small as 200 nm shows very little degradation compared to the unprocessed NbN film. Moreover, they exhibited excellent uniformity both in their critical temperature and resistance.

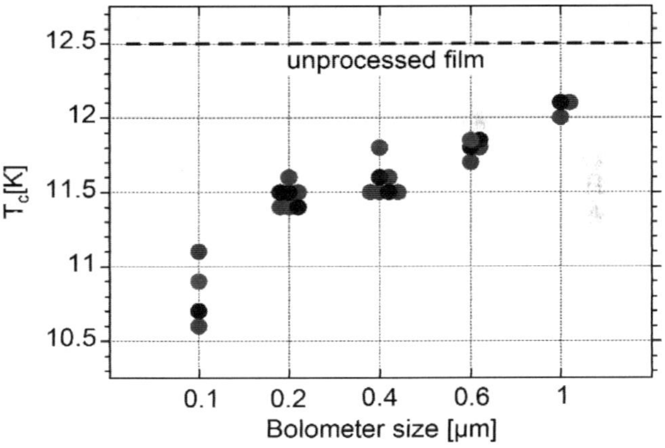

Fig. 3: Electrical characterization of the HEBs with different bridge dimensions in comparison with the unprocessed film prior patterning and thinning down of the membranes.

Once etched, the HEB mixer chips were sorted, identified and one pair was subsequently placed inside the waveguide channel. The inspection under the SEM revealed the high quality of the mixer chips and compliance of the RF structure dimensions to the designed values. An enlarged view on the E-probe and RF structure as well as the alignment of the entire chip can be seen in Fig. 4.

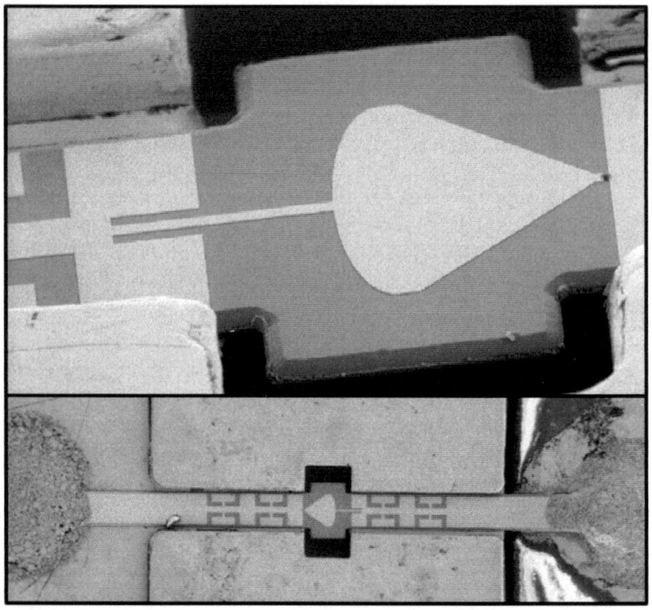

Fig. 4: SEM image of the aligned HEB mixer chip with electrical contacts and the enlarged view onto the E-probe and high impedance line for providing the DC bias.

The HEB mixer was cooled down in a cryostat to 4K and its critical current re-measured and compared to the values obtained before etching of the chip membrane. As depicted in Fig. 5, the critical current is largely identical and amounts to 135 μA for a device with 200 nm and 95 Ohm normal state resistance. The difference of the IV response beyond the critical current is due to different bias circuits.

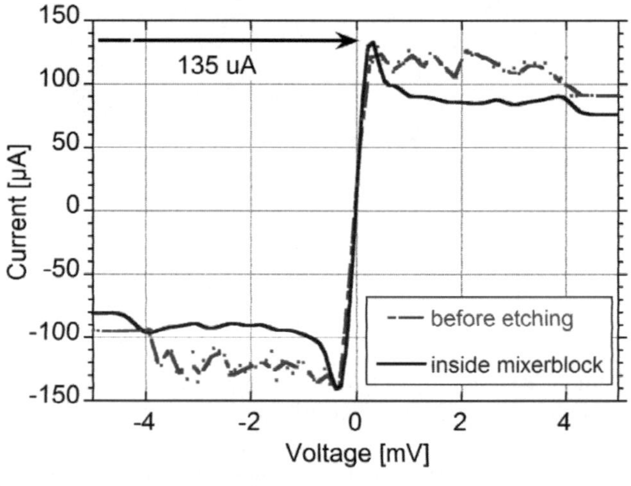

Fig. 5: IV curves of one particular HEB mixer with 200 nm bridge length before the membrane patterning and after patterning and mounting inside the waveguide channel. The critical current does not show any degradation

Conclusion

This manuscript presented the design, implementation and first DC characterization of a wideband balanced waveguide NbN HEB mixer using a GaN substrate. The latter is believed to increase the IF bandwidth as recent studies predict. The optimized E-probe and high-impedance line enable the wideband operation from approximately 1 to 1.6 THz as demonstrated by HFSS simulations, thus covering important absorptions lines in 3 different atmospheric windows with a single HEB mixer. Moreover, the balanced receiver scheme promises higher stability by employing balanced layout to suppress LO AM noise and replaces the inefficient beam splitter with its significant waste of LO power. The fabrication of the presented design was successfully implemented with only very little degradation of the NbN film's critical temperature from 12.5 K of the unprocessed film to 11.5 K for a HEB mixer with 200 nm bridge length. DC measurements have proven that even the most crucial process steps such as the GaN patterning and removal of bulk Si from the backside of the chip had not deteriorated the critical current of the HEB mixer.

Further studies will be focusing on an extended RF characterization including noise temperature and bandwidth measurements.

References

[1] D. Leisawitz, "Scientific motivationand technology requirements for the SPIRIT and SPECS far-infrared/submillimeter space interferometers," *Proc. SPIE,* vol. 4013, pp. 36-46, 2000.

[2] D. Meledin, A. Pavolotsky, V. Desmaris, I. Lapkin, C. Risacher, V. Perez, D. Henke, O. Nyström, E. Sundin, D. Dochev, M. Pantaleev, M. Fredrixon, M. Strandberg, B. Voronov, G. Gol'tsman and V. Belitsky, "A 1.3-THz Balanced Waveguide HEB Mixer for the APEX Telescope," *IEEE Transactions on Microwave Theory and Techniques, vol. 57, no. 1,* 2009.

[3] J. Pardo, E. Serabyn and J. Cernicharo, "Submillimeter atmospheric transmission measurements on mauna kea during extremely dry El Nino conditions: Implications for broadband opacity contributions," *J. Quant. Spectrosc. Radiat. Transf.,* vol. 68, pp. 419-433, 2001.

[4] E. Gershenzon, G. Gol'tsman, I. Gogidze, A. Elant'ev, B. Karasik and A. Semenov, "Millimeter and submillimeter range mixer based on electronic heating of supercondcuting films in the resistive state," *Sov. Phys. Superconductivity,* vol. 3, p. 1582, 1990.

[5] K. Il'in, I. Milostnaya, A. Verevkin, G. Gol'tsman, E. Gershenzon and R. Sobolewski, "Picosecond hot-electron energy relaxation in NbN superconducting photodetectors," *Applied Physics Letter,* vol. 76, no. 19, 2000.

[6] P. Pütz, D. Büchel, K. Jacobs, M. Schultz, C. E. Honingh and J. Stutzki, "Waveguide hot electron bolometer mixer development for upGREAT," in *39th International Conference on Infrared, Millimeter, and Terahertz waves (IRMMW-THz),* Tucson, AZ, 2014.

[7] S. Cherednichenko, V. Drakinskiy, J. Baubert, J. M. Krieg, B. Voronov, G. Gol'tsman and V. Desmaris, "Gain bandwidth of NbN hot-electron bolometer terahertz mixers on 1.5 μm Si3N4/SiO2 membranes," *Journal of Applied Physics,* vol. 101, no. 6, p. 124508, 2007.

[8] D. Meledin, E. Tong, R. Blundell, N. Kaurova, K. Smirnov, B. Voronov and G. Gol'tsman, "Study of the IF Bandwidth of NbN HEB Mixers," *Applied Superconductivity,* vol. 13, no. 2, pp. 164-167, 2003.

[9] J. Kooi, J. Baselmans, M. Hajenius, J. Gao, T. Klapwijk, P. Dieleman, A. Baryshev and G. de Lange, "IF Impedance and Mixer Gain of NbN Hot-Electron Bolometers," *Journal of Applied Physics,* vol. 101, p. 044511, 2007.

[10] S. Miki, Y. Uzawa, A. Kawakami and Z. Wang, "IF bandwidth and Noise Temperature Measurements of NbN HEB Mixers on MgO Substrates," *Applied Superconductivity,* vol. 11, no. 1, p. 175, 2001.

[11] D. Dochev, V. Desmaris, A. Pavolotsky, D. Meledin, Z. Lai, A. Henry, E. Janzen, E. Pippel, J. Woltersdorf and V. Belitsky, "Growth and characterization of epitaxial ultra-thin NbN films on 3C-SiC/Si substrates for terahertz applications," *Supercond. Sci. Technol.,* 2011.

[12] S. Krause, D. Meledin, V. Desmaris, A. Pavolotsky, V. Belitsky, M. Rudzinski and E. Pippel, "Epitaxial growth of ulta-thin NbN films on

AlxGa1-xN buffer-layers," *Supercond. Sci. Technol.*, pp. vol. 27, no. 6, Apr 2014.

[13] S. Krause, V. Afanas'ev, V. Desmaris, D. Meledin, A. Pavolotsky, V. Belitsky, A. Lubenschenko, A. Batrakov, M. Rudzinski and E. Pippel, "Ambient temperature growth of mono- and polycrystalline NbN nanofilms and their surface and composition analysis," *Applied Superconductivity,* 2016.

[14] S. Krause, V. Mityashkin, S. Antipov, G. Gol'tsman, D. Meledin, V. Desmaris, V. Belitsky and M. Rudzinski, "Reduction of Phonon Escape Time for NbN Hot Electron Bolometers by Using a GaN Buffer Layers," *IEEE Transactions on Terahertz Science and Technology,* 2017.

[15] V. Desmaris, D. Meledin, A. Pavolotsky, R. Monje and V. Belitsky, "All-metal micromachining for the fabrication of sub-millimetre and THz waveguide components and circuits," *Journal of Micromechanics and Microengineering,* vol. 18, no. 9, p. 095004, 2008.

[16] V. Desmaris, D. Meledin, D. Dochev, A. Pavolotsky and V. Belitsky, "Terahertz components packaging using integrated waveguide technology," *IEEE MTT-S International Microwave Workshop on Millimeter Wave Integration Technologies,* no. 2011, pp. 81-84, 2011.

[17] D. Dochev, V. Desmaris, A. Pavolotsky, D. Meledin and V. Belitsky, "A Technology Demonstrator for 1.6-2 THz Waveguide HEB Receiver with a Novel Mixer Layout," *Journal of Infrared, Millimeter and Terahertz Waves,* vol. 32, p. 451, 2011.

[18] S. Cherednichenko, V. Drakinskiy, T. Berg, P. Khosropanah and E. Kollberg, "Hot-electron bolometer terahertz mixers for the Herschel Space Observatory," *Review of Scientific Instruments,* vol. 79, p. 034501, 2008.

[19] S. B. Kaplan, "Acoustic Matching of Superconducting Films to Substrates," *Journal of Low Temperature Physics,* vol. 37, pp. 343-364, 1979.

[20] R. Nebosis, A. Semenov, Y. Gousev and K. Renk, "Analysis of a Superconducting Hot-Electron Bolometer Mixer: Theory and Comparison with Experiment," *7th International Symposium on Space and Terahertz Technology,* 1996.

[21] K. Ilin, R. Schneider, D. Gerthsen, A. Engel, H. Bartolf, A. Schilling, A. Semenov, H.-W. Huebers, B. Freitag and M. Siegel, "Ultra-thin NbN films on Si: crystalline and superconducting," *Journal of Physics: Conference Series,* vol. 97, p. 012045, 2008.

[22] J. Gao, M. Hajenius, F. Tichelaar, T. Klapwijk, B. Voronov, E. Grishin, G. Gol´tsman, C. Zorman and M. Mehregany, "Monocrystalline NbN nanofilms on a 3C-SiC/Si substrate," *Applied Physics Letters,* vol. 91, p. 062504, 2007.

[23] H. Rashid, D. Meledin, V. Desmaris, A. Pavolotsky and V. Belitsky, "Superconducting 4-8 GHz Hybrid Assembly for 2SB Cryogenic THz Receivers," *IEEE Transactionson Terahertz Science and Technology,* vol. 4, no. 2, pp. 193-200, 2014.

[24] I. K., M. Lindgren, M. Currie, A. Semenov, G. Gol'tsman and R. Sobolewski, "Picosecond hot-electron energy relaxation in NbN superconducting photodetectors," *Applied Physics Leters,* vol. 76, no. 19, 2000.

MgB_2 THz HEB mixer operation from 5K till 20K

Evgenii Novoselov*, Sergey Cherednichenko

Chalmers University of Technology, SE-41296, Gothenburg, Sweden
*Contact: evgenii@chalmers.se

Abstract—It has been already demonstrated that MgB_2 hot-electron bolometer (HEB) mixers can provide simultaneously both low noise sensitivity and wide bandwidth operation (see also "MgB_2 HEB mixer with an 11GHz bandwidth" on this conference). However, a receiver noise of MgB_2 HEB mixers increases if a bath temperature is raised, especially above ½ of a device critical temperature (T_c). In order to study the origin of such behavior HEB mixers should be characterized at various conditions (bath temperatures, local oscillator (LO) frequencies, LO powers, biasing) and intrinsic mixer parameter should be extracted.

Here we present the detailed study of one of the devices which has demonstrated good receiver performance. Studied device was fabricated from a 5-7nm thick MgB_2 film deposited with a custom in-house built Hybrid Physical Chemical Vapor Deposition (HPCVD) system. The HEB size is 1x1μm^2 and has a T_c of 30K. A room temperature resistance R_{295} is 230Ohm, and a critical current I_c is 1mA at 4.2K. Device has been tested with two LOs (0.7THz and 1.6THz) using Y-factor and U-factor techniques and has demonstrated high sensitivity (about 1000K at 5K bath temperature and wide noise bandwidth (11GHz) at both frequencies. The HEB mixers was also characterized at 15K and 20K bath temperatures and showed that the increase of the noise temperature corresponds to the reduction of the mixer conversion gain, while the output mixer noise stays constant (210K). The HEB noise performance was studied at various bias points on the IV-plane at 0.7THz and 1.6THz. The bias point region providing the minimum receiver noise is wide (5-10mV and 0.2-0.4mA). The noise characterization was supplemented by independent heterodyne mixing measurements at 0.7THz in order to define mixer gain bandwidth (GBW) and study mixer gain behavior. The GBW is appeared to be at least 8GHz.

Since the MgB_2 HEB mixers has both a higher conversion gain (-9.5dB at 5K) and output noise temperature compare to a typical NbN HEB mixer (-12dB, 60K) the intermediate frequency (IF) chain is less critical and increase of the IF chain noise should result in a smaller increase of the receiver noise temperature. The device was also tested with a room temperature readout which led to increase of receiver noise temperature by about factor of 2 while the noise bandwidth was kept the same.

The wSMA receivers – a new wideband receiver system for the Submillimeter Array

Paul K. Grimes[1*], Raymond Blundell[1], Scott Paine[1], John Test[1], C-Y. Edward Tong[1], Robert W. Wilson[1], and Lingzhen Zeng[1]

[1]*Harvard-Smithsonian Center for Astrophysics, Cambridge, MA 02138, USA*
[*]Contact: pgrimes@cfa.harvard.edu

***Abstract*—** The current Submillimeter Array (SMA) receiver systems were designed in the mid-1990s and have been operating for more than fifteen years. With regular upgrades to the receiver inserts installed within each antenna's receiver cryostat, the deployment of the SWARM correlator, expansion of the IF signal transport bandwidth via improvements to the analog IF signal processing hardware, and many other enhancements, the SMA currently greatly outperforms its original specifications in terms of sensitivity and instantaneous bandwidth. It also supports additional observing modes such as full-Stokes polarization at 0.86 mm and 1.3 mm and dual polarized phased array operation for mm-VLBI.

Further significant upgrades to the SMA's performance will require major changes to the receivers and IF signal transport systems, as well as further expansion of the correlator. We have begun development of a program of major upgrades to these systems with the aim of significantly increasing the SMA's instantaneous bandwidth, which will result in a significantly upgraded instrument called the wSMA (wideband Submillimeter Array). This program will include completely replacing the cryostats, receiver inserts and several other elements of the receiver systems, building additional segments of the SWARM correlator and upgrading the IF signal optical fiber transport to >20 GHz analog bandwidth on multiple channels using a commercial 1550nm DWDM system.

The new receiver system will operate with an IF of 4 – 18 GHz. Since the correlator of the interferometer separates the two sidebands of the DSB mixers, the wSMA will deliver 14 GHz bandwidth per polarization per sideband, for a total of 56 GHz of processed on-sky bandwidth. The new receiver system also features simplified receiver optics. Instead of the current use of heterogeneous single polarized receivers combined using a wire grid polarizer, the polarization performance of the wSMA will be enhanced by the use of a single dual-polarized feedhorn and an ortho-mode transducer feeding two identical SIS mixers to receive the two polarizations. The further simplification of the ambient temperature optics enabled by the use of waveguide LO coupling within the cooled receiver and use of cooled receiver selection optics will reduce ambient temperature optical losses and thus improve sensitivity. This upgrade will also have the significant operational benefit of replacing the current aging and commercially obsolete GM/JT cryogenic coolers with modern pulse-tube coolers having significantly reduced maintenance requirements.

In this paper we will describe the proposed receiver systems, including the new receiver optics and cryostats, the dual-polarization double sideband front-ends, wide band isolators and IF LNAs, and new LO systems, as well as required enhancements to the SMA's IF transport and signal processing capabilities to fully utilize the increased bandwidth of the front-end receivers. We will also discuss the current and future capabilities of the SMA, and the potential for further upgrades and guest/PI instrumentation that will be enabled by this upgrade.

NOEMA receivers: Upgrade for simultaneous dual-band observations

A-L. Fontana*, D. Maier, J. Reverdy, M. Parioleau, M. Rival and V. Jubard[1]

[1]*IRAM, 300 rue de la piscine, 38400 Saint Martin d'Hères, France*
*Contact: fontana@iram.fr

Abstract—NOEMA (Northern Extended Millimeter Array) is the successor of the IRAM interferometer located on the Plateau de Bure. Currently composed of eight 15-m diameter antennas, it is still under construction and will consist in twelve antennas in the next future. All the antennas are equipped with state of the art wide band SIS heterodyne receivers.

Each NOEMA receiver comprises, in a single cryostat, four dual-linear polarization side band separating SIS modules covering the frequency bands 72-116GHz, 127-179GHz, 200-276GHz and 275-373GHz and reaching excellent noise performances. Each of these modules delivers four 4-12GHz IF signals.

The construction of the receivers for the 12 antennas is almost finished, and IRAM is already working on the future upgrade of these receivers. The goal of this upgrade is to observe simultaneously the same point of the sky with two different frequency bands (currently, the four dual polarized beams of each receiver are separated by 4 to 6 arc minutes on the sky, depending on the bands considered). To co-align the different receiver bands, several design options are envisaged. The band combination can be made at room temperature, or at cryogenic temperature. For this last option we are developing dual band, dual polarization cryogenic side band separating SIS receiver modules. In each of these modules, a common (for two frequency bands) optics, composed of two focusing mirrors, is used to have a frequency independent illumination of the sub-reflector. Inside the module, the two bands are diplexed by a dichroic filter (developed by QMC Instruments, Cardiff University) operating at 4K. Each beam is then re-directed into a dual polarization feed horn coupled to a waveguide Ortho Mode Transducer (OMT) used to diplex the two polarizations of the band considered. Each OMTs output is then connected to a side band separating SIS mixer, delivering two 4-12GHz IF signals.

The different options envisaged for the dual band upgrade will be presented, and in particular, the design and first tests results of a 72-116GHz/127-179GHz and of a 72-116GHz/200-276GHz cryogenic dual band module will be detailed. The future perspectives of this work, in particular the way to offer several band combinations, will also be discussed.

A Cartridge-type Multi-pixel Receiver for the 1.5 THz Frequency Band of GLT

Yen-Ru Huang, Chuang-Ping Chiu, Hsian-Hong Chang, Chun-Lun Wang, Hiao-Wen Chang, and Ming-Jye Wang
Institute of Astronomy and Astrophysics, Academia Sinica, Taipei 10617, Taiwan
*Contact: yenruhuang@asiaa.sinica.edu.tw, phone: +886-2-2366-5309

Abstract—A prototype of 2 × 2 NbN hot electron bolometer (HEB) mixer array receiver, operating at 1.5 THz, has been designed and demonstrated for the Greenland Telescope (GLT). In this design, the LO beam splits into four equal-power sub-beams with a spacing of 18 mm by a power distributor module, then these sub-beams arrives at a four-pixel silicon lens array and are coupled by the twin slot antenna (TSA) of HEB mixer through a large-area beam splitter made of a Mylar film with 13 μm thickness. An additional HDPE lens is located at 120 mm in front of the individual silicon lens to increase the size of beam waist for fitting to the aperture parameter of GLT sub-reflector, thus this design can match the optical parameters between Si/TSA and GLT antenna. The pixel number can be further extended to 9 based on the high output power of LO module.

Introduction

The National Science Foundation (NSF) in the United States awarded the 12-m ALMA-NA Vertex Prototype Telescope to a team, led by the ASIAA, in 2011 to deploy the telescope on the summit of Greenland ice sheet, named as Greenland Telescope (GLT), for pursuing Submillimeter Very Long Baseline Interferometry (submm-VLBI) observations on the northern sky [1, 2]. Besides, due to the excellent weather conditions, the GLT could be also applied for single dish observations up to 1.5 THz window. The major advantage of GLT operating at THz frequencies is the higher angular resolution, as compared to space or airborne THz telescopes. Thus, high resolution and high speed mapping will be the important subjects for GLT.

Multi-pixel array receiver is more preferred for enhancing the mapping speed of telescope [3, 4]. In recent years, we devoted on developing a 1.5 THz multi-pixel cartridge receiver based on hot-electron-bolometer (HEB) mixers for GLT [5, 6]. The ALMA-type receiver cartridge was chosen in our design, which has many advantages, such as high cooling efficiency and good modularity. The local oscillator (LO) source in our receiver cartridge can provide an output power of 14 μW at 1.5 THz, and up to 35 μW while the frequency multipliers are cooled. The boosted LO power can further extend the pixel number to 9 in the future.

Performance Testing of the Receiver Cartridge

The engineering model of 1.5 THz 2 × 2 HEB mixer array receiver cartridge has been designed and assembled basically, and we tested the single-pixel configuration first to check the performance of cartridge. The single and four pixel configurations are identical, except of the power distributor module and the mixer block.

In our cartridge design, the quasi-optical path mainly comprises two parts: the LO beam path and the radio frequency (RF) beam path. For the LO beam path, the amplified Q-band frequency signal is fed through the 300 K base-plate to reduce power losses, and then passes through the cryogenic two-doubler module and two-tripler module to get the LO frequency at 1.5 THz. The horn of the LO module and a 90 degree off-axis parabolic mirror with an effective focal length of 30 mm are mounted on the 85 K plate to make the Rayleigh length long enough. Then the LO beam is divided into four sub-beams with a spacing of 18 mm through the power distributor module which consists of TE and TM mode polarizing beam splitters made of quartz and silicon slabs respectively. After that the LO sub-beams and RF beams are combined by a large-area beam splitter made of 13 μm Mylar film, and then coupled by a four-pixel silicon lens module with the twin-slot antenna (TSA) of individual HEB mixer chip. The calculated transmittance of the Mylar beam splitter is about 80%. For the RF beam path, an additional four-pixel HDPE lens model is located at 120 mm in front of the silicon lens to increase the beam waist size for matching with the aperture parameter of GLT sub-reflector.

After the engineering model assembled, the LO output power measurement was carried out through the single-pixel power distributor by using the bolometer from QMC instrument. The single-pixel power distributor module simply consists of two reflectors to align the LO signal to HEB mixer. It is noticed that the Mylar beam splitter on 4 K stage was mounted perpendicularly to the original attitude for guiding the signals to outside. A large-area 90 degree off-axis parabolic mirror was placed in front of the vacuum window to focus LO signals into the bolometer. An optical chopper and a lock-in amplifier were applied here to measure the modulated output peak-to-peak voltage. The measurement result is shown in Fig. 1, where the red curve is the output amplitudes at room temperature. The variation of voltage (or output power) with frequency perfectly meets the specifications provided by VDI. The black curve is the output amplitude of power detector as the frequency two-doubler and two-tripler modules cooled, which is about 2.5 times higher than that at room temperature.

A FULL VERSION OF THIS PROCEEDINGS ARTICLE HAS BEEN SUBMITTED TO THE IEEE TRANSACTIONS OF TERAHERTZ SCIENCE AND TECHNOLOGY (TST) FOR PEER-REVIEW, PLEASE CHECK THE WEBSITE: HTTPS://WWW.MTT.ORG/TERAHERTZ

The LO beam profile is shown in Fig. 2, which is measured by the same setup but with an additional 2-mm pinhole mounted on an x-y linear motorized stage to define the measurement pixel on the scanning frame. The profile of beam pattern is close to the Gaussian function with negligible distortion, and the full width at half maximum (FWHM) is about 4.479 mm.

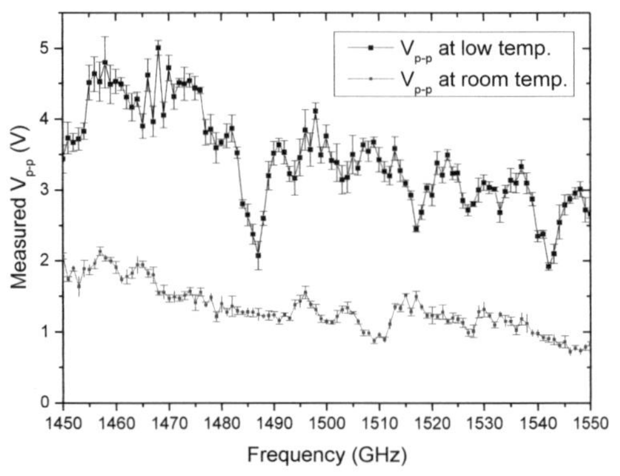

Fig. 1 The output power measurement of 1.5 THz LO module at room and low temperatures.

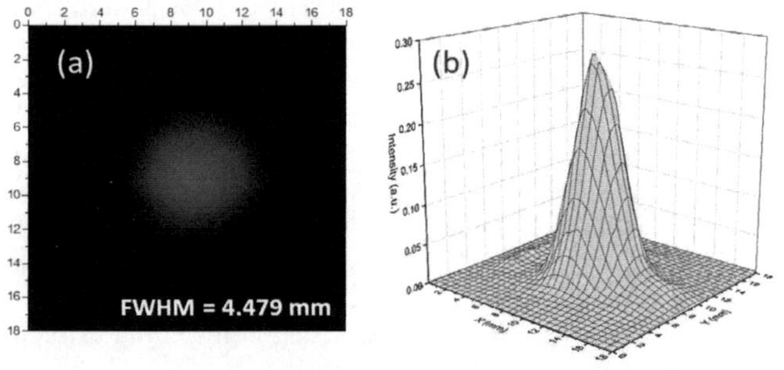

Fig. 2 The beam profile of LO module.

The Temperature of Cold Cartridge

The Sumitomo RDK-3ST three-stage cryocooler is applied in our testing cryostat, which has cooling powers of 1.0 W at 4.4 K for the 4 K stage, 8 W at 18 K for the 15 K stage, and 33 W at 85 K for the 85 K stage. The cooling test had been carried out and the result shows that the balanced temperature is 3.2 K, 15.8 K, and 83.5 K on the three temperature stages of cartridge respectively. The DC biases for cooled two-doubler module of the LO system were applied before cooling. On the other hand, in order to understand the thermal loading as the LO system works normally (synthesizer of LO module turned on and the LO signal amplified), we mounted six sensors on different positions to measure the temperature distribution. As LO system working, the two-doubler module is heated from 95.3 K to 108.1 K and the temperature of two-tripler module increases about 3.2 K. However, the 4K plate and the mixer block (~ 3.6 K) only have a temperature increase less than 10 mK.

Fig. 3 shows the temperature variation at mixer block with a resolution of 0.2 second while the LO system was turned on. The temperature variation is just 8 mK within 17.5 hours and therefore we can expect that the HEB mixer performance won't be affected during whole system in operation.

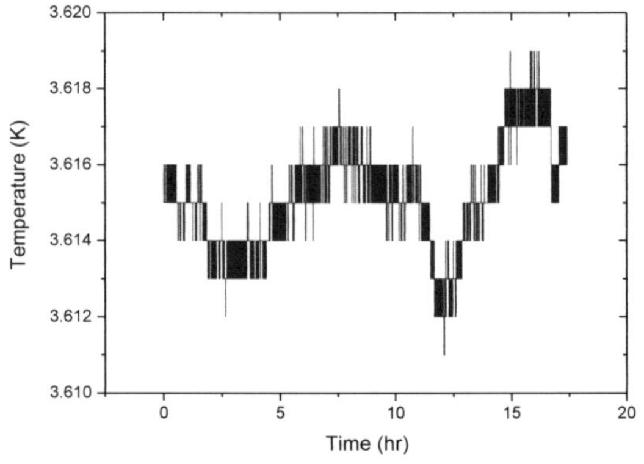

Fig. 3 The temperature stability of mixer block.

RF Performance Testing of HEB Mixers

The RF performance of single-pixel receiver cartridge with HEB mixer was tested. The HEB mixer is made of NbN film with a thickness of 17 nm, and the 1.5 THz twin-slot antenna was applied here [6]. To match the characteristic impedance of the lumped port in the antenna design, the gap length and micro-bridge width were defined to 0.55 μm and 1.57 μm respectively. The DC testing results of HEB mixer shows that the critical temperature (T_C) and normal-state resistance (R_n) is about 10.4 K and 61 Ohm respectively. For the RF performance test, the user controlled attenuation (UCA) port in the LO module was used for optimizing the LO power. It is noted that there is no stabilization mechanism of LO pumping power presently. The current-voltage (I-V) characteristics of HEB mixer was measured under various LO pumping levels, as shown in Fig. 4(a). The optimal pumped situation, with UVA bias voltage of 1.55 V, was chosen for receiver noise temperature (T_{rec}) measurement. The intermediate frequency (IF) power under hot/cold load and the T_{rec} of cartridge receiver are indicated in Fig. 4(b). The T_{rec} is about 2200 K as the HEB is biased at 0.3 mV. The test result is quite convincing which is similar to that tested in wet dewar before. It should be noted that the Si lens has no anti-reflection (AR) coating. The performance of cartridge receiver could be further improved by using Si lens with AR coating and a better HEB mixer.

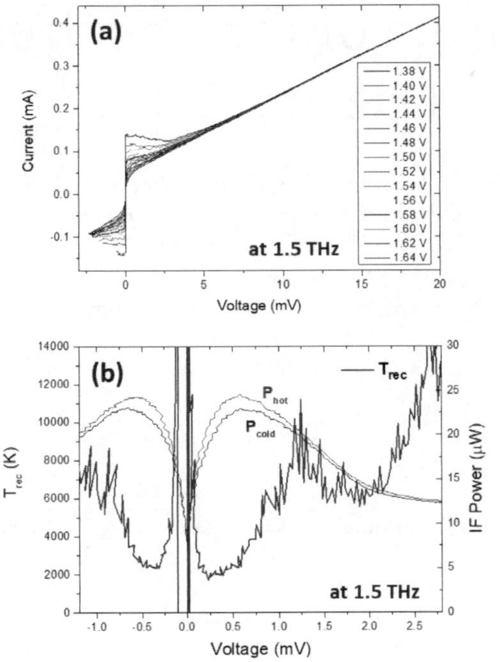

Fig. 4 (a) The I-V curves with different LO pumping power. (b) The result of hot-cold load measurement.

SUMMARY

We have successfully designed and assembled an engineering model of 1.5 THz single/four -pixel cartridge-type receiver based on HEB mixers. At present, the performance testing of single-pixel design has been completed, including the vacuum and cooling test, the output power and beam profile of LO source, and receiver noise measurement. From our results, the LO output power is large enough for applying more pixels in the future and the temperature variation of mixer block is quite small (~ 8 mK) within a long monitoring period. In addition, the RF performance testing of HEB mixers has also been carried out, and the optimal pumped situation can be achieved. In the future, we will deploy HEB mixers with better performance on the cartridge and characterize the four-pixel cartridge receiver cartridge.

ACKNOWLEDGMENT

The authors would like to thank Dr. Sheng-Cai Shi in Purple Mountain Observatory, Nanjing, China, for the help on HEB mixer design.

REFERENCES

[1] H. Hirashita, P. M. Koch, S. Matsushita, S. Takakuwa, M. Nakamura, et al. "First-generation science cases for ground-based terahertz telescopes," *Publ. Astron. Soc. Jpn.*, vol. 68, R1, Feb. 2016.

[2] P. K. Grimes, K. Asada, R. Blundell, R. Burgos, H-H. Chang, et al. "Instrumentation for single-dish observations with the Greenland Telescope", in *Proc. of SPIE*, 2014, vol. 9153, 91531V.

[3] F. Rodriguez-Morales, K. S. Yngvesson, E. Gerecht, N. Wadefalk, J. Nicholson, et al. "A terahertz focal plane array using HEB superconducting mixers and MMIC IF amplifiers," *IEEE Microw. Wireless Compon.*, vol. 15, pp. 199-201, April 2005.

[4] U. U. Graf, C. E. Honingh, K. Jacobs, and J. Stutzki, "Terahertz heterodyne array receivers for astronomy," *J. Infrared Millim. Te.*, vol. 36, pp. 896-921, June 2015.

[5] Y.-R. Huang, C.-P. Chiu, H.-H. Chang, Y.-Y. Chiang, and M.-J. Wang, "Development of 1.5 THz Cartridge-type Multi-pixel Receiver Based on HEB Mixers," in *Proceedings of the 27th International Symposium on Space Terahertz Technology (ISSTT 2016)*, Apr. 2016, Nanjing, China, paper T4-P2.

[6] Y.-R. Huang, C.-P. Chiu, W.-C. Lu, H.-H. Chang, Y.-Y. Chiang, and M.-J. Wang, "Development of 1.5 THz Cartridge-type Multi-pixel Receiver Based on HEB Mixers," in *Proceedings of 2016 URSI Asia-Pacific Radio Science Conference (AP-RASC 2016)*, Aug. 2016, Seoul, Korea, pp. 569-571.

4GREAT: A Multiband Extension of GREAT from 490 GHz to 2.7 THz

Carlos A. Durán[1], Rolf Güsten[1], Christophe Risacher[1], Oliver Ricken[1], Stefan Heyminck[1], Urs U. Graf[2], Cornelia E. Honingh[2], Karl Jacobs[2], Patrick Pütz[2], Gert de Lange[4], Darek Lis[3], Yan Delorme[3], Hans-Joachim Wunsch[1], Andrej Görlitz[1], Andreas Schmitz[1], Bernd Klein[1], Andreas Bell[1]*

[1]Max Planck Institut für Radioastronomie, Auf dem Hügel 69, 53121 Bonn, Germany
[2]KOSMA, I. Physikalisches Institut, Universität zu Köln, Zülpicher Straße 77, 50937 Köln, Germany
[3]LERMA Observatoire de Paris, 61 avenue de l'Observatoire, 75014 Paris, France
[4]SRON Netherlands Institute for Space Research, Zernike Building, PO Box 800, NL-9700 AV Groningen, The Netherlands
**Contact: cduran@mpifr-bonn.mpg.de*

Abstract— The German REceiver for Astronomy at Terahertz frequencies (GREAT) has been in successful service onboard SOFIA since 2011. GREAT, with its modular approach, is composed of a group of cryostats containing detectors for different frequency bands (until now, between 1.25 and 4.7 THz). At any time, GREAT can carry two cryostats.

4GREAT (4G), a new member of the GREAT constellation, is a 4-color single-pixel module. Two channels, 4G-1 and 4G-2, are implemented using spare flight mixers developed for Herschel's *Heterodyne Instrument for the Far-Infrared,* namely HIFI band 1 and band 4. The third channel, 4G-3, makes use of the current GREAT L1 detector (1.2–1.5 THz), while 4G-4 covers the frequency range of GREAT $M_{a,b}$ (2.5-2.7 THz), using a newly developed mixer, (similar in design to the upGREAT HFA mixers). The four channels, co-aligned on sky, are operated in a single closed-cycle cooled cryostat.

4GREAT, scheduled for commissioning in July 2017, will be used simultaneously with the upGREAT-HFA (an array of 7 pixels working at 4.745 THz), allowing multiple frequency observations of astrophysically important species including among many others, the ground-state transitions of many hydrides (HDO, HCl, CH, ammonia NH_3, isotopic water $H_2{}^{18}O$, hydroxyl OH), as well as mid-J transitions of carbon monoxide.

INTRODUCTION

With the successful addition of the upGREAT LFA [2] and HFA [3] to the highly modular GREAT instrument [1], greatly enhanced scientific opportunities have been offered to the interested SOFIA communities. However, while these instruments increase the scientific yield of SOFIA by spatial multiplexing (14+7 pixels operating at the same time), there is also a strong demand for a complementary instrument that does span a wider range in sky frequencies. 4GREAT has been developed in response to these needs. The instrument will operate simultaneously four state-of-the-art detectors at science-defined frequencies between 0.5 and 2.7 THz.

Since Herschel ceased operations in 2012 [7], astronomers are lacking access to those parts of the sub-Terahertz spectrum that are not visible from ground-based observatories. In 4GREAT, we make use of the spare mixers developed for HIFI band 1 (4GREAT channel 4G-1) and for band 4 (4G-2). Much of their frequency ranges are blocked by Earth's atmosphere, even at high dry sites like the Chajnantor Plateau in Chile, the sites of APEX and ALMA [4,5].

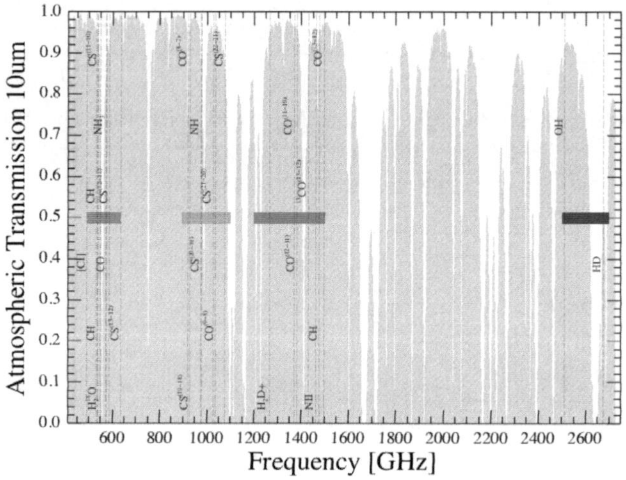

Fig. 1. The atmospheric transmission at PWV of 10 μm, for a SOFIA flight altitude of 43.000 ft. Astrophysical lines of interest and the frequency bands of the 4GREAT channels are marked.

4GREAT channel 3 reuses the existing GREAT-L1 (1.25 – 1.5 THz) mixer, while the channel 4 mixer (2.5 – 2.7 THz) is a newly developed detector, similar in design to the upGREAT-HFA mixers.

Fig. 1 shows the frequency coverages of the different channels plotted along with some scientific lines of interest and the atmospheric transmission for a high-altitude flight of SOFIA [6].

The use of closed-cycle cooled cryostats instead of cryogenics LHe/LN2 (wet) cryostats is a step forward in terms of operation and reliability of the GREAT receivers, as they do not require daily (cryo) services.

Design Description

The main requirement on the 4GREAT design is to allocate 4 different frequency detectors inside its closed-cycle cooled cryostat. The detectors shall operate simultaneously, and their beams must co-align on the sky. With this in mind and with tight constraints on the available volume for the mixers and cold optics inside the cryostat, a maximum weight budget for the overall system, the 4GREAT opto-mechanical design was optimized. The design is modular again. Special attention was paid to good maintainability.

TABLE I
4GREAT main components and subcomponents characteristics

Channel	CH1	CH2	CH3	CH4
RF Bandwidth (GHz)	492 - 630	892 - 1100	1200-1500	2490 - 2700
IF Bandwidth (GHz)	4 - 8	4 - 8	0 - 4	0 - 4
Mixer	SIS - Herschel HIFI - 1 (LERMA)	SIS-Herschel HIFI - 4 (SRON)	HEB - GREAT - L1 (KOSMA)	HEB - GREAT - M-HD (KOSMA)
LNA / Warm Amplifier	LNF-LNC4_8C (LNF)	LNF-LNC4_8C (LNF)	CITLF4 (CMT)	CITLF4 (CMT)
	AFS3-00100800 (Miteq)	AFS3-00100800 (Miteq)	AFS3-00100800 (Miteq)	AFS3-00100800 (Miteq)
Local Oscillator	S.S.Chain AMC563@LO-U (200uW)	S.S. Chain AMC581@LO-U (150uW)	S.S. Chain AMC627@LO-D (30uW)	S.S. Chain AMC616@LO-D (3.5 uW)
LO Coupling	Wiregrid Splitter	Wiregrid Splitter	Wiregrid Splitter	Diplexer
Optics	Common optics plate + Cold Tower + LO-U Optics	Common optics plate + Cold Tower + LO-U Optics	Common optics plate + Cold Tower + LO-D Optics	Common optics plate + Cold Tower + LO-D Optics
Trec (K) - DSB	100	300	800	1500
IF Processor	IFX x 1. High Order BPF 4-8 GHz	IFX x 1. High Order BPF 4-8 GHz	IFX x 1. High Order BPF 0-4 GHz	IFX x 1. High Order BPF 0-4 GHz
Backend	FFTS4G. Nyquist Band 4-8	FFTS4G. Nyquist Band 4-8	dFFTS4G x 1ch	dFFTS4G x 1ch
Taper (dB)	11.86 - 16.54	12.25 - 16.09	13.29 - 14.78	14.35 - 13.68

A. Mixers

The mixers for 4GREAT originate from different technologies and sources (Tables I, II). The 4G-1 and -2 mixers use Superconducting Insulator Superconducting (SIS) junctions developed by LERMA and SRON [7] for HIFI-Herschel. 4G-3 and -4 operate Hot Electron Bolometers (HEB), designed and built by KOSMA.

TABLE II
4GREAT mixer specifications

Band	Technology	Manufacturer	Remark
CH1	SIS	LERMA	HIFI-1 Flight Spare Mixer
CH2	SIS	SRON	HIFI-4 Special Qualification Model Mixer
CH3	HEB (NbTiN)	KOSMA	GREAT L1 Spare Mixer
CH4	HEB (NbN)	KOSMA	GREAT M-HD Mixer

B. Local Oscillators

4GREAT utilizes four solid-state local oscillator units or “LO chains”, built by Virginia Diodes Inc. [13]. They consist of a reference oscillator and several multiplication stages along with high power amplifiers. Given the large response widths of the 4GREAT mixers (Q > 20%), three of the LO chains consist of two sub-chains each, which are combined to drive the last multiplier stages. Table I summarizes the band coverage and average output power over each band.

Space for the local oscillator units is very limited and thus LO chains were grouped in couples and split into two modules (Fig. 2): the lower LO unit (LO-D) houses the LO chains for the channels 4G-3 and -4 and uses one of the standard GREAT LO compartments, while the upper LO unit (LO-U) for 4G-1 and -2 attaches to the flange for the auxiliary calibration unit [1].

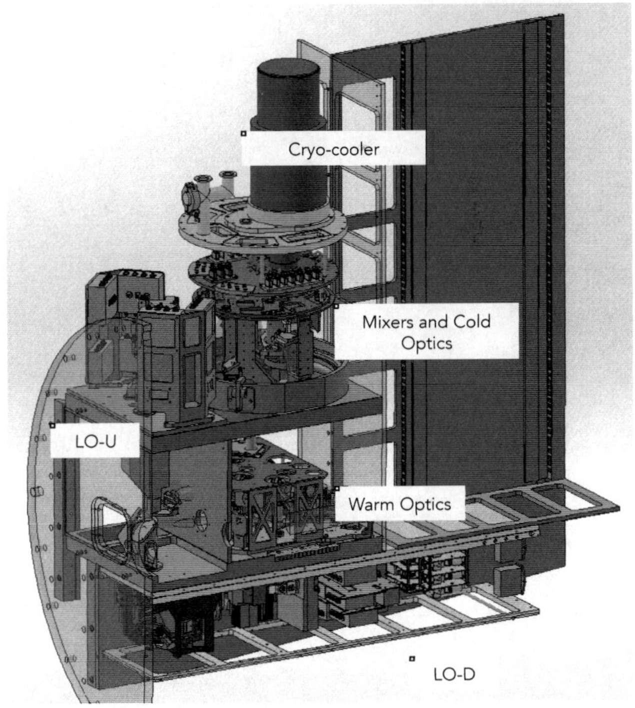

Fig. 2. 4GREAT subcomponents mounted on the GREAT SI structure: The cryostat (mixers, cooler and cold optics), the lower local oscillator unit (LO-D), the upper local oscillator unit (LO-U) and the warm optics are shown.

C. Optics

The 4GREAT optics consist of *Cold Optics* inside the cryostat, the *Optics Plate* where the splitting of the telescope beam into the 4 signal beams and superposition with the LO beams is done, and dedicated *LO Optics* for each of the LO chains. In general, a multiple *Gaussian telescope* approach was taken to make the system as frequency independent as possible, with the only exception of the LO for 4G-3. For every mirror and optical component, a 5*w* beam criterion was used, with a designed edge-taper of 14dB at the central frequency of each band. RF windows for the cryostat were manufactured by QMC [11] for 4G-1 and -2 (Quartz with A/R coating). Tydex [12] provided the windows (Silicon coated with A/R) for 4G-3 and 4G-4.

1) The Cold Optics

Each of the 4 mixers is mounted on a block that also contains part of the channel optics and the low noise amplifiers (LNA). Each of these blocks (*cold towers*) contains a paraboloidal mirror, located in front of each mixer horn, a flat mirror and an ellipsoidal mirror. The ellipsoidal mirror M1 then forms a Gaussian telescope with the first active mirror (M2) mounted outside the cryostat, on the optics plate, just in front of each window (Fig. 4).

Fig. 3 shows the four *cold towers* installed. Because of the low frequency and hence large beam size of 4G-1, the *cold tower* for this channel is much larger than the others.

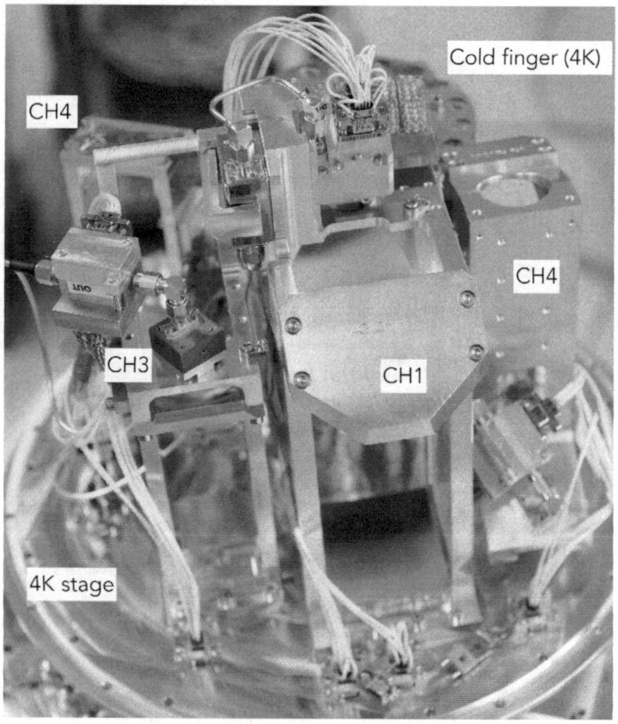

Fig. 3. Each *cold tower* mounts a paraboloidal (M0), a flat and an ellipsoidal mirror (M1), along with the corresponding mixer and cold LNA amplifier.

2) The Optics Plate

The optics plate or "*warm optics*" is placed just below the cryostat (Fig. 2). Fig. 4 shows the schematic of the optics for a single channel. Splitting the telescope beam into four separate signal beams is done using a wire grid and two dichroic filters. The wire grid is placed after the active mirror M4 that is common to all channels. Each of the transmitted (combined signal beams of 4G-2 and 4G-4) and reflected beams (combined signal beam of 4G-1 and 4G-3) is directed to a low-pass dichroic filter positioned between the next two active mirrors (M3-13 / M3-24 and M2n, n=1-4). Consequently, there are two M3 and four M2 mirrors on the optics plate. Between each M2 and before entering the cryostat, the signal beams of channels 1, 2 and 3 are superimposed with the corresponding LO beams by a wire grid (LO in reflection). Because of the relatively weak LO source (Table I) channel 4 uses a Martin-Puplett-Diplexer.

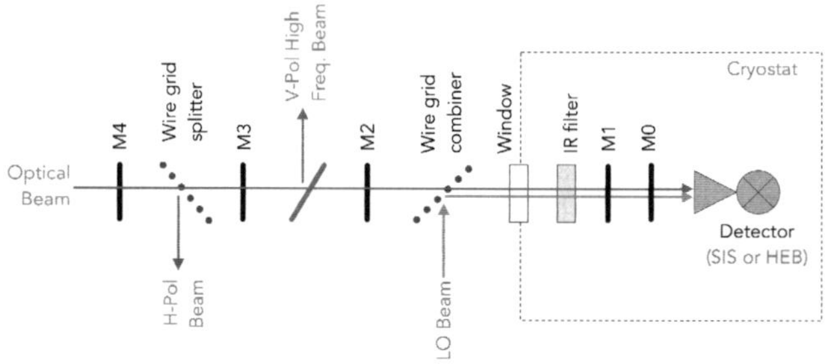

Fig. 4. Diagram showing the main optics components. 4G-4 uses a diplexer instead of a wire grid as combiner. The beam optics comprises two Gaussian telescopes: M4-M3 (at the Optics Plate) and M2 (Optics Plate) with M1 (Cold Tower).

On the optics plate, optical elements are positioned on two levels, spaced by 50mm. The lower level consists mainly of signal beam-related elements while the upper one is mostly used for the LO signal (Fig. 5).

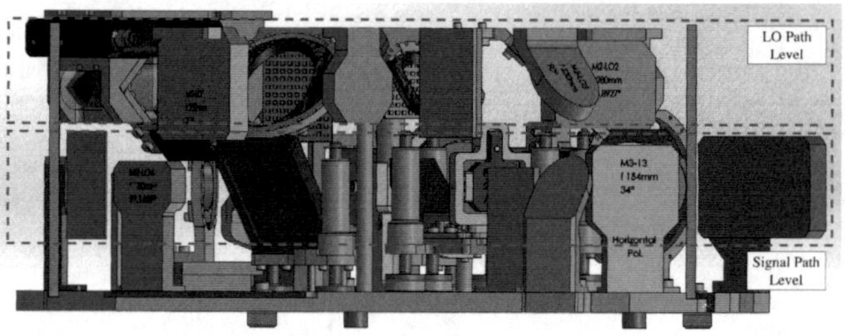

Fig. 5. Side view of the optics plate. Red dashed lines frame the two planes. Some non-critical elements have been omitted in the figure for better visualization.

3) The Optics of the upper LO Unit (LO-U)

The LO-U optics has a very particular design as it hosts the two "*low*" frequency channels, which require larger optical elements. Because of the location of the LO-U, its components are placed on an extension that sticks into the SI frame and then couples to the upper level of the optics plate (Figs. 5, 6). Each of the LO output beams is mapped to its respective mixer input beam by two Gaussian telescopes (4 active mirrors). The first Gaussian telescope is located inside/at the LO-U. The second Gaussian telescope is formed by an active mirror on the upper optics plate level and with M1 inside the cryostat.

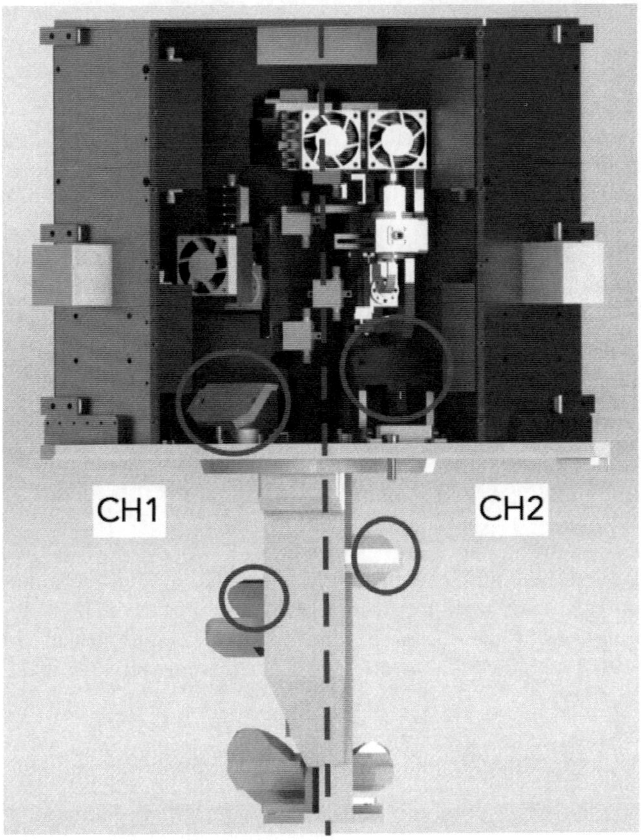

Fig. 6. The upper LO unit (serving 4G-1 and4G-2). The LO chains and optics components are visible in the image. The two active mirrors for each LO are encircled. The dashed line indicates marks the optical and electrical separation between the two channels.

4) *The Optics of the lower LO Unit (LO-D)*

For the upGREAT-LFA [2] the requirement of fitting two LO chains in the LO compartment was solved by tilting one of the LO beams. The design of 4GREAT LO-D with the LOs for 4G-3 and 4G-4 follows the same approach and uses some of the optical components already installed in the SI structure. This constraint requires that for the optics of the 4G-3 LO three active mirrors have to be used, which introduces a minor frequency dependence of the approach. The LO optics for 4G-4 uses four mirrors again as for the lower frequency channels, with a coupling wire grid, which is part of the diplexer optics on the optics plate, in between two of the active mirrors (M2 and M1).

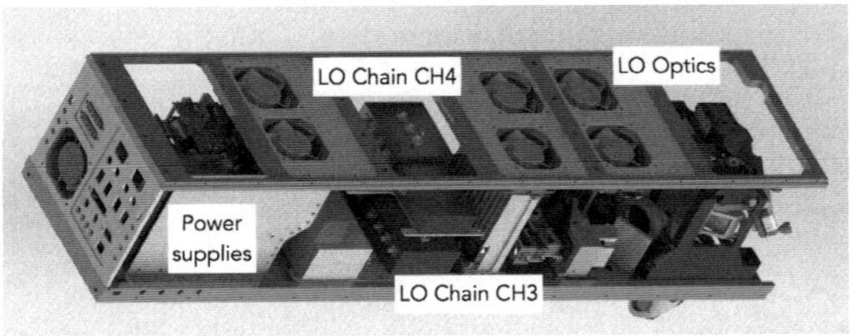

Fig. 7. The lower LO unit (serving 4G3 and 4G-4) is a very compact unit that includes its power supplies, the LO chains and the LO fore-optics.

5) *Cryostat and Closed Cycle Cooler*

As for the upGREAT LFA and HFA, 4GREAT uses a Pulse Tube closed-cycle cooler PTD-406C from Transmit GmbH [9] in a cryostat manufactured by Cryovac GmbH [10]. The cooler load map is shown in Fig. 8.

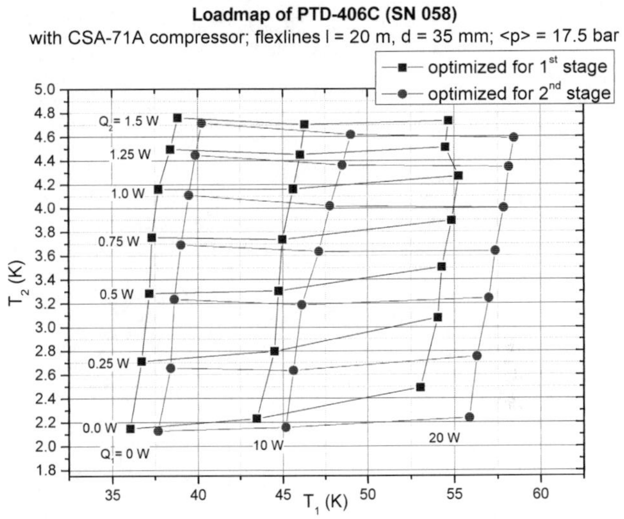

Fig. 8. Thermal load map for the Pulse Tube cooler used by 4GREAT. The operation temperatures are 3K (second stage) and 55K (fist stage).

6) *Intermediate Frequency Unit and Backend*

The Intermediate Frequency (IF) units for 4G-3 and 4G-4 make use of the 0 - 4 GHz IF processor modules already in use for the upGREAT LFA and HFA, while for 4G-1 and -2 new IF processor modules were developed to accommodate the 4 to 8 GHz range (given by the SIS IF bands).

A new version of the high-resolution digital back-end has been recently introduced to GREAT/upGREAT. dFFTS4G [8] is a dual 0-4 GHz bandwidth 32k channels spectrometer. A single card then serves the signals from both 4G-3 and -4 chains, while for 4G-1 and 4G-2 one card each of the FFTS4G spectrometer is employed. The latter operates directly in the 2nd Nyquist band (4-8 GHz).

LABORATORY RESULTS AND COMMISSIONING

4GREAT has been integrated at MPIfR during the months of April and May 2017. The preliminary results presented here have been derived from tests performed with the final aircraft configuration in the AFRC laboratories in Palmdale during May - June. Instrument commissioning is scheduled to take place during the New Zealand SOFIA deployment in July 2017.

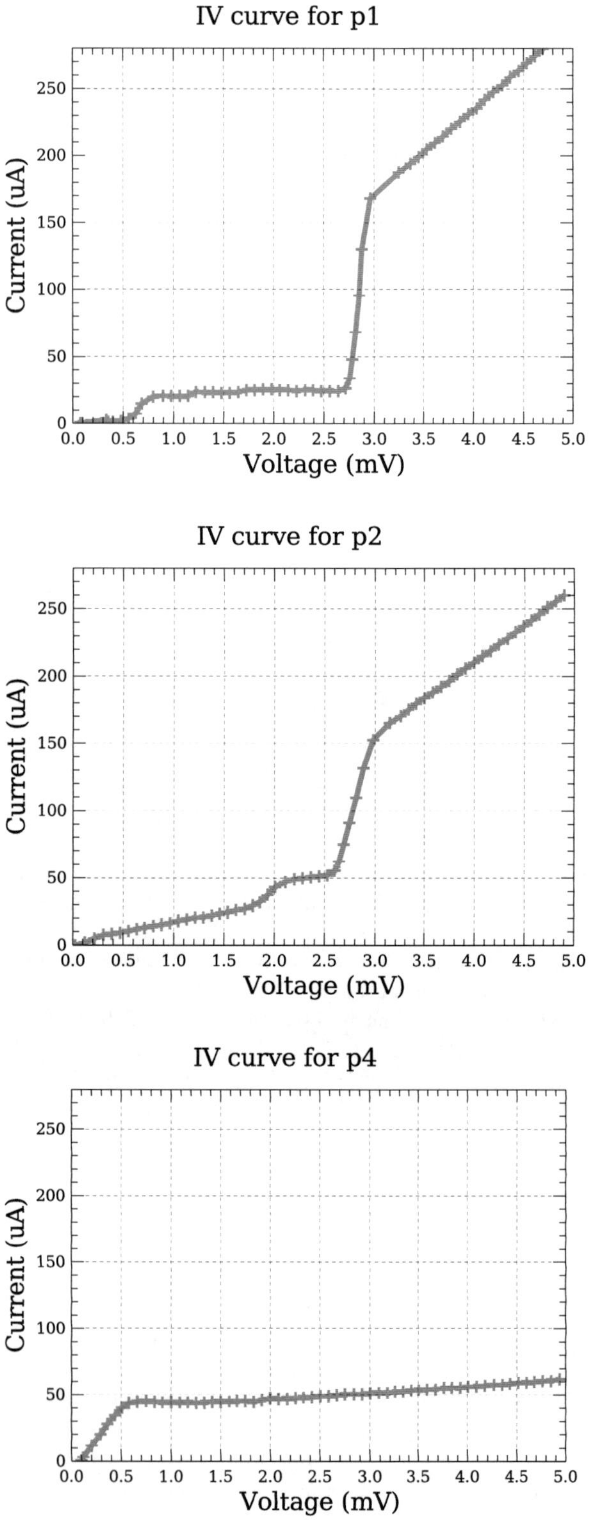

Fig. 9. Pumped I-V curves of the channels 4G-1, 4G-2 and 4G-4 mixers at 3.5 K. Curves for channels 1 and 2 were obtained with magnetic field applied.

CONCLUSIONS

The frequency-multiplexing of 4GREAT will bring new scientific opportunities to the SOFIA communities. The instrument will allow making more efficient use of the limited observing time by performing simultaneous observations at 4 different frequencies.

Though with extended science capabilities, the number of instrument configurations is now limited to two (LFA & HFA, 4GREAT & HFA) which will reduce significantly the technical overheads. All-closed-cycle operation does eliminate the need for daily cryogen service before flight and makes system operation more robust.

ACKNOWLEDGMENT

We thank the MPIfR mechanical workshop for their commitment and excellent work, the KOSMA mechanical workshop for their invaluable help, and in particular all the members of the Division for Submm Technologies at the MPIfR who devotedly worked to make this project a reality.

REFERENCES

[1] S. Heyminck, U. U. Graf, R. Güsten, J. Stutzki, H. W. Hübers and P. Hartogh, "GREAT: the SOFIA high-frequency heterodyne instrument", *Astron. and Astrophys*., Vol. 542, L1, June 2012.
[2] C. Risacher, R. Güsten, J. Stutzki et al. "First Supra-THz Heterodyne Array Receivers for Astronomy With the SOFIA Observatory", *IEEE Transactions On Terahertz Science And Technology*, Vol. 6, No. 2, March 2016.
[3] C. Risacher et al. "The upGREAT High Frequency Array: 7 Pixels at 4.7 GHz", *in preparation*
[4] C. Leinz, M. Caris, T. Klein et al. "A 1THz Receiver System at APEX", 21st International Symposium on Space Terahertz Technology, held March 23-25, 2010 at Oxford University's Said Business Center and the STFC Rutherford Appleton Laboratory, Oxford, UK. National Radio Astronomy Observatory (NRAO), 2010., p.130-135
[5] Atmospheric transmission calculator for Llano de Chajnantor, Available: http://www.apex-telescope.org/sites/chajnantor/atmosphere/
[6] Lord, S. D., 1992, NASA Technical Memorandum 103957. Available: https://atran.sofia.usra.edu/cgi-bin/atran/atran.cgi
[7] T de Graauw et al. The Herschel-Heterodyne Instrument for the Far-Infrared (HIFI): instrument and pre-launch testing", *Proc. SPIE* 7010, Space Telescopes and Instrumentation 2008: Optical, Infrared, and Millimetre, 701004 (July 12, 2008)
[8] B. Klein. "A new Fast Fourier Transform Spectrometer for upGREAT and 4GREAT". *Spectroscopy with SOFIA: new results & future opportunities*. Schloss Ringberg, Germany. (5-8 March 2017)
[9] Transmit GmbH: http://www.transmit.de
[10] Cryovac GmbH & Co KG: http://www.cryovac.de
[11] QMC Instruments Ltd & Thomas Keating Ltd http://www.terahertz.co.uk
[12] Tydex Optics: http://www.tydexoptics.com/pdf/THz_Windows.pdf
[13] Virginia Diodes Inc., Available: https://www.vadiodes.com

Results from the Kilopixel Array Pathfinder Project (KAPPa): a 6mm × 6mm 650 GHz Heterodyne Mixer Pixel with Integrated SiGe LNA and Permanent Magnet

Christopher Groppi[*,1], Caleb Wheeler[1,2], Hamdi Mani[1], Kristina Davis[1], Matthew Underhill[1], Marko Neric[1], Sander Weinreb[3], Damon Russell[4], Jacob Kooi[4], Arthur Lichtenberger[5], Craig Kulesa[6]

[1]*Arizona State University School of Earth and Space Exploration, 85287, Tempe AZ, USA*
[2]*Fisk University Department of Physics, 37208, Nashville TN, USA*
[3]*California Institute of Technology, 91125, Pasadena CA, USA*
[4]*NASA Jet Propulsion Laboratory, 91109, Pasadena CA, USA*
[5]*University of Virginia Department of Electrical and Computer Engineering, 22904, Charlottesville VA, USA*
[6]*University of Arizona Steward Observatory, 85721, Tucson AZ, USA*
[*]Contact: cgroppi@asu.edu

Abstract—We present the results of the Kilopixel Array Pathfinder Project (KAPPa) instrument, a technology development project aimed at addressing several key technological issues confronting the expansion of heterodyne arrays to formats approaching ~1000 pixels. The KAPPa receiver was built to test an integrated 6mm × 6mm heterodyne pixel cell operating from 600-700 GHz. The pixel cell contains a single ended waveguide SIS mixer, IF matching network, low power dissipation SiGe LNA, IF output and permanent magnet all contained beneath the 6mm aperture of the waveguide feedhorn. This design allows the pixel cell to be tiled to arbitrary array size without mechanical interference from neighboring pixels. The KAPPa test receiver can house a tunable electromagnet, used to optimize the applied magnetic field and also a permanent magnet that applies a fixed field. Our permanent magnet design uses off-the-shelf neodymium permanent magnets and then reshapes the magnetic field using machined steel concentrators. These concentrators allow the use of an unmodified commercial permanent magnet in the back of the detector block while two small posts provide the required magnetic field across the SIS junction in the detector cavity. Performance of the test mixer meets the requirements for ALMA band 9 for most of the 600-720 GHz band, while retaining the simplicity and reliability needed for use in large format heterodyne array applications.

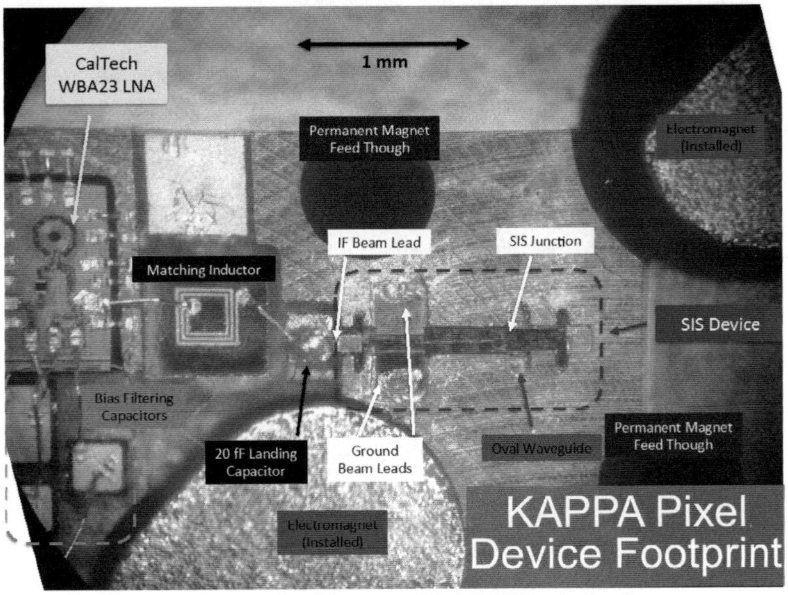

A micrograph of the assembled KAPPa pixel cell.

Development of a 16-pixel monolithic 1.9 THz superconducting waveguide HEB mixer

Jonathan Kawamura[1,*], Jenna Kloosterman[1], Jose Siles[1], Faouzi Boussaha[1], Bruce Bumble[1], Victor White[1], Imran Mehdi[1], Choonsup Lee[1], Damon Russell[1], Christopher Groppi[2], and Kristina Davis[2]

[1]*Jet Propulsion Laboratory, California Institute of Technology, Pasadena, CA 91109*
[2]*SESE, Arizona State University, Tempe, AZ, 85287*
[*]Contact: kawamura@jpl.nasa.gov

Abstract—Compact heterodyne imaging receivers are needed for future science instruments for aircraft, balloon-borne and space instruments that have limited observing time. Obviously, increasing the number of pixels affords a proportional increase in mapping speed provided that the sensitivity remains the same. In recent years, we have worked towards this goal by building modular linear 4-pixel waveguide mixer arrays that can be stacked with little difficulty to fill a focal plane. At the same time, we have built local oscillators in the same manner, and successfully built several local oscillator chains with 4 output pixels. Similarly, we have designed and built a 4 channel low-noise amplifier. With these components in hand, it is a straightforward exercise to expand the array. In the current configuration the mixer and local oscillator arrays are simply imaged through a beam splitter as in a conventional heterodyne receiver.

Currently, our efforts involve expanding the system to a 4 x 4 array receiver by conceptually stacking the linear components. In this talk, we focus our discussion on the mixer: in contrast to the sparsely spatially populated linear array we have previously built, we have designed and produced a compact 16-pixel mixer block more suitable for practical operation by direct drilling an array of feed horns into a block (Figure 1). Noise performance and beam measurements will be presented.

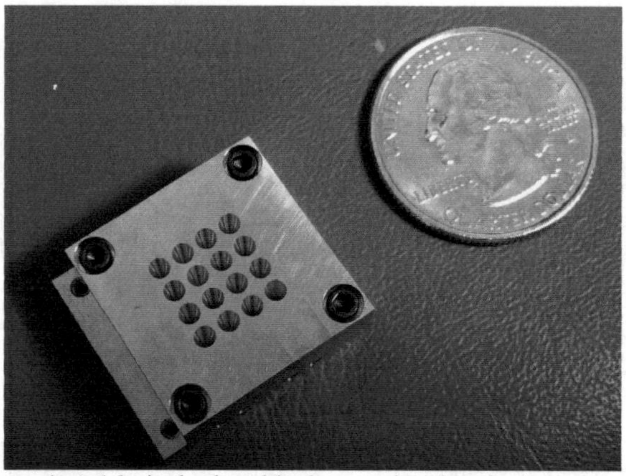

Figure 1. A 16-pixel mixer block next showing the horn array. The horn spacing is 2.5 mm.

4.7 THz flight mixers for upGREAT

P. Pütz*, K. Jacobs, M. Schultz, S. Fathi, M. Justen, C. E. Honingh, and J. Stutzki

I. Physikalisches Institut, Universität zu Köln, 50937 Köln, Germany
*Contact: puetz@ph1.uni-koeln.de

Abstract—We summarize our laboratory results for the flight mixers that we delivered to the 7-pixel high frequency array (HFA) focal plane extension of the GREAT receiver in operation on SOFIA. The HFA targets the [O I] fine structure transition at 4745 GHz (63 μm) of mainly galactic sources. The HFA went into commissioning in November 2016 and greatly increases mapping speed over its single pixel receiver predecessor GREAT H (see presentation from C. Risacher et al. on the receiver performance on SOFIA).

Based on the performance of the single-pixel receiver we selected waveguide NbN hot electron bolometer devices with a similar LO power consumption. The quantum-cascade laser (QCL) based LOs used at 4.7 THz demonstrate significantly higher output power levels (> 1 mW) than the solid-state multiplier sources used e.g. for the 1.9 THz low frequency array (LFA). This significantly relaxes the maximum allowable LO pump power limit during mixer selection.

We compare the heterodyne characterization results at 4.7 THz for these mixers using one common lab setup with a QCL based LO and single shot 0-5 GHz IF processing capability. All measurements were performed with an evacuated signal path using a 3 μm thick Mylar beamsplitter as diplexer. We measure uncorrected Trec noise temperatures less than 1500 K for all mixers averaged over an 1-2 GHz IF band in our lab setup with 3 dB IF noise roll-offs above 3 GHz.

4-Pixel Heterodyne Receiver at 1.9 THz using a CMOS Spectrometer

Jenna L. Kloosterman[1,*], Jonathan H. Kawamura[1], Adrian J. Tang[1,2], Rod Kim[2], Jose Siles[1], Fahouzi Boussaha[3], Bruce Bumble[1], Choonsup Lee[1], Alejandro Peralta[1], Robert Lin[1], Imran Mehdi[1]
[1]*Jet Propulsion Laboratory, California Institute of Technology, 91106 Pasadena, CA, USA*
[2]*University of California, Los Angeles, Los Angeles, CA 90095, USA*
[3]*Observatoire de Paris, 75014 Paris, France*
*Contact: Jenna.L.Kloosterman@jpl.nasa.gov

Abstract— This paper reports on engineering results from a high-sensitivity 4x1-pixel 1.9 THz heterodyne array for the astrophysically important [CII] spectral line. A 4-pixel multiplier chain at 1.9 THz, with a compact spacing of 5 mm, is used as an LO. Superconducting Hot Electron Bolometers (HEB) are used as mixers. Receiver performance is verified by y-factor measurements with measured sensitivities of 900 K uncorrected. Spectral measurements were obtained by mixing a coherent source near the target frequency of 1.9 THz and down converting the IF to baseband. Spectra were recorded with both an IBOB and a newly available CMOS-based spectrometer for comparison. This paper is the first to report on spectra obtained by CMOS spectrometer technology and this spectrometer provides several advantages over more traditional FFT spectrometers because of its compact size and low power consumption. This 4x1 receiver prototype is easily scalable to other frequencies and larger focal plane arrays. Furthermore, because of its compact configuration, it can be easily packaged for orbital or sub-orbital missions.

I. Introduction

Large spectroscopic THz arrays are needed for surveys that can resolve large-scale motions within giant molecular clouds (GMCs) to formulate a more complete understanding of the star formation process and the lifecycle of the Interstellar Medium (ISM). Because of atmospheric attenuation, most important atomic and molecular species are only visible from near space or space, the most important being 1.9 THz (158 μm) fine structure line of [CII]. The *Herschel Space Observatory* [1], launched in 2009, carried the first 1.9 THz heterodyne receiver as part of the Heterodyne Instrument for the Far Infrared (HIFI) [2]. The list of receivers at 1.9 THz that have flown since HIFI is short. The German Receiver for Astronomy at Terahertz Frequencies (GREAT) [3-4] followed by upGREAT [5-6] have operated in flight on the Stratospheric Observatory for Infrared Astronomy (SOFIA) [7]. upGREAT contains two 7-pixel arrays, one for each polarization. The Stratospheric Terahertz Observatory (STO) [8] and the Stratospheric Terahertz Observatory – 2 (STO-2) [9] contained multiple 1.9 THz pixels and the Galactic/Extragalactic ULDB (ultra-long duration balloon) Stratospheric Terahertz Observatory (GUSTO) was just selected under the NASA Explorer's Program for a launch in 2021.

Airborne and balloon flights provide for comparatively large mass and power margins. For the next generation of THz heterodyne arrays to be appealing to a broader range of platforms such as a CubeSat or a satellite, receivers with more compact configurations and lower power consumption are needed. This paper presents lab measurements from a compact 4-pixel array receiver. Heterodyne performance is verified with simultaneous measurements of spectra using a second 1.9 THz source and an IBOB spectrometer. Finally, measurements from the first use of a CMOS-based spectrometer are presented.

II. Instrument Description

A. Local Oscillator

The LO is a frequency multiplied source with 4 individual pixels spaced 5 mm apart. The compact design is only 18 x 10 x 12 cm and uses around 24 W of power. The multiplier chain, shown in Figure 1, uses a QuickSyn synthesizer to drive the AMC. The signal is then amplified by a GaN power amplifier and split 4 ways. Each pixel has its own set of Schottky diode triplers chained together to output power at 220 GHz, 650, GHz and 1.9 THz. Furthermore, each of these triplers can be biased individually to maximize power output for each pixel. Measured powers for each pixel on the LOat 1.9 THz were in the 20-40 μW range. This also provides a mechanism to tune each pixel individually, which is necessary to optimize pump power for each HEB mixer.

B. Optics

For lab testing we used two 30 deg off axis parabolic mirrors with a focal length of ~100 mm. The first mirror focused the LO power and signal through the center of the window. The second mirror, attached to the cold plate inside the cryostat, redirected the LO power and signal to the mixer block via a folding mirror. These optics were a cheaper alternative than machining a larger cryostat window to couple LO from all 4 pixels simultaneously.

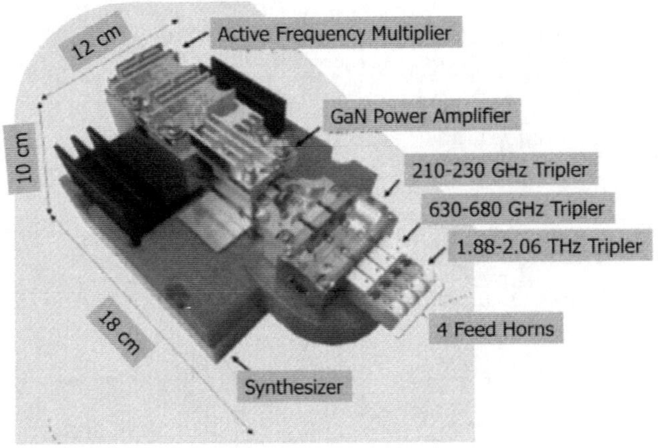

Figure 1. Compact 4-pixel 1.9 THz frequency multiplier chain.

C. HEB Mixer Module

The 4-pixel HEB mixer block uses a matching diagonal feed horn array to that of the LO. The pixel spacing for each is only 5 mm. The compact mixer block measures 30 x 10 x 10 mm when assembled as seen in Figure 2. The front half contains diagonal feed horns and a very short 1.9 THz waveguide. To machine diagonal feed horns, an E-plane split through the middle of the block is necessary. The back half contains a pocket for the mixer chip and feed thrus to GPO connectors for IF signals.

The mixers are made from silicon-on-insulator (SOI) chip technology. The HEB sits in the center of a bowtie antenna on the chip. The SOI chips are then inserted and subsequently wirebonded into a 2 mm x 2 mm backshell constructed from photolithograpy and micro-plating techniques. The backshells ensure that the mixer is precisely aligned to the front piece of the mixer block 1.9 THz waveguide. A ground connection between the mixer block and the HEB device is made when the backshell contacts the front feed horn array [10].

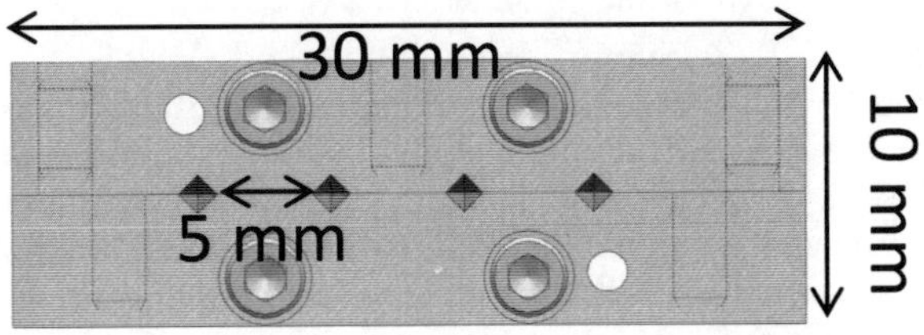

Figure 2. Photograph (below) and model (above) of a compact 4-pixel HEB mixer block.

The optics, mixer bock, and LNAs for all 4 pixels can been seen installed inside the cryostat in Figure 3. The cryostat was originally designed for one pixel and has been subsequently retrofitted for two and then four pixel systems.

Figure 3. 4-pixel HEB system installed inside the cryostat.

D. IF Electronics

A bias-tee is attached to the IF output of every pixel and the bias circuitry is contained inside the cryostat, controlled by a bias box and NI-DAQ on the outside. The IF LNAs are clamped to the 4 K cold plate. Outside of the cryostat, the receiver is completed by room temperature amplifiers and filters. For y-factor measurements the IF is readout by an Agilent power meter. For spectral measurements, the IF goes through a second down conversion to baseband.

E. Spectrometers

This paper demonstrates the receiver's functionality using two different spectrometers, a more traditional FPGA IBOB spectrometer and a new CMOS-based spectrometer. The CMOS-based spectrometer uses a UCLA-JPL developed system-on-chip (SOC) technology containing the digitizers, FFT processors, and USB readout circuitry to provide highly integrated low-power spectral processing. The entire SOC is about the size of a credit card and draws an order of magnitude less power over its FPGA predecessor. This spectrometer design is described in more detail in Ref. [11].

III. Measurements And Results

A. Receiver Performance

Initial optical alignment was completed by shining a laser through the optics before the cryostat was closed and cooled. The window was removed during the laser alignment. The LO was mounted to a 3-axis stage. This setup is very similar to the one shown in Figure 6, although y-factor measurements did not use the 1.9 THz source, only the 4-pixel LO. Once cold the HEB currents were monitored until all 4 pixels were maximally pumped at the same time. I-V curves taken simultaneously were recorded with a LabView program and replotted in Figure 4. Room temperature resistances and critical currents from unpumped I-V curves (not shown) are summarized in TABLE 1.

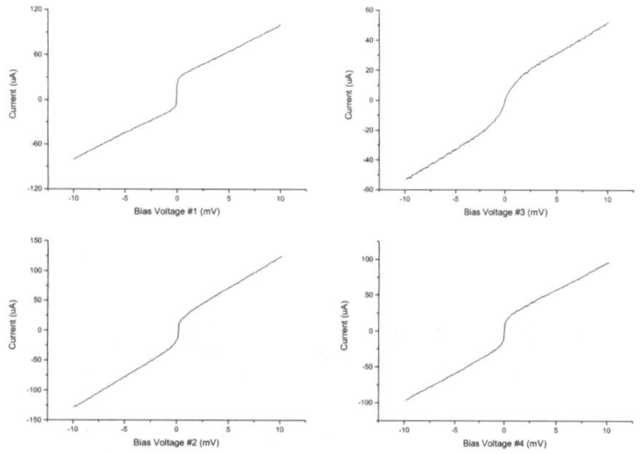

Figure 4. The 4-pixel LO is simultaneously pumping all 4 HEB mixers.

Each pixel was then measured for sensitivity using the y-factor method. A wire grid was used to inject warm and cold blackbody loads into the cryostat while sweeping the HEB in bias voltage. The measurements and calculated DSB noise temperature measurements are shown in Figure 5. The warm load at 290 K, is shown in red, and the liquid nitrogen load, 80 K, is shown in blue. The resulting DSB noise temperature is shown in black. The two best pixels were pixels 1 and 4 with double sideband noise temperatures of ~900 K uncorrected. Pixels 2 and 3 had devices that were slightly below and above the target room temperature resistance of 100 Ω, respectively. This creates an impedance mismatch with the IF circuity when cold thus degrading the overall performance of the mixer.

TABLE 1. ROOM TEMPERATURE AND COLD TEMPERATURE RESISTANCES, CRITICAL CURRENTS, AND NOISE TEMPERATURES FOR ALL 4 PIXELS.

Pixel	Ω_{RT}	Ω_{100K}	I_C (μA)	$T_{RX,DSB}$ (K)
1	109	126	130	900
2	82	93	165	1800
3	130	179	30	1700
4	105	120	185	900

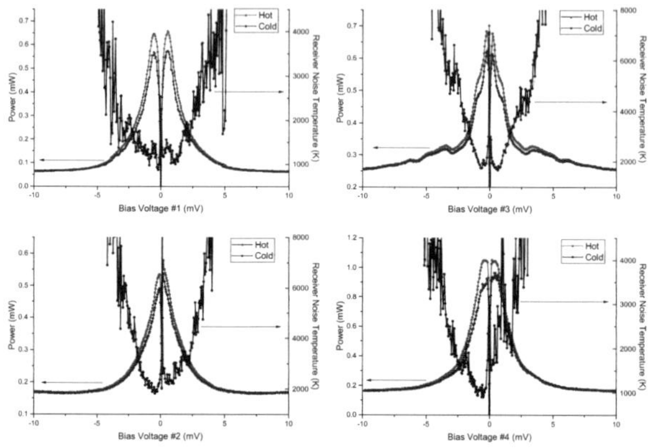

Figure 5. Sensitivity measurements for 4-pixel HEB array. The warm load (290 K) is in red. The cold load (80 K) is in blue. The calculated noise temperature is in black.

B. Spectral Measurements

1. IBOB Spectrometer: After verifying the performance of the receiver, a tone at 1.8977976 THz was quasi-optically injected via the polarizer grid and observed at baseband by all 4 pixels simultaneously with IBOB spectrometers. A picture of the setup can be seen in Figure 6 and the results of the LSB measurements can be seen in Figure 7. The tone moved through the band pass when it was adjusted in frequency. To illustrate this observation, a second frequency is shown at 1.8978300 THz, which is ~34 MHz from the initial tone. The measurements have been normalized to 1.

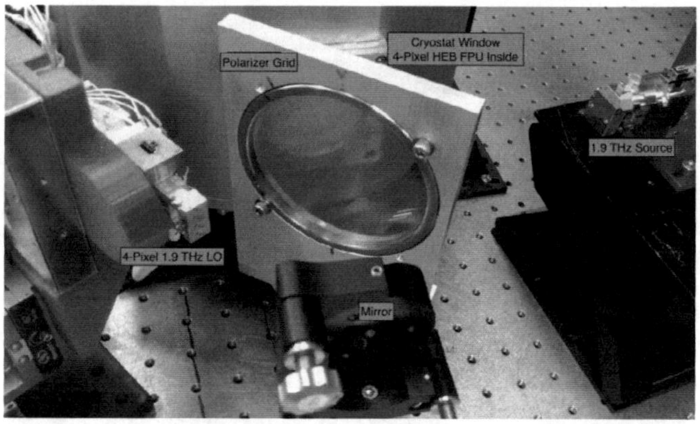

Figure 6. Measurement setup for spectral measurements at 1.9 THz.

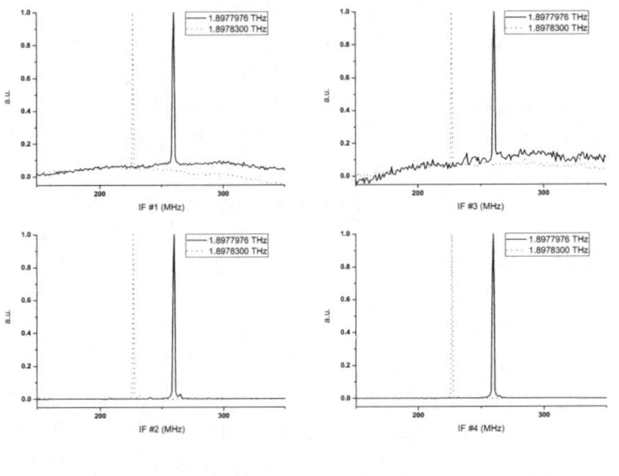

Figure 7. Spectral measurements using an IBOB spectrometer for two different tones near 1.9 THz. The measurements from each tone were taken simultaneously and only the LSB is shown. Tones have been normalized to 1.

2. CMOS-Based 2.0 GHz Complex FFT Spectrometer: Finally, following the verification of heterodyne performance of the receiver using an IBOB spectrometer, the use of a new CMOS spectrometer was demonstrated. At the time of the 4-pixel measurements, a single 2.0 GHz Complex FFT spectrometer with 256 channels was available. The 2 GHz naming refers to the maximum possible input clock frequency, which determines the maximum spectrometer bandwidth. Power consumption for this spectrometer typically runs about 500 mW. It was designed with Earth and planetary science applications in mind, and subsequently had options for other Stokes' Parameters to compute a complex FFT. For galactic observations of [CII], only the real FFT would be used and power consumption would be lowered with the same number of channels.

The first measurement of this new spectrometer was an Allan Variance to determine stability. In order to take the measurement, noise from the receiver baseband amplifiers was injected into the spectrometer. This measurement, shown in Figure 8, was taken over a night lasting 15.5 hours. Two channels of the spectrometer were chosen at random to compute the spectroscopic Allan Variance time, which is nearly ~10,000 s. However, the error bars on this measurement grow larger as fewer data points were available for averaging.

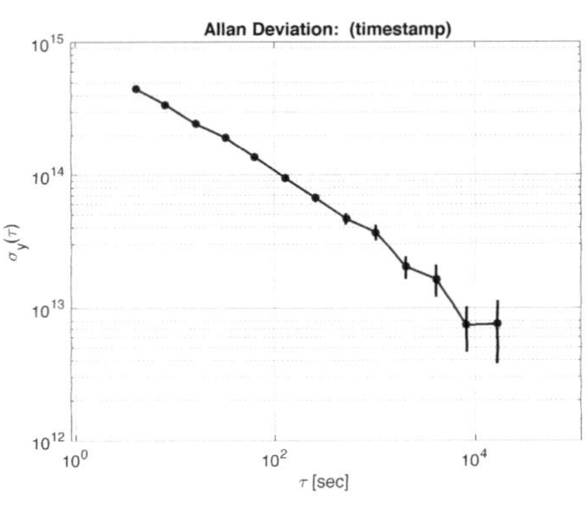

Figure 8. Allan variance measurement for the CMOS 2.0 GHz spectrometer.

Next, the spectral measurements at 1.9 THz with the receiver were repeated with this spectrometer. Due to the limitation of only one available spectrometer, one signal was recorded at a time, but the setup shown in Figure 6 did not otherwise change and the baseband configuration remained the same. The resulting spectra are shown for the LSB in Figure 9 for 1 GHz of bandwidth.

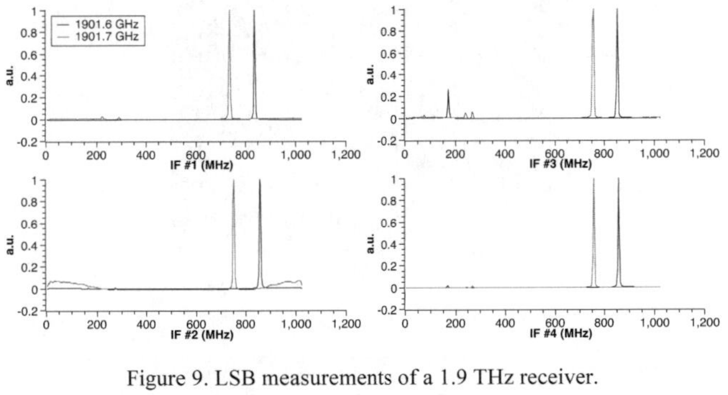

Figure 9. LSB measurements of a 1.9 THz receiver.

A final Allan Variance test was done for the end-to-end receiver using Pixel 4. For this test, the frontend of the receiver was stabilized through a PID loop between the 660 GHz tripler LO bias voltage and the HEB mixer current. Once again two spectrometer channels were chosen at random to calculate the spectroscopic Allan Variance time. Shown in Figure 10, this measurement resulted in a spectroscopic receiver Allan Time of ~5 seconds.

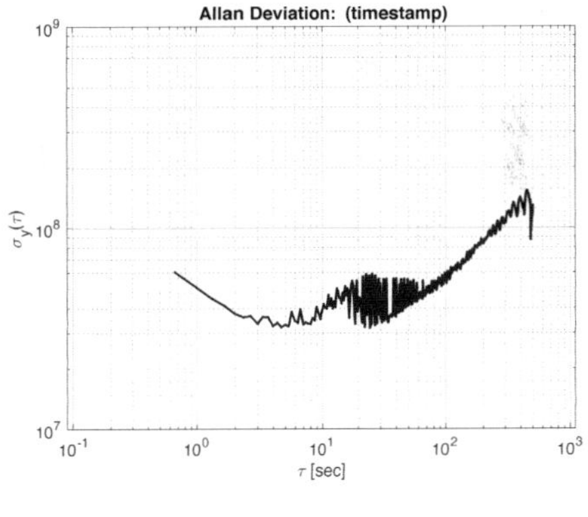

Figure 10. The stabilized end-to-end receiver spectroscopic Allan Variance for Pixel 4.

3. 2.6 GHz Real FFT Spectrometer: A newer version of the CMOS spectrometer with a maximum 2.6 GHz clock frequency became available and was tested using the 557 GHz water line. This version was designed for only real inputs and subsequently drew about the same amount of power as the previous version, despite having 4 times as many channels for a much higher spectral resolution. The measurement setup included a gas cell containing water vapor at 1 mTorr against a 77 K reference load and a 500-600 GHz room temperature Schottky receiver. The spectrometer was placed in a small red 3-D printed box. The box, shown in Figure 11, provides a scale of the size of an SOC. The spectrum recorded at 557 GHz from this setup is shown in Figure 12.

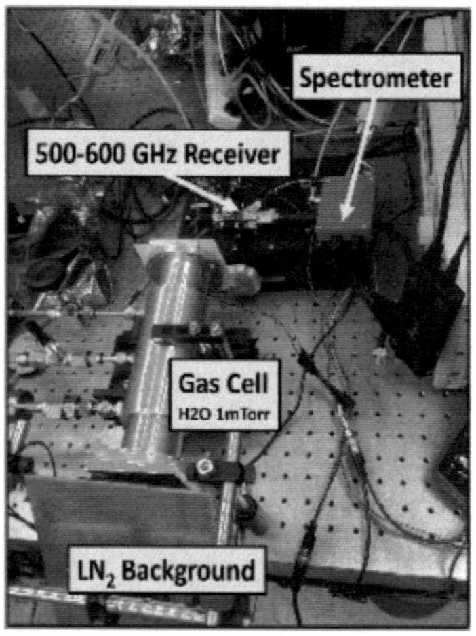

Figure 11. The measurement setup for the demonstration of a 557 GHz receiver using the new 2.6 GHz Real FFT SOC spectrometer.

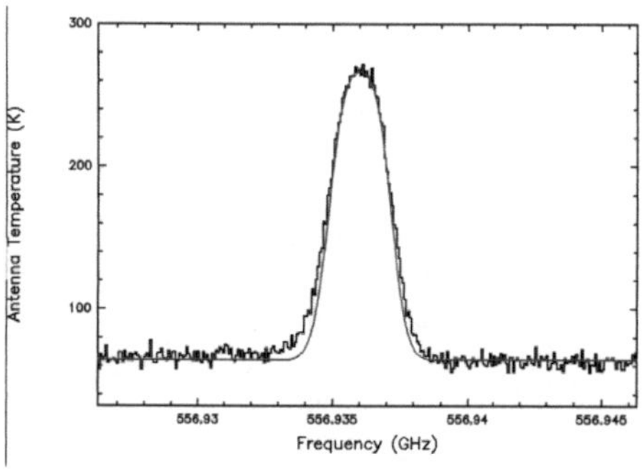

Figure 12. Spectrum of water vapor recorded at 557 GHz during a simulated 1800-second observation. The receiver, which is operated in ambient air, stares through a 25 cm column of room temperature water vapor gas held at 1 mTorr against a 77 K reference load. The red curve is the model brightness temperature from Ref. [12].

IV. Discussion

A. Lesson Learned from the Mixer Block Design

The 4-pixel mixer block used in this paper had continual problems with wire bonds from the IF signal lifting from the post of the GPO connector. This lead to a number of the better matched HEB devices breaking as the mixer block was opened and closed a number of times to re-bond the wires. It was determined that issue with the wire bonds was a result of the post of the GPO connector being too close to the HEB backshell to make a good loop. Subsequently, designs of future blocks have doubled the space left between these and should improve the mixer block assembly process.

B. Larger-Format HEB Focal Plane Arrays

After completing a 4-pixel 1.9 THz receiver, the next steps were to design and fabricate a 16-pixel LO [13] and a 16-pixel mixer block [14]. The new design not only corrects for the problem identified in Section IV-A, it also uses Picket-Potter feed horns with a circular to rectangular waveguide transition, which can be directly drilled without an E-plane split that is needed for a traditional diagonal feed horn [15]. Furthermore, these feed horns also have the added bonus of a lower cross-polarization.

B. FPGA and CMOS Spectrometer Comparison

CMOS spectrometers will have distinct advantages over the more traditional FPGA-based spectrometers, especially for space-based applications due to their compact size and lower power consumption. The IBOB used for this paper is the size of a record player and draws an average of 25-30 W. On the other hand, the new CMOS-based technology has shown it can match the spectral resolution and bandwidth of an FPGA-based FFT spectrometer at a fraction of the size and power consumption.

V. Summary and Conclusion

The 4-pixel HEB receiver has been demonstrated end-to-end using a 4-pixel LO and a 4-pixel mixer module. The 4-pixel LO draws only 24 W of power, a factor of 2 less than its 4-pixel predecessor and in a more compact configuration [16]. The receiver demonstrates high sensitivity with 2 pixels at 900 K DSB uncorrected. Finally, spectra using an FPGA-based IBOB spectrometer were compared to spectra using new CMOS technology. This new technology has demonstrated that the CMOS technology is a compact, low-powered viable alternative to the power hungry and large FPGA FFT spectrometers. With advances in frequency multiplier chains, mixer block design and fabrication, and CMOS technology, this receiver demonstrates a new compact, low-powered technology for space-based THz applications.

Acknowledgment

The research was carried out at the Jet Propulsion Laboratory, California Institute of Technology, under a contract with the National Aeronautics and Space Administration.

References

[1] G. L. Pilbratt, J. R. Riedinger, T. Passvogel, G. Crone, D. Doyle, U. Gageur, A. M. Heras et al., "Herschel Space Observatory-An ESA facility for far-infrared and submillimetre astronomy," *Astronomy & Astrophysics,* vol. 518, p. L1, July 2010.

[2] T. De Graauw, F. P. Helmich, T. G. Phillips, J. Stutzki, E. Caux, N. D. Whyborn, P. Dieleman et al., "The Herschel-heterodyne instrument for the far-infrared (HIFI)," *Astronomy & Astrophysics,* vol. 518, p. L6, July 2010.

[3] S. Heyminck, U. U. Graf, R. Güsten, Jürgen Stutzki, H. W. Hübers, and P. Hartogh, "GREAT: the SOFIA high-frequency heterodyne instrument," *Astronomy & Astrophysics,* vol. 542, p. L1, June 2012.

[4] P. Pütz, C. E. Honingh, K. Jacobs, M. Justen, M. Schultz, and J. Stutzki, "Terahertz hot electron bolometer waveguide mixers for GREAT." *Astronomy & Astrophysics* vol. 542, p. L2, June 2012.

[5] C. Risacher, R. Güsten, J. Stutzki, H-W. Hübers, A. Bell, C. Buchbender, D. Büchel et al., "The upGREAT 1.9 THz multi-pixel high resolution spectrometer for the SOFIA Observatory," *Astronomy & Astrophysics,* vol. 595, p. A34, Nov. 2016.

[6] C. Risacher *et al.*, "First Supra-THz Heterodyne Array Receivers for Astronomy With the SOFIA Observatory," in *IEEE Transactions on Terahertz Science and Technology*, vol. 6, no. 2, pp. 199-211, March 2016.

[7] E. T. Young, E. E. Becklin, P. M. Marcum, T. L. Roellig, J. M. De Buizer, Terry L. Herter, R. Güsten et al., "Early science with SOFIA, the stratospheric observatory for infrared astronomy," *The Astrophysical Journal Letters,* vol. 749, no. 2, p. L17, March 2012.

[8] C. Walker, C. Kulesa, P. Bernasconi, H. Eaton, N. Rolander, C. Groppi, J. Kloosterman *et al.*, "The Stratospheric THz Observatory (STO)," in *SPIE Astronomical Telescopes+ Instrumentation*, 2010, pp. 77330N-77330N.

[9] A. Young, C. Walker, C. Kulesa, P. Bernasconi, R. Dominguez, J. Siles, et al., "Stratospheric Terahertz Observatory 2016, Sub-orbital flight from

McMurdo, Antarctica," in *Proc. 28th Int. Symp. Space Terahertz Technol.*, 2017.
[10] F. M. Boussaha, J. H. Kawamura, J. A. Stern, A. Skalare and V. White, "A Low Noise 2.7 THz Waveguide-Based Superconducting Mixer," in *IEEE Transactions on Terahertz Science and Technology*, vol. 2, no. 3, pp. 284-289, May 2012.
[11] A. Tang, T. Reck and G. Chattopadhyay, "CMOS system-on-chip techniques in millimeter-wave/THz instruments and communications for planetary exploration," in *IEEE Communications Magazine*, vol. 54, no. 10, pp. 176-182, October 2016.
[12] S. Paine, private communication, 2016.
[13] J. V. Siles, J. Kawamura, R. Lin, C. Lee, and I. Mehdi, "Ultra-Compact THZ Multi-Pixel Local Oscillator Systems for Balloon-borne, Airborne and Space Instruments," in *Proc. 28th Int. Symp. Space Terahertz Technol.*, 2017.
[14] J. Kawamura, J. Kloosterman, J. Siles, F. Boussaha, B. Bumble, V. White, *et al.*, "Development of a 16-pixel monolithic 1.9 THz superconducting waveguide HEB mixer," *Proc. 28th Int. Symp. Space Terahertz Technol.*, 2017.
[15] K. K. Davis, J. Kloosterman, C. Groppi, J. Kawamura, and M. Underhill, "Micro-Machined Integrated Waveguide Transformers in THz Pickett-Potter Feedhorn Blocks," *Proc. 28th Int. Symp. Space Terahertz Technol.*, 2017.
[16] J.V. Siles, I. Mehdi, C. Lee, R. Lin, J.Kawamura, E. Schlecht, *et al.*, "A multi-pixel room-temperature local oscillator subsystem for array receivers at 1.9 THz," in *Proc. of SPIE Astronomical Telescopes+ Instrumentation*, 2014 pp. 914777-914777.

Performance of NbN and NbTiN HEB waveguide mixers for GREAT and upGREAT

D. Büchel*, K. Jacobs, P. Pütz, M. Schultz, C. E. Honingh, and J. Stutzki
1. Physikalisches Institut, Universität zu Köln,
Zülpicher Str. 77, 50937 Köln, Germany
*Contact: buechel@ph1.uni-koeln.de

Abstract—We present results and analysis of heterodyne measurements at 1.9 THz using hot electron bolometer (HEB) waveguide mixers comparing NbN micro-bridge on a 2 μm thick Si membrane with NbTiN on a 3 μm thick SiN membrane. The mixers are designed and fabricated in Cologne for the GREAT/upGREAT instrument, which is in operation on the Stratospheric Observatory for Infrared Astronomy (SOFIA).

All mixers are measured in the same heterodyne setup in a liquid helium cryostat at optimum bias conditions (lowest noise temperature), over a 0.5 – 5 GHz IF range. For the NbN HEB mixers the average mixer gain over a 1-2 GHz IF bandwidth is (-6 +/- 1) dB with an average mixer noise of (450 +/- 50) K and a noise bandwidth of (4.0 +/- 0.4) GHz. For the NbTiN HEB mixer the mixer gain is around (-10 +/- 1) dB with a mixer noise of about (500 +/- 50) K and a noise bandwidth of (2.3 +/- 0.4) GHz.

For both HEB mixer types the gain and noise are also calculated from the lumped-element model using the broken-line transition model. The moderate agreement between calculated and the experimental data will be discussed.

Silicon Micromachined Integrated 4-Pixel Heterodyne Receiver at 1.9 THz

Goutam Chattopadhyay[1*], Maria Alonso-delPino, Theodore Reck[1], Cecile Jung-Kubiak[1], and Choonsup Lee[1]

[1]*Jet Propulsion Laboratory, California Institute of Technology, Pasadena, CA 91109*
[*]Contact: goutam@jpl.nasa.gov

Abstract—Deep reactive ion etching (DRIE) based silicon micromachining technology is proving to be one of the most suitable option for the fabrication of next generation of terahertz multi-pixel heterodyne instruments. Silicon micromachining provides high precision fabrication capabilities and high level of integration within the receiver. Moreover, it enables the fabrication of hundreds or thousands of pixels from a single wafer, which reduces cost, saves time, and allows to develop multi-pixel arrays in a fast parallel path instead of serial nature of current generation of multi-pixel instrument architecture.

We are developing a highly compact 4-pixel heterodyne receiver at 1.9 THz with integrated orthomode transducer (OMT) for dual-polarization capability and a balanced receiver architecture fabricated with micromachined silicon waveguide packaging. Fig. 1 shows the block diagram of the proposed receiver architecture and the waveguide implementation of the receiver. The receiver has a dual-polarized antenna which couples the incoming radiation to an OMT where the signal is separated into two linear polarizations. Each signal goes into a 90-degree waveguide hybrid coupler that divides the RF signal equally between the two output ports with 90-degrees phase difference. The LO signal is injected through the isolated port of the hybrid and the combined RF and LO signals feed two sets of HEB mixers in a balanced configuration. The IF outputs are then combined in a 180- degree hybrid obtaining the corresponding polarization signal, as shown in Fig. 1.

One of the challenges of the development of multi-pixel heterodyne receivers is the antenna array. For single-pixel instruments, horns are widely used because of their good performance and bandwidth. However, for multi-pixel instruments, these antennas struggle in terms of fabrication –higher frequency requires higher fabrication tolerances – and integration with the rest of the detector because of their volume, mass, and the difficult integration of silicon and metal at cryogenic temperatures. Thus, having an antenna that can be fabricated in silicon micromachining techniques is essential for these receivers. We developed a silicon microlens based array antenna that can be integrated with the silicon micromachined receiver. In this design, the gold plated micromachined silicon wafers are vertically stacked to reduce loss and provide a highly integrated receiver system. The key advantage of having this vertical integration architecture is that we can easily transition to a two-dimensionally arrayed multi-pixel system.

In this paper, we will describe the design of the 1.9 THz silicon micromachined receiver system, the microlens antenna design, fabrication techniques with DRIE based silicon micromachining, assembly of the components, and preliminary test results.

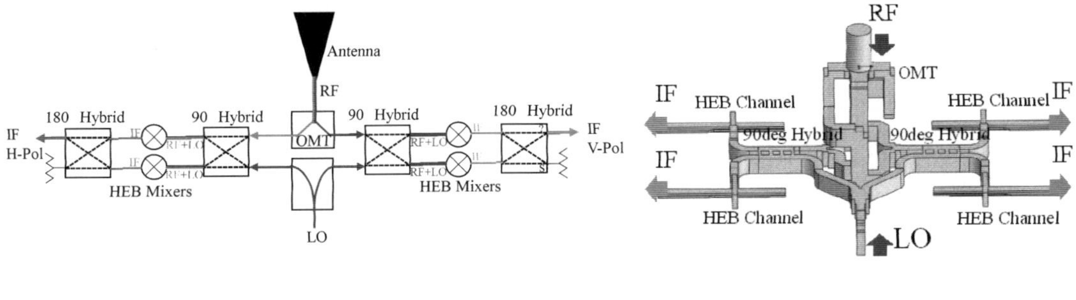

Figure 1: (a) Block diagram and (b) actual waveguide design of the dual-polarized, balanced receiver front-end at 1.9 THz fabricated with silicon micro-machining technology.

The work described here was carried out at Jet Propulsion Laboratory, California Institute of Technology, Pasadena, California, USA under a contract with National Aeronautics and Space Administration.

Performance and surface wave reduction in large monolithic kinetic inductance detector arrays

Andrey M. Baryshev[*,1,2], Stephen J. C. Yates[1], Ozan Yurduseven[4], Juan Bueno[1], Akira Endo[4,5], Lorenza Ferrari[1], Willem Jellema[1], Nuria Llombart[4], Vignesh Murugesan[3], Kristina K. Davis[6], David J. Thoen[4], Stefan Heyminck[7], Rolf Guesten[7], Jochem J. A. Baselmans[3,4]

[1] *Netherlands Institute for Space Research (SRON), 9747 AD, Groningen, The Netherlands*
[2] *Kapteyn Astronomical Institute, University of Groningen, 9700 AV, Groningen, The Netherlands*
[3] *Netherlands Institute for Space Research (SRON), 3584 CA, Utrecht, The Netherlands*
[4] *Terahertz Sensing Group, Faculty of Electrical Engineering, Mathematics and Computer Science, Delft University of Technology,2628 CD Delft, The Netherlands*
[5] *Kavli Institute of NanoScience, Faculty of Applied Sciences, Delft University of Technology, 2628 CJ, Delft, The Netherlands*
[6] *Arizona State University, 781 Terrace Rd., Tempe, AZ, U.S.A*
[7] *Max Planck Institute for Radioastronomy, D-53010, Bonn, Germany*

[*]Contact: andrey@astro.rug.nl

Abstract— Microwave kinetic inductance detector (MKID) technology is quickly becoming the main choice for very large focal plane arrays for radioastronomy due to combination of high quantum efficiency, low noise and large multiplexing rations (>1000) due to inherent frequency domain readout. We are developing MKID arrays for APEX MKID instrument, a dual frequency camera with total pixel count of 25000 and 1 f λ sampled field view of 16 x 16 arcmin.

The low frequency band of AMKID covers 350 GHz atmospheric window It consists of four monolithic detector chip tiled in the instrument cold focal plane. Each chip consists of monolithic silicon lens array, glued to a silicon substrate containing 880 MKIDs which are coupled to printed double slot type planar antenna. This design demonstrated high optical efficiency and low noise performance but it contains naturally a parasitic cross talk path, where radiation can travel from one array pixel to another through silicon substrate and monolithic lens array. The latter gives rise to a surface wave and produces spurious response at spatial position where detector is not present. For monolithic detector arrays, this effect has been found to be significant to affect the astronomical observations.

In this contribution we report the MKID array design, analyze the origins and effect of surface wave. We will present laboratory measurements and analysis clearly demonstrating existing effect. We were able to reduce the surface wave significantly by implementing carefully designed absorption mesh layer in the detector structure. We will report on mesh design and laboratory measurements. Finally, we will present laboratory sensitivities and yields of latest generation MKID chips.

A Fast Machine Learning Based Algorithm for MKID Readout Power Tuning

Rupert H. Dodkins[1,*], Kieran O'Brien[1], Niranjan Thatte[1], Sumedh Mahashabde[1], Neelay Fruitwala[2], Seth Meeker[2], Alex Walter[2], Paul Szypryt[2] and Ben Mazin[2]
[1]*Department of Astronomy, University of Oxford, Oxford, OX1 3RH, UK*
[2]*Department of Physics, University California, Santa Barbara, California 93106, USA*
*Contact: `rupert.dodkins@physics.ox.ac.uk`

***Abstract*— As high pixel count Microwave Kinetic Inductance Detector (MKID) arrays become widely adopted, there is a growing demand for automated device readout calibration. These calibrations include ascertaining the optimal driving power for best pixel sensitivity, which, because of large variations in MKID behavior, is typically performed by manual inspection. This process takes roughly 1 hour per 1000 MKIDs, making the manual characterization of ten-kilopixel scale arrays unfeasible. We propose the concept of using a machine-learning algorithm, based on a convolution neural network (CNN) architecture, which should reliably tune ten-kilopixel scale MKID arrays on the order of several minutes.**

INTRODUCTION

Microwave Kinetic Inductance Detectors, or MKIDs, are superconducting detectors that sense photons by measuring the change in ac surface impedance produced by the separation of Cooper pairs [1]. Each MKID device is a superconducting thin film lithographically patterned into an array of high quality factor microwave resonators. Through passive frequency domain multiplexing, arbitrarily large arrays can be fabricated and thousands of pixels can be read out per feed-line using room temperature electronics. The current generation of MKID devices are kilopixel arrays [2] and in the UVOIR (UV, optical and IR) regime, the current generation are many tens of thousands of pixels [3].

While fabrication of large arrays is comparatively straightforward, maintaining consistency across each array is challenging. Ideally a wide frequency sweep through the device should yield transmission dips at equal spacing with uniform, high quality factors. In reality there is a distribution in the separation between resonances and the quality factors, and therefore the power handling ability. Due to this variety in behavior, the probe signal power allocations have been, until now, mostly decided by manual human inspection.

MKID DIGITAL READOUT

For readouts in both the THz [4] and UVOIR regime [5], a software defined radio transceiver is used to generate and monitor a comb of microwave frequencies to drive each of the resonators. Since the quality factors of the resonators are high,

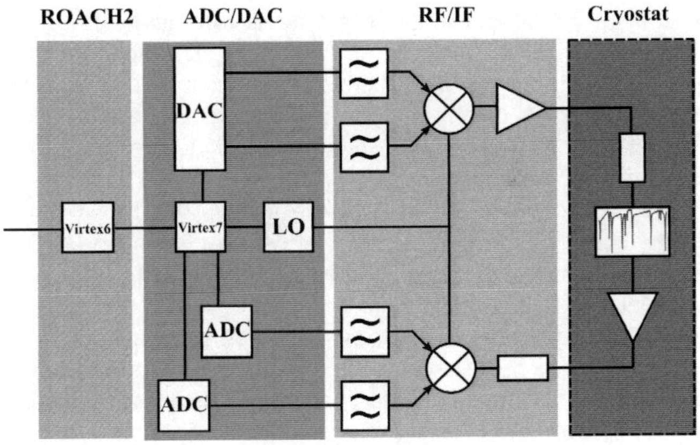

Fig. 1: A block diagram showing the general functions of the hardware that make up a single feed-line of the second generation UVOIR Software Defined Radio digital readout [5]. The CASPER ROACH2 boards [6] contain firmware, that produces the baseband frequencies to drive the resonators; as well as channelizes the incoming signal, performs optimal filtering, triggering, and creates photon packets to send to the data acquisition machine.

the carrier frequency for a particular pixel does not interact with other pixels while propagating through the array. The complexity of device readout is transferred to the digital backend, which demultiplexes the sum of probe tones using custom firmware. The number of resonators a feed-line can probe is limited by bandwidth of the analogue-to-digital converter (ADC); therefore, many feed-line readout systems operate in parallel, each probing a fraction of the array. Fig. 1 shows the readout architecture of a single feed-line for the UVOIR second-generation readout [5].

In order to detect the phase (or amplitude) modulations of each resonator during observations, the readout uses I/Q carrier signals, where I and Q represent the magnitude of the real and imaginary signal components respectively. I is a sine wave at the resonator frequency and Q is the same waveform offset in phase by one-quarter cycle. The phase between the readout tone and the resonant frequency is calculated using the equation $\Phi = \arctan(Q/I)$. During observations, the readout electronics constantly monitor the phase of each pixel and if that phase crosses a threshold then a photon packet is created and sent to the data acquisition machine.

Drive Signal Power Tuning

Readout tone powers need to be evaluated and programmed into the readout in order to probe the resonators under optical illumination. The cryogenic amplifier (usually a HEMT amplifier) noise and Two-Level System Noise [7] place limits on the sensitivity of the device. One technique to reduce these contributions is to drive each resonator with as much power as possible. However, in the high-power regime resonators exhibit a nonlinear response. If the power is sufficiently high, the resonator can occupy two stable states, and the resonator undergoes bifurcation. This has been attributed to an inherent nonlinearity in the kinetic inductance [8] and readout power heating [9]. The transmission profile distorts away from the Lorentzian profile (Fig. 2), which can be detrimental to the phase measurements and therefore sensitivity. The goal in readout tuning is therefore to identify the power for the onset of bifurcation and step down a few dB – which is essentially a classification problem.

De Visser et al. [10] have shown that, in the THz regime, even though the readout photons are below the Cooper pair binding energy, the readout photons will heat the quasiparticle population, which leads to an increase in generation-recombination noise. This study assumes that optimum sensitivity is achieved by driving the resonators at the highest power prior to bifurcation, and describes an optimization to that method. In theory, the algorithm could be optimized to any criteria so long as the training data exists.

Presently, the optimal power for each resonator is evaluated through manual visual inspection. The user will study each resonator, observing the resonance loops, the derivative of the transmission spectra with respect to frequency, termed v_{IQ}, and the relative change of these parameters at increasing power. With this information at hand they can ascertain if, and why, a resonator is showing non-ideal behavior, and make an informed estimate of the optimal power.

For the second generation UVOIR readout [5], the input data is created by sweeping the digital readout frequency across the resonators and then stepping the power of the digital readout in 1 dB intervals. The power limits are chosen such that each resonator is sampled in the low power regime and after it has transitioned into the bifurcation regime. This creates a datacube for each resonator consisting of I and Q magnitudes at each frequency and power. Hierarchical Data Format (HDF5) [11] is used to store each of these datacubes for every pixel on a feed-line.

Fig. 2 shows two example resonator datacubes. The first resonator in Fig. 2 (a) shows a resonator with near-ideal transmission spectra and power handling ability. The resonator clearly bifurcates at the power proceeding lime-green (power index 12 from the lowest power).

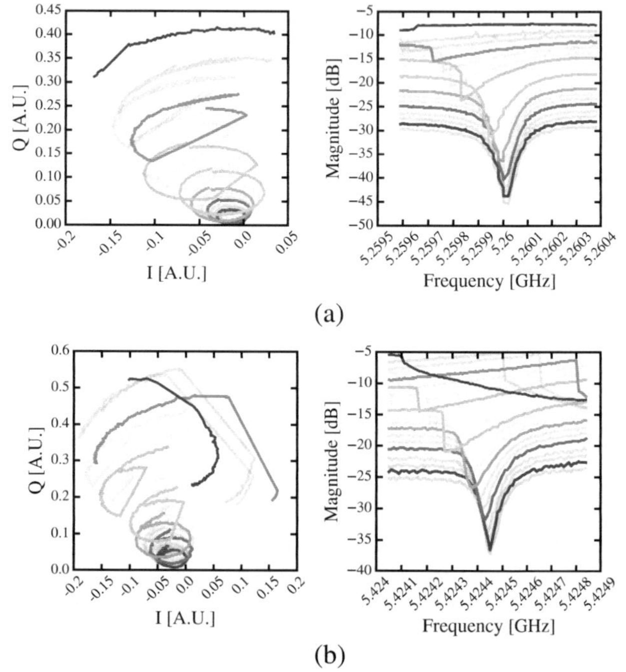

Fig. 2: (a) The resonance loop and transmission spectrum of an ideal resonator sampled at a range of powers. (b) The resonance loop and transmission spectrum of two near-colliding resonators sampled at a range of powers. The red and blue trends are sampled at the highest and lowest powers respectively; high saturation colors are separated by 3dB. The units of I and Q are uncalibrated raw data produced by the Analog to Digital converter in the digital readout.

The second resonator, Fig. 2b, shows a resonator datacube where the sampling window becomes contaminated at high powers with an adjacent resonator from higher frequencies and the initial resonator simultaneously translates out of the sampling window at low frequencies. Depending on the sampling window, it will sometimes appear as though an unbifurcated resonator disappears and a bifurcated resonator reappears, or vice versa. Sometimes, two resonator profiles will merge, and it is non-trivial to analytically fit the merger. Other times a well-separated resonator will show non-ideal power-handling behavior, for example the resonator in Fig. 2a shows multiple discontinuities at the penultimate power sample. Or simply, a resonator could have low signal to noise ratio.

When preparing an MKID device for observations it is often preferable to include as many resonators as possible, to ensure sufficient pixel count. However, there are many ways resonators can show non-ideal behavior and it is challenging to tune analytical algorithms to accurately account for each type of behavior.

Deep Learning Classification

Deep learning is a machine learning method used to discover and model high level abstractions in data [12], and the

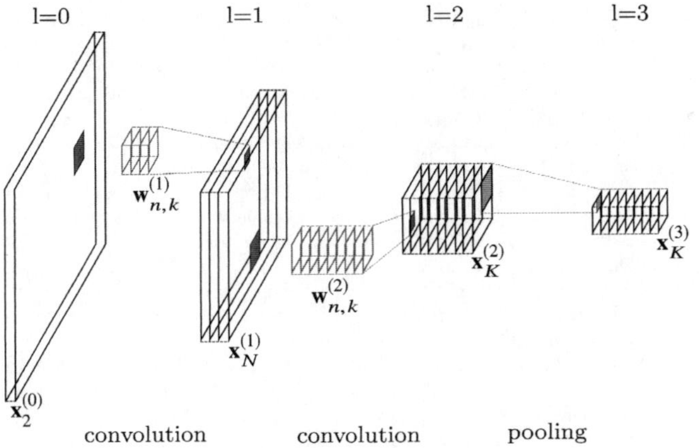

Fig. 3: The first four layers of a generic convolution neural network. The n and k labels for the feature maps are displayed in the reference frame of layer 2. For clarity, the n dimension of the weight tensors are not displayed, nor are any of the bias vectors. The input image has two spatial dimensions (frequency and power) and an effective depth of two (amplitudes of I and Q). The filled squares in each layer represent a projection to all depths. For a given weight, the filled square of the previous feature map displays the receptive field and, on the current feature map, the destination of the weighted sums along the n dimension. Between layers 2 & 3 is a 1/4 subsampling or pooling

objective in classification is simply to create some model which can take an input vector and assign the correct label to it. In deep learning this is accomplished with an interconnected network of multiple layers each consisting of many transfer functions termed neurons. Each neuron takes the input from the previous layer, transforms it according to a learnable parameter and then passes the output through a nonlinear activation function. During training, these parameters are tuned in successive iterations such that the outputs of the final layer converge on the labels of the input images for that batch. For resonator power tuning, the target label would be the optimal power index, or in this study, the highest power prior to bifurcation.

The more layers in the neural network, the more complex the features the model can discern. However, a excessive number of parameters in the neural network compared to the amount of training data, increases the chances of overfitting, meaning the model does not generalize well to unseen input data. Image recognition algorithms typically use millions of training images. For resonator power tuning, using manual inspection data as the reference training set, this magnitude of training data will not be available. To maintain the model sophistication in this instance, more training data could be created through label preserving transformations, or regularization techniques such as dropout [13], early stopping and batch normalization [14] could be utilized.

In a fully connected neural network (FCNN), each neuron computes a weighted sum of all the neurons in the layer that precedes it. These networks suffer from the 'curse of dimensionality', making them computationally expensive and, in some cases, easy to overfit. The goal of this study is to create a fast and rigorous algorithm. Given the limited scope of the training data, an alternative neural network architecture should be more effective.

Convolution neural networks (CNN) conversely are locally connected networks. Smaller weight matrices are convolved across the preceding layer, probing smaller receptive fields and exploiting spatially local correlations. To increase the number of free parameters, a four-dimensional weight tensor is convolved across the previous layer producing a tensor, $\boldsymbol{x}$, with depth K for an input depth N.

$$\boldsymbol{x}_k^{(l)} = a\left(\sum_n^N \boldsymbol{w}_{n,k}^{(l)} * \boldsymbol{x}_n^{(l-1)} + \boldsymbol{b}_k^{(l)}\right)$$

where $n = 0,1 \ldots N$ and $k = 0,1 \ldots K$. The weights, $\boldsymbol{w}^{(l)}$, act as filters extracting two-dimensional features from each layer in the n dimension. The biases $\boldsymbol{b}^{(l)}$ are also learnable parameters. The area of each successive layer in a CNN architecture decreases (assuming no padding) which means the learnable parameters of the successive layers extract successively larger scale features.

For resonator power tuning, the power-sweep datacubes are three-dimensional, and the spatial information between all the dimensions could be best exploited with a CNN architecture. Fig. 3 shows an arbitrary implementation of the first few layers of a CNN-based algorithm for resonator power tuning. The benefits of both FCNN and CNN architectures could be exploited by using several convolution and pooling layers, followed by fully connected layers.

CONCLUSIONS

It has been shown that a simple machine learning algorithm is an interesting new method for characterizing resonator arrays for readout. The advantages are that it can potentially characterize an array of many thousands of resonators in just a few seconds – compared to manual inspection, which takes many human hours. This advancement is vitally important as more kilopixel resonator arrays come online and as 10^4 pixel scale array are developed. Dodkins et al. (2017) [15] details the implementation of a CNN algorithm for resonator classification and provides a comparison with conventional analytical algorithms in order to quantify the accuracy.

ACKNOWLEDGMENT

The authors would like to thank the Science Technology Facilities Council for supporting this research through their studentship and Long Term Attachment programs.

REFERENCES

[1] Day, P. K., LeDuc, H. G., Mazin, B. A., Vayonakis, A., & Zmuidzinas, J. "A broadband superconducting detector suitable for use in large arrays," *Nature*, vol. *425*, pp. 817-821, 2003.

[2] Catalano, A., Adam, R., Ade, P., André, P., Aussel, H., Beelen, A., ... & Calvo, M. "The NIKA2 commissioning campaign: performance and first results". arXiv preprint arXiv:1605.08628. 2016.

[3] Meeker, S., Mazin, B., Jensen-Clem, R., Walter, A., Szypryt, P., Strader, M., & Bockstiegel, C. "Design and development status of mkid integral

field spectrographs for high contrast imaging," *in Proc. Adaptive Optics for Extremely Large Telescopes 4–Conference Proceedings* 2015 paper 1.1

[4] van Rantwijk, J., Grim, M., van Loon, D., Yates, S., Baryshev, A., & Baselmans, J. "Multiplexed readout for 1000-pixel arrays of microwave kinetic inductance detectors," *IEEE Transactions on Microwave Theory and Techniques*, vol. 64, pp. 1876-1883, 2016.

[5] Strader, M. J. "*Digitial readout for microwave kinetic inductance detectors and applications in high time resolution astronomy,"* (Doctoral dissertation, University of California, Santa Barbara). 2016.

[6] CASPER Group Collaboration of Astrophysical Signal Processing and Electronics Research. [online] Available: https://casper.berkeley.edu/ 2009.

[7] Gao, J. "*The physics of superconducting microwave resonators"*. California Institute of Technology. 2008.

[8] Swenson, L. J., Day, P. K., Eom, B. H., Leduc, H. G., Llombart, N., McKenney, C. M., ... & Zmuidzinas, J. "Operation of a titanium nitride superconducting microresonator detector in the nonlinear regime," *Journal of Applied Physics*, vol. 113, pp. 104501, 2013

[9] Thomas, C. N., Withington, S., & Goldie, D. J. "Electrothermal model of kinetic inductance detectors," *Superconductor Science and Technology*, vol. *28*, pp. 045012. 2015.

[10] De Visser, P. J., Withington, S., & Goldie, D. J. "Readout-power heating and hysteretic switching between thermal quasiparticle states in kinetic inductance detectors," *Journal of Applied Physics*, vol. *108*, 114504. 2010.

[11] The HDF Group Hierarchical data format version 5. [online] Available: http://www.hdfgroup.org/HDF5 2000.

[12] Bengio, Y., & LeCun, Y. "Scaling learning algorithms towards AI," *Large-scale kernel machines*, vol. *34*, pp. 1-41 2007.

[13] Srivastava, N., Hinton, G. E., Krizhevsky, A., Sutskever, I., & Salakhutdinov, R. "Dropout: a simple way to prevent neural networks from overfitting," *Journal of Machine Learning Research*, vol. 15, pp. 1929-1958, 2014.

[14] Ioffe, S., & Szegedy, C. "Batch normalization: Accelerating deep network training by reducing internal covariate shift," *International Conference on Machine Learning* pp. 448-456 Jun 2015

[15] Dodkins, R., Fruitwala N., O'Brien K., Thatte N., Mahashabde S., Meeker S., Walter A., Szypryt P., Mazin B., in prep. "MKID Digital Readout Tuning with Deep Learning". Monthly Notices of the Royal Astronomical Society 2017.

The effects of changes in bath temperature on Kinetic Inductance Detectors

Tejas Guruswamy*[1], Chris N. Thomas[1], Stafford Withington[1], David J. Goldie[1]

[1]*Quantum Sensors Group, Cavendish Laboratory, Cambridge CB3 0HE*
*Contact: tg307@mrao.cam.ac.uk

Abstract—We present calculations of the effects on responsivity and noise of changes in bath temperature and bath temperature noise on a typical Kinetic Inductance Detector (KID), and how these might be affected by electrothermal feedback, or thermal isolation of the device. KIDs are a leading ultra low noise, large format superconducting detector technology for astrophysics at THz frequencies and above, with performance strongly determined by the superconductor quasiparticle effective temperature. Particularly for space deployments, which often require background limited detectors but have challenging constraints on cryosystem design, knowing exactly how fluctuations in fridge temperature affect measured noise is crucial.

Our calculations are derived from a general electrothermal model [1] applicable to KIDs and similar superconducting microresonator-based devices. We are able to calculate both the large-signal – steady-state operating point – and small-signal – responsivity and noise-equivalent power (NEP) limits – behaviour of an electrical resonator in a complex thermal circuit with a number of power flows. In particular, our model includes the effects of readout power heating of the superconducting quasiparticles – absorbed readout power affects quasiparticle effective temperature which changes readout power absorbed, the phenomenon we describe as electrothermal feedback in KIDs – using recent work on the detailed functional form of quasiparticle-phonon cooling in superconductors [2,3].

Using this model, we have explored how a changing bath temperature affects both the large and small-signal behaviour of a typical KID. We find when readout power heating at the steady state operating point has increased the effective quasiparticle temperature significantly above the bath temperature, this effective quasiparticle temperature is no longer sensitive to small changes in the bath temperature or thermal conductivity from the device to the bath. Consequently, the responsivity to a signal of interest and the intrinsic generation-recombination noise of the device are unchanged as bath temperature varies. In general, KID performance is significantly reduced by an increased quasiparticle effective temperature, but this result suggests allowing some heating may reduce the noise contribution from an unstable fridge temperature.

1. Thomas, C. N., Withington, S. & Goldie, D. J. Electrothermal model of kinetic inductance detectors. Supercond. Sci. Technol. 28, 45012 (2015).
2. Goldie, D. J. & Withington, S. Non-equilibrium superconductivity in quantum-sensing superconducting resonators. Supercond. Sci. Technol. 26, 15004 (2013).
3. Guruswamy, T., Goldie, D. J. & Withington, S. Nonequilibrium superconducting thin films with sub-gap and pair-breaking photon illumination. Supercond. Sci. Technol. 28, 54002 (2015).

Photon-Counting with KID Resonators for THz/Submillimeter Space Spectroscopy

O. Noroozian*[1, 2, 3], E. M. Barrentine[1], T. R. Stevenson[1], A. D. Brown[1], V. Mikula[1], E. J. Wollack[1], K. U-Yen[1], S. H. Moseley[1]

[1]*NASA Goddard Space Flight Center, 20771, Greenbelt, MD, USA*
[2]*University of Maryland, 20742, College Park, MD, USA*
National Radio Astronomy Observatory, 22903, Charlottesville, VA,
*Contact: o.noroozian@gmail.com

Abstract— Photon-counting direct detectors are highly desirable for reaching the ~ 10^{-20} W/Hz$^{1/2}$ power sensitivity permitted by the Origins Space Telescope (OST), a notional cryogenic facility inspired by NASA's Astrophysics Roadmap. We are developing Kinetic Inductance Detectors (KIDs) with photon counting capability in the far-infrared/THz combined with integrated spectrometers suitable for the OST facility. To reach the required sensitivity we are experimenting with single-layer superconducting resonators made from aluminum films (10 – 25 nm thick) on single-crystal Si substrates. Small-volume inductors made from such thin Al films have the potential to become ultra-sensitive to single pair-braking far-IR photons (>90 GHz) under the right conditions. Understanding the physics of these superconducting films and superconducting-dielectric systems is critical to achieving detector performance with ultra low-loss and low-noise substrates.

In our measurements of these resonators, we have achieved very high internal quality factors ($Q_i \sim 7 \times 10^6$ for 25 nm Al, and 1.1×10^6 for 10 nm Al) at $\sim 10^6$ microwave photon drive power. At single-photon drive powers both films remarkably maintain a very high $Q_i \sim 0.5 \times 10^6$, by far the highest value for such thin films reported in literature to the best of our knowledge. In addition we have obtained quasi-particle (QP) lifetimes of ~ 1.0 ms for 100 nm Al on Si resonators, another critical material parameter for reaching photon-counting sensitivity. Our cryogenic testbed was optimized for ultra low stray radiation, which was confirmed by measurement cross comparisons. To realize a practical device, we are integrating these films with our Silicon-on-Insulator (SOI) process to form parallel-plate capacitors on single-crystal dielectric to minimize two-level system frequency noise and loss. We have designed KID geometries with small-volume inductors that allow us to probe the quasi-particle dynamics at these never before explored volume limits, and to measure quality factor and QP lifetime as a function of microwave readout power. Based on a detailed physical model, we simulated the detector output time stream for a given design when illuminated with random photon events, and using an optimal linear filter were able to show that photon counting with >95% efficiency at 0.5 and 1.0 THz is possible.

We report on these developments and discuss our plans to implement these devices into optically coupled ultrasensitive KIDs suitable for photon counting in space.

Compact diffractive optics for THz imaging

L. Minkevičius[1*], S. Indrišiūnas[1], R. Šniaukas[1], G. Račiukaitis[1], V. Janonis[1], V. Tamošiūnas[1], I. Kašalynas[1], G. Valušis[1]
[1]*Center for Physical Sciences and Technology, Saulėtekio Ave. 3, LT-10257 Vilnius, Lithuania*
*Contacts: linas.minkevicius@ftmc.lt,

Abstract— we presented a compact diffractive silicon-based multilevel phase Fresnel lens (MPFL) with up to 50 mm in diameter and numerical aperture (NA) up to 0.86 designed and fabricated for compact terahertz imaging systems. The laser direct writing (LDW) technology based on picosecond laser was used to fabricate the diffractive optics on silicon with different number of phase quantization levels *P* reaching almost kinoform spherical surface needed for efficient THz beam focusing. Focusing performance was investigated by measuring Gaussian beam intensity distribution in the focal plane and along the optical axis of lens. The influence of phase quantization number to the focused beam amplitude was estimated, and power transmission efficiency reaching more than 90 % was demonstrated. THz imaging resolution less than 1 mm using robust 50 mm diameter multilevel THz lens was achieved and demonstrated at frequency 580 GHz.

INTRODUCTION

Suitability of terahertz (THz) imaging for nondestructive testing [1], biomedical research [2] or food and pharmaceutical industry [3] stimulates a search for compact practically convenient solutions. As a particular important issue can be assumed development of compact THz diffractive optic components, which enable to replace massive parabolic mirrors into much more attractive flat compact optics elements[4], [5].

In this letter, multilevel phase Fresnel lens (MPFL) designs, fabrication and focusing performance are studied. The design of MPFLs starts from two-phase quantization levels *P* and extends up to continuous kinoform shape. Samples were developed for the focusing of 580 GHz frequency beam.

Silicon wafer of 0.46 mm thickness was patterned using industrial-scale-compatible LDW system based on 1064 nm wavelength 13 ps pulse duration, 1 MHz repetition rate, 60 μJ peak energy laser (Atlantic 60 from Ekspla LTd.) [6]. The LDW technology enabled us to manufacture different complexity multilevel zone-phase elements in the same process with the opportunity to modify parameters in time.

Two groups of the MPFL samples with outside diameter of 17.5 mm with the focal distances of 10 mm and 5 mm and one sample with 50 mm diameter and the focal distance of 30 mm were designed and fabricated. The group of 17.5 mm diameter MPFL contained five samples with different number of phase quantization levels for each focal distance allowing to figure out the optimal phase quantization level [5]. The radii and depths of the subzones were calculated according to methodology presented in [7].

The SEM images of the 5 mm focal length MPFL with 4 subzones are shown in Fig. 1 a-b. The step-profile scanned across the center of each sample is given in Fig. 1c. The photo of flat 50 mm diameter MPFL with P=16 quantization levels is depicted in Fig. 1d. As one can see, step-like structure of individual zones start to merge into a continuous kinoform shape for the case of 8 phase quantization levels. Such transition from step like to kinoform shape occurred due to the limited size of the ablation spot of 44 μm. Fig. 1b shows the SEM picture of the surface microstructure at each area of 4 steps MPFL. The laser-processed surface of the silicon contained randomly distributed columnar structures the scale of whose was increased with higher material exposure by laser. Such increase of surface roughness and growth of columnar morphology with number of laser pulses was well known phenomena in materials laser processing [8], [9].

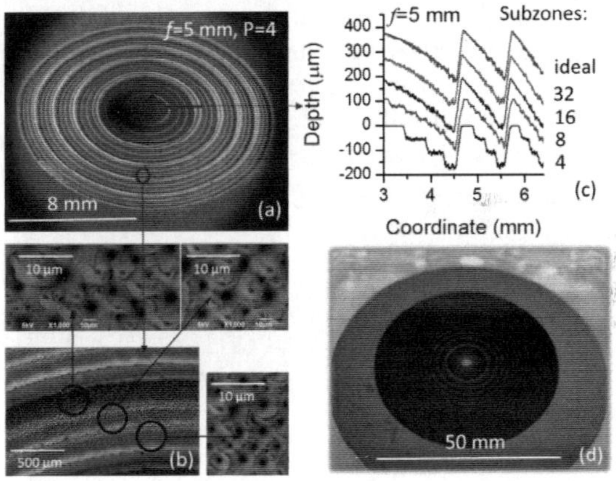

Fig. 1. SEM image of the MPFL of focal distance f=5 mm consisting of 4 phase quantization levels (a), zoomed area of each subzone (b). The cross section of 5 mm focal length MPFL measured with a step profiler (c). Each following profile line was moved by 100 μm [5]. Photo of the 50 mm diameter, 16 subzones MPFL with the focal lengths *f*=30 mm (d).

The multilevel phase Fresnel lens focusing performance was investigated by measuring Gaussian beam intensity distribution in the focal plane and along the optical axis at the 0.58 THz frequency. The peak signal dependency on phase quantization number for 5 mm and 10 mm focal length MPFLs is shown in Fig. 2. It is clearly seen that peak signal saturates at 8 phase quantization levels. This result is in a good agreement with theoretical diffraction efficiency distribution [10]. Detailed investigation was performed aiming to explain the signal deviation in case of 16 and >1000 phase quantization levels. Absolute value of the focused beam intensity was found to be depended of the THz reflection and absorption losses in the

A FULL VERSION OF THIS PROCEEDINGS ARTICLE HAS BEEN SUBMITTED TO THE IEEE TRANSACTIONS OF TERAHERTZ SCIENCE AND TECHNOLOGY (TST) FOR PEER-REVIEW, PLEASE CHECK THE WEBSITE: HTTPS://WWW.MTT.ORG/TERAHERTZ

laser affected silicon. It is worth noting that laser ablation changes silicon transmittance [11], [12].

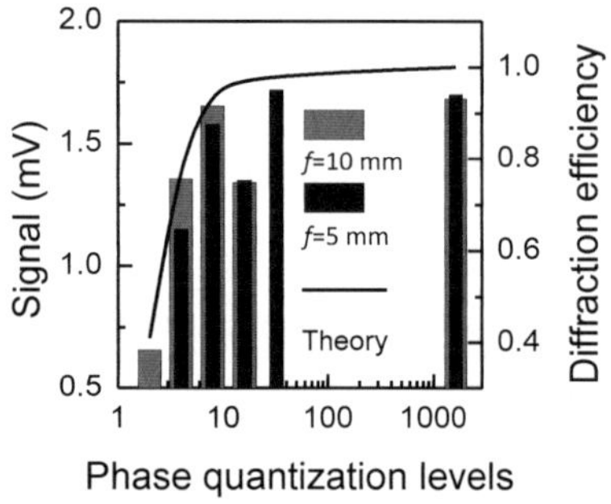

Fig.3. Peak signal of the THz detector dependence on phase quantization number *P*. Theory - diffractive efficiency of the Si-lens depending on the number of phase quantization levels [5].

The comparison of metal zone plate and commercial parabolic mirror focusing performance was presented in [4]. It was shown that the imaging system with the 5 mm focal length zone plate improves the special resolution up to 25% in comparison with that of commercial parabolic mirror. Using the same experimental set up, presented in [4], the resolution target imaging using 50 mm diameter THz MPFL is recorded. Imaging result of the resolution target is shown in Fig. 7. As one can see, periodic stripes were distinguishable if the period was not smaller than 1 mm. Such result is comparable with the parabolic mirror performance, but at the same time it has an advantage of compact size and absence of Fabry-Perrot oscillations as it was observed in the case with metal zone plate [13], [14].

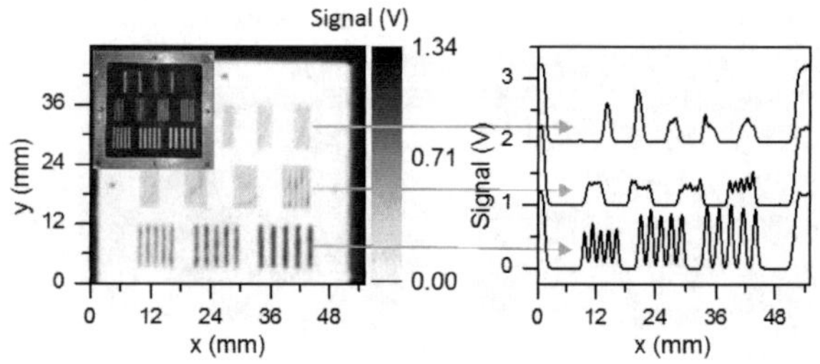

Fig. 7. THz image of the resolution target at 580 GHz frequency obtained by 3 cm focal length diffractive lens (left). Dark color in the THz images corresponds to the transmittance maxima. The cross-section of resolution target different period stripes line (right). Vertical scale is shifted by 1 V. The photo of resolution target is shown in the inset. Detailed target parameters are described in [4].

Conclusions

The laser direct writing technology was found to be powerful and convenient tool to fabricate diffractive optics for THz frequencies. The multilevel phase Fresnel lenses for 580 GHz reaching 0.86 numerical apertures have been fabricated with a different number of phase quantization levels. The effect of phase quantization number to the focused beam amplitude was determined, and radiation power transmission efficiency reaching more than 90 % was demonstrated. It was shown that imaging resolution less than 1 mm can be obtained employing robust 50 mm diameter multilevel THz lens at 580 GHz frequency.

Acknowledgment

This work was supported by the Research Council of Lithuania (LAT 04/2016).

References

[1] P. Lopato, G. Psuj, and B. Szymanik, "Nondestructive Inspection of Thin Basalt Fiber Reinforced Composites Using Combined Terahertz Imaging and Infrared Thermography," *Adv. Mater. Sci. Eng.*, vol. 2016, pp. 1–13, Jul. 2016.

[2] K. W. Kim, K.-S. Kim, H. Kim, S. H. Lee, J.-H. Park, J.-H. Han, S.-H. Seok, J. Park, Y. Choi, Y. Il Kim, J. K. Han, and J.-H. Son, "Terahertz dynamic imaging of skin drug absorption.," *Opt. Express*, vol. 20, no. 9, pp. 9476–84, Apr. 2012.

[3] J. A. Zeitler, P. F. Taday, D. A. Newnham, M. Pepper, K. C. Gordon, and T. Rades, "Terahertz pulsed spectroscopy and imaging in the pharmaceutical setting - a review," *J. Pharm. Pharmacol.*, vol. 59, no. 2, pp. 209–223, Feb. 2007.

[4] I. Kašalynas, R. Venckevičius, L. Minkevičius, A. Sešek, F. Wahaia, V. Tamošiūnas, B. Voisiat, D. Seliuta, G. Valušis, A. Švigelj, and J. Trontelj, "Spectroscopic Terahertz Imaging at Room Temperature Employing Microbolometer Terahertz Sensors and Its Application to the Study of Carcinoma Tissues," *Sensors*, vol. 16, no. 4, p. 432, 2016.

[5] L. Minkevičius, S. Indrišiūnas, R. Šniaukas, B. Voisiat, V. Janonis, V. Tamošiūnas, I. Kašalynas, G. Račiukaitis, and G. Valušis, "Terahertz multilevel phase Fresnel lenses fabricated by laser patterning of silicon," *Opt. Lett.*, vol. 42, no. 10, p. 1875, May 2017.

[6] B. Voisiat, S. Indrišiūnas, R. Šniaukas, L. Minkevičius, I. Kašalynas, and G. Račiukaitis, "Laser processing for precise fabrication of the THz optics," *Proc. SPIE*, vol. 10091, no. 10.1117/12.2253634, p. 100910F, Feb. 2017.

[7] W. G. J.M. Rodríguez, Hristo D. Hristov, "Fresnel zone plate and ordinary lens antennas: Comparative study at microwave and terahertz frequencies," *41st Eur. Microw. Conf.*, pp. 894–897, 2011.

[8] J. Zhu, G. Yin, M. Zhao, D. Chen, and L. Zhao, "Evolution of silicon surface microstructures by picosecond and femtosecond laser irradiations," *Appl. Surf. Sci.*, vol. 245, no. 1, pp. 102–108, 2005.

[9] C. A. Zuhlke, T. P. Anderson, and D. R. Alexander, "Formation of multiscale surface structures on nickel via above surface growth and below surface growth mechanisms using femtosecond laser pulses," *Opt. Express*, vol. 21, no. 7, p. 8460, Apr. 2013.

[10] E. D. Walsby, S. Wang, J. Xu, T. Yuan, R. Blaikie, S. M. Durbin, X.-C. Zhang, and D. R. S. Cumming, "Multilevel silicon diffractive optics for terahertz waves," *J. Vac. Sci. Technol. B Microelectron. Nanom. Struct.*, vol. 20, no. 6, p. 2780, Dec. 2002.

[11] S. Indrišiūnas, B. Voisiat, A. Rėza, I. Šimkienė, R. Mažeikienė, A. Selskis, and G. Račiukaitis, "Effect of laser-induced conversion of silicon nitride to silicon oxy-nitride on antireflective properties of passivation layer in polysilicon solar cells," *Opt. Mater. Express*, vol. 5, no. 7, p. 1532, Jul. 2015.

[12] M. E. Shaheen, J. E. Gagnon, and B. J. Fryer, "Femtosecond laser ablation behavior of gold, crystalline silicon, and fused silica: a comparative study," *Laser Phys.*, vol. 24, no. 10, p. 106102, Oct. 2014.

[13] L. Minkevičius, B. Voisiat, A. Mekys, R. Venckevičius, I. Kašalynas, D. Seliuta, G. Valušis, G. Račiukaitis, and V. Tamošiūnas, "Terahertz zone plates with integrated laser-ablated bandpass filters," *Electron. Lett.*, vol. 49, no. 1, pp. 49–50, 2013.

[14] L. Minkevičius, K. Madeikis, B. Voisiat, I. Kašalynas, R. Venckevičius, G. Račiukaitis, V. Tamošiūnas, and G. Valušis, "Focusing Performance of Terahertz Zone Plates with Integrated Cross-shape Apertures," *J. Infrared, Millimeter, Terahertz Waves*, vol. 35, no. 9, pp. 699–702, Jul. 2014.

Complex Beam Mapping of Large MKID Focal Plane Arrays

Kristina K. Davis[1*], Stephen J. C. Yates[2], Willem Jellema[2,3], Ronald Hesper[2], Jochem J. A. Baselmans[4,5], Christopher Groppi[1], Lorenza Ferrari[2], Kotaro Kohno[6], and Andrey Baryshev[2,3]

[1] *School of Earth and Space Exploration, Arizona State University, Tempe, AZ, USA*

[2] *Netherlands Institute for Space Research, Groningen, NL*

[3] *Kapteyn Institute, University of Groningen, Groniningen, NL*

[4] *Netherlands Institute for Space Research, Utrecht, NL*

[5] *Terahertz Sensing Group, Delft University, Delft, NL*

[6] *Institute for Astronomy and Research Centre for the Early Universe, University of Tokyo, Tokyo, JP*

*Contact: Kristina.davis.1@asu.edu

Abstract—Complex radiation pattern measurements of cameras for astronomical observatories are advantageous over total power radiation pattern measurements because of the wider range of analysis they allow. Probing both the amplitude and phase structure of the camera allows direct comparison between the measured field and electromagnetic simulation data, and allows propagation from the measurement plane to parallel planes along the optical axis. Initial results of laboratory characterization of complex radiation pattern measurements of a 350 GHz, 880 pixel microwave kinetic inductance focal plane array are presented here. The multiplexing readout allow the whole array to be measured with just two beam pattern scans. This allows rapid characterization of a large field camera, comparable to that required by current, planned, and future instruments.

Introduction

In just the last two decades, astronomical microwave kinetic inductance detectors (MKID) instruments have grown from single pixel prototypes to large focal plane arrays (FPAs) of several hundreds of pixels. Until recently, the only option available to measure the radiation pattern of these detectors was with amplitude-only measurements, as direct detectors do not intrinsically record phase information from coherent sources. With the introduction of a phase reference system demonstrated in [1], it is possible to record the full complex field structure of direct detector arrays. We have expanded on the previous work for a single array pixel to a large MKID array consisting of 880 pixels.

A complex field analysis is more data intensive than the analysis of simpler power patterns, but opens the door to a much more complete set of analysis techniques. In addition to constraining beam fitting routines in more degrees of freedom, complex field analysis allows spatial filtering of stray light from off-axis sources, full beam reconstruction by propagation to arbitrary planes along the optical axis, determination of pointing offsets from a single measurement scan plane, and full wave front error analysis [2]. The beam fitting routine we adopt corrects for misalignment between the scanning plane and optical system of the measurement set-up, reducing the complexity of the measurement system verification process.

We point out that though MKIDs were analyzed in this research, the measurement technique we use here is applicable to other direct detector devices, in the full range of the frequency spectrum. In particular, the methods we employ for rapid analysis of large datasets can be adopted for large FPA analysis.

Methods

The MKID FPA analyzed here consists of high-Q, $\lambda/4$ resonators [3] on a silicon substrate, each with a resonant frequency in the range 4-8 GHz. Optical coupling is achieved with a twin-slot antennae [4], combined with elliptical lens [5]. The twin-slot is integrated into the MKID, where the ground plane is sputtered NbTiN 500 nm thick and the central line is sputtered Al 55 nm thick. The Al section acts as the active area of the device, as its kinetic inductance—and hence the MKID resonant frequency—are modified by the optical loading. The antenna is optimized for 350 GHz, while an on-chip stray light mesh absorber of 40 nm Ta is integrated on the back of the detector chip [6]. A laser-machined silicon lens array is aligned on the array using alignment markers on the back of the detector chip. The array has 880 pixels with hexagonal packing and a pixel pitch of 2 mm. All MKIDs are coupled to single throughline, allowing all 880 pixels to be read out with a single co-axial readout line. The readout system [7] has only 2 GHz of instantaneous bandwidth, so two separate scans of the array are required to cover the full 4-8 GHz readout spectrum. In the presented measurements, the array is mounted in a large field-of-view optical system [6], allowing measurements to be made with sources in the reimaged focal plane.

We measure the complex beam pattern of the incoherent detector array by implementing an optical scheme involving 2 sources in a heterodyne configuration. The amplitude and phase signal of each pixel is determined by the relative signal strength of the two monochromatic sources, which are offset by a small frequency difference such that they modulate the MKID at the difference frequency. As the source is scanned across the image

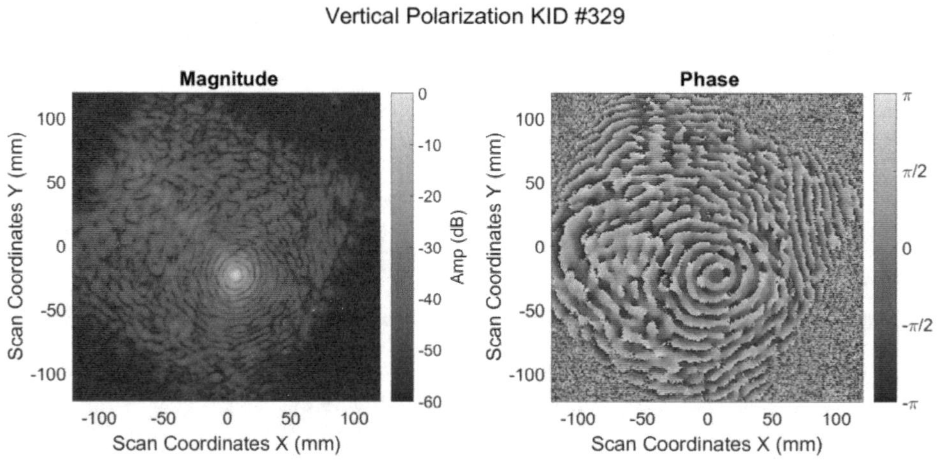

Fig. 1. Magnitude (right) and phase (left) of a central pixel of the array. The rotation of the image plane with respect to the scan plane is obvious. The phase structure shows surprising coverage across the image plane given the mechanical and thermal instabilities of the system.

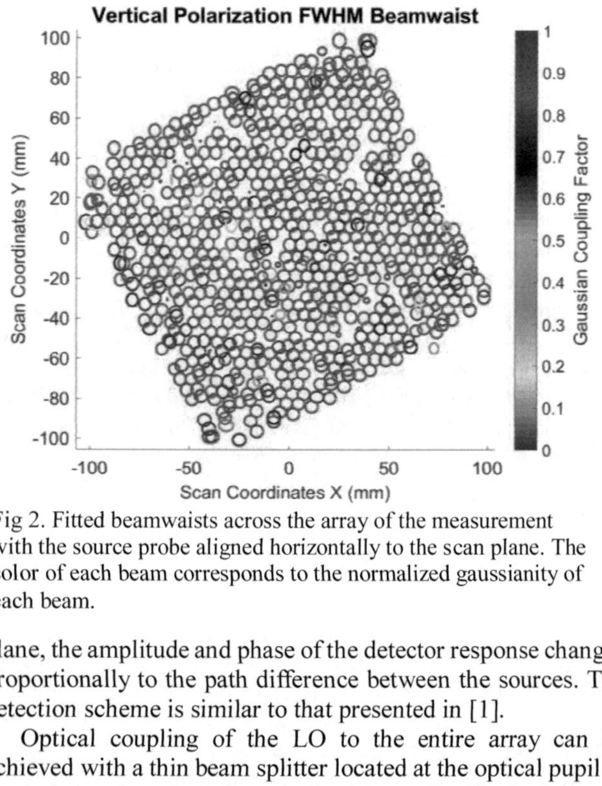

Fig 2. Fitted beamwaists across the array of the measurement with the source probe aligned horizontally to the scan plane. The color of each beam corresponds to the normalized gaussianity of each beam.

plane, the amplitude and phase of the detector response changes proportionally to the path difference between the sources. The detection scheme is similar to that presented in [1].

Optical coupling of the LO to the entire array can be achieved with a thin beam splitter located at the optical pupil in the warm reimaging system. At the pupil, all pixel beams overlap spatially but can be distinguished by their beam pointing angle. To improve coupling for all pixels at larger angles, the LO is deliberately defocused. In this optical system the pupil is directly available with little perturbation to the optics.

Results

Timeseries signals from each pixel in the FPA were converted to a complex field map. The amplitude and phase of the entire scan range for a representative pixel is shown in Figure 1. There is a strong peak signal in the amplitude maps but a large dispersed signal at the ~-40 dB level present throughout the entire map. We have attributed this signal to diffraction and imperfect optics; in detector substrate and lens array stray light is suppressed by the on-chip absorber as discussed in [6]. The phase map shows a spherical phase center, near the amplitude center, and the ring-like structure is caused by the phase 'wrapping' from $-\pi$ to π. For the subsequent analysis, we use only a portion of the field map corresponding to an area of 8x the expected beamwidth in the near field of the system. This area was chosen to include at least 2 phase wraps in the selected area to determine whether these points were actual wraps caused by the spherical phase roll-off or from nulls in the amplitude map where phase jumps due to the sign change in the complex field map.

We then take the selected area from each pixel in the complex field maps and use a Gaussian-Hermite fitting algorithm to determine the Gaussian beam efficiency and waist parameters for each pixel in both polarizations, as described in [1]. This routine fits for the center position of the beam as well as the beam width independently in X and Y. Fig. 2 shows the fitted beam locations and full width, half maximum (FWHM) elliptical beamshapes for across the entire FPA in a single polarization of the source probe.

Conclusion

Analysis of this large dataset is still ongoing, but initial results confirm that complex field parameters can be measured easily for a large focal plane MKID array. The range of analysis shown here exceeds that would be available for more standard, power pattern measurements. These results show that to first order, the beams appear consistent across the array and agree with the designed parameters.

Acknowledgment

The authors would like to thank David Thoen and Vignesh Murugesan for the fabrication of the focal plane array. The authros also thank Jan Barkoff for letting us borrow testing equipment. K.D. would like to thank Andrey Khudchenko for assisting with operation of the testing equipment.

References

[1] K.K. Davis, W. Jellema, S.J.C. Yates, L. Ferrari, J.J.A Baselmans, K. Khono, D. Thoen, V. Muregesan, A. Baryshev, "Proof of Concept Demonstration of Vector Beam Pattern Measurements of Kinetic Inductance Detectors". IEEE Thz Sci. and Tech. Jan. 2017. doi: 10.1109/TTHZ.2016.2617869

[2] W. Jellema, "Optical Design and Performance Verification of Herschel-HIFI," PhD. Thesis, 2015, University of Groningen, the Netherlands

[3] P. K. Day, H. G. LeDuc, B. A. Mazin, A. Vayonkias, J. Zmuidzinas, "A Braodbans Superconducting Detector Suitable for Use in Large Arrays," *Nature.* vol. 425. pp 817-821. October, 2013. doi:10.1038/nature02037

[4] J. Zmuidzinas and H. G. LeDuc, "Quasi-optical slot antenna SIS mixers," *IEEE Trans. on Microw. Theory Techn.,* vol. 40, no. 9, pp. 1797–1804, Sep 1992. [Online]. Available: http://dx.doi.org/10.1109/22.156607

[5] D. F. Filipovic, S. S. Gearhart, and G. M. Rebeiz, "Double-slot antennas on extended hemispherical and elliptical silicon dielectric lenses," *IEEE Trans. on Microw. Theory Techn.*, vol. 41, no. 10, pp. 1738–1749, Oct 1993. [Online]. Available: http://dx.doi.org/10.1109/22.247919

[6] S. J. C. Yates, A. M. Baryshev, O. Yurduseven, J. Bueno, K. K. Davis, L. Ferrari, W. Jellema, N. Llombart, V. Murugesan, D. J. Thoen, J. J. A. Baselmans. "Surface wave control for large arrays of microwave kinetic inductance detectors", accepted for publication in *IEEE Trans. THz Sci. Tech.*

[7] J. van Rantwijk, M. Grim, D. van Loon, S. Yates, A. Baryshev,and J. Baselmans, "Multiplexed readout for 1000-pixel arrays of microwave kinetic inductance detectors," *IEEE Trans. Microw. Theory Techn.,* vol. 64, no. 6, pp. 1876–1883, June 2016. [Online]. Available: http://dx.doi.org/10.1109/TMTT.2016.2544303

A spline-profile Diagonal horn with low cross polarization and sidelobes, suitable for THz split-block machining

Hugh J. Gibson*, Bertrand Thomas[2], Marc Warnke[2], Alain Maestrini[3], Peter de Maagt[4]
[1]*Gibson Microwave Design , 92160, Antony, France*
[2]*RPG Radiometer Physics GmbH, 53340, Meckenheim, Germany*
[3]*Observatoire de Paris, LERMA, 75014, Paris, France*
[4]*ESA/ESTEC, Noordwijk, The Netherlands*
*Contact: hugh.gibson@radiometer-physics.de

Abstract— This paper presents a novel diagonal horn design with improved performance over a standard diagonal horn. The cross polarization is reduced from -10dB to better than -20dB and co-polar side-lobes are also reduced. Power coupling to a single linear polarization fundamental Gaussian mode is improved from 84% to approximately 97%, similar to smooth-walled spline horns and only slightly worse than a conventional corrugated horn. The horn can be directly machined in split-block up to several THz with relative ease using standard milling techniques.

Introduction

Horn antennas which can synthesize a Gaussian beam are needed for coupling to Gaussian optics. Corrugated horns have traditionally been the preferred horn choice with best performance but are impractical to produce above about 1.5THz due to difficulty in machining very thin corrugations. Modern "dual-mode" horns including smooth walled spline-curve conical horns [1] [2] and multi-taper conical horn designs [3] have excellent properties but are still relatively difficult to manufacture. The only horn that is uniquely suited to easy direct machining in split-block form (with a Gaussian power beam-profile) is the straight-walled diagonal horn although the performance has up till now been significantly worse than that of other types of horns, especially in cross-polar losses.

Background

The diagonal horn was first described by Love [4]. Johansson and Whyborn [5] suggested a transition from rectangular waveguide and an optimal value for the beam size at the aperture. Withington and Murphy [6] published a mode analysis. The diagonal horn is directly machined by cutting a deepening "V" groove into each half of a waveguide split-block. It is in widespread use and has excellent beam-coupling properties and good circular symmetry but performance is compromised by a relatively high cross-polarization component (-15dB) and sharp -16dB secondary lobes in the 45° radiation axis. Improvements have already been suggested [7] but most are not compatible with THz frequencies.

Improvements to the Design

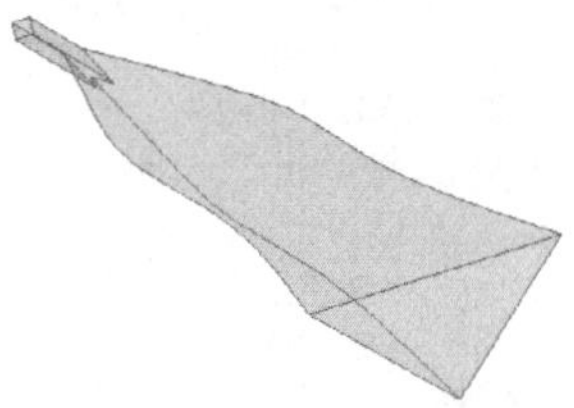

A smooth natural spline curve was added to the vertical axis (depth) of the "V" cutting tool. The horn antenna was optimised for the 1080 GHz to 1280 GHz frequency range for the Submillimeter Wave Instrument (SWI) onboard JUICE ESA Jupiter mission. The horn-antenna gain is about 23dBi which is compatible with similar horns. The cross-polarization is improved to < -20dB and the unwanted shoulders in the beam-shape (45° plane) reduced from -16dB to < -20dB. The horn is only slightly longer than the original diagonal horn for similar gain. Figure 1 shows the beam shape at three different frequencies, with cross-polar component, at 45° (green), and co-polar 45° plot (brown). The far-field phase of the antenna is flatter over a larger radiation angle, which improves antenna coupling efficiency as there is less de-focusing off-axis. All plots are using a spherical scan, far-field, with the origin set to optimised phase centre[8].

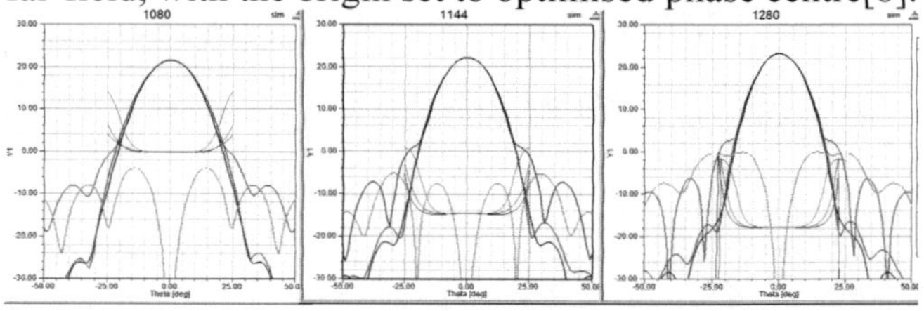

Fig. 1 Improved diagonal horn simulation for three frequencies. X-pol (45°) is in green. Ludwigs 3rd definition of cross-polarization[8] is used.

The estimated integrated power coupling to a fundamental mode, single polarization Gaussian is better than 95% (including cross-polar losses) which is a significant improvement with respect to the standard diagonal design which has approximately 84% coupling (with 9.5% loss in cross-polarization alone, or 0.43dB). We calculate the 2D coupling to a fitted Gaussian to be 98% for 0 and 90 degree cuts, and 92% for the 45 degree cut, including cross-polar losses.

EASE OF MANUFACTURE AT THZ FREQUENCIES

Because the spline curve has no abrupt corners, a surprisingly large diameter milling tool can be used to create the gentle curve. It is estimated that a horn for 10THz can be directly machined with an end-mill tool with a diameter of 0.1 mm with relative ease. The horn is also extremely tolerant to machining errors [9].

WIDE BANDWIDTH TESTS

The diagonal horn using the Whyborn [5] direct-transition from waveguide to horn suffers from poor match at extreme low frequencies (bottom end of a complete waveguide band). Good performance can be regained using a slightly modified flared-waveguide transition to the horn. More details are given in [9].

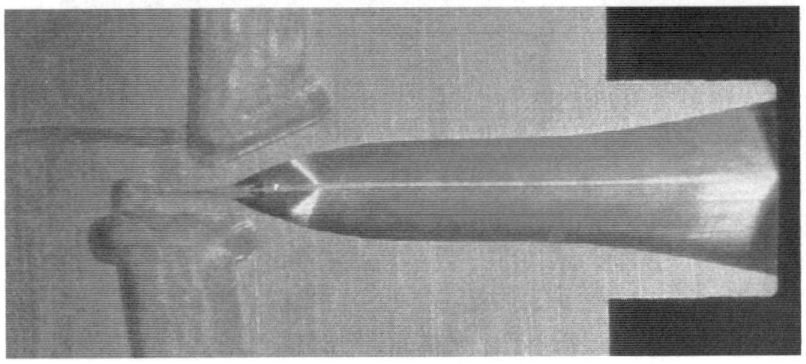

Fig. 2 1.2THz Sub-harmonic mixer split-block including spline diagonal horn antenna (design by GMD, machined by RPG Radiometer Physics GmbH)

CONCLUSION

A spline diagonal horn has been built as part of 1.2THz mixer developments and it is planned to extend this design up to 2THz as part of ESA contract AO/1-8271 in collaboration with Observatoire de Paris and RPG. The traditional diagonal horn design has been modified using a spline-curve "s"shape modulation in the depth of the "V" groove. This modification is similar in shape to smooth-wall spline horns and improves cross-polarization purity, side-lobe level and the spherical wave front of the beam. The result is a high quality horn which can be directly machined into split-block, exactly as the original diagonal horn. The technology is extendable to 10THz and promises to be an easy way to realise high quality horns at multi-THz frequencies with conventional split-block technology. Integrating the horn directly in the mixer block also solves the difficult problem of the alignment of waveguide flanges at THz frequencies. The reduction of waveguide losses due to a very short connecting waveguide is also very advantageous. Final measurement tests are on-going. Complete details of the new design and comparisons with other horns are to be found in [9].

ACKNOWLEDGEMENTS

This project was funded under ESA contract AO/1-8271. This project would not be possible without the skill of expert micro-machinists at RPG including Benjamin Elsner and Stefan Rajtschan.

REFERENCES

[1] C. Granet, G.L. James, R. Bolton, G. Moorey "A Smooth-Walled Spline-Profile Horn as an Alternative to the Corrugated Horn for Wide Band Millimeter-Wave Applications" *IEEE trans. Antennas and Propagation* , vol. 52, no. 3, march 2004

[2] L. Zeng, C.L. Bennett, D.T. Chuss, E. J. Wollack "A Low Cross-Polarization Smooth-Walled Horn with Improved Bandwidth" *IEEE trans. Antennas and Propagation* , vol. 58, no. 4, April 2010

[3] G. Yassin, P. Kittara, A. Jiralucksanawong, S. Wangsaya, J. Leech, M. Jones "A High Performance Horn for Large Format Focal Plane Arrays" 18th ISSTT 2007

[4] A.W. Love "The diagonal horn antenna" *Microwave Journal* vol V. pp 117-112, 1962

[5] J.K. Johansson N.K. Whyborn "Antennas for sub-millimetre wave receivers" second ISSTT 1991

[6] S. Withington J.A. Murphy "Analysis of diagonal horns through Gaussian-Hermite modes" *IEEE Trans. Antennas and Propagation* vol 40, no.2 pp 198-206 Feb 1992

[7] O. Xu "Diagonal horn Gaussian efficiency enhancement by dielectric loading for submillimeter wave aookication at 150GHz" Progress in Electronics Research Vol. 114, 117-192 2011

[8] A. Ludwig, "The definition of cross polarization" IEEE Transactions on Antennas and Propagation, vol. 21, no. 1, pp. 116–119, 1973

[9] H.J. Gibson, B. Thomas, M Warnke, A. Maestrini, P. De Maagt "A spline-profile Diagonal horn with low cross polarization and sidelobes, suitable for split-block machining" submitted to *IEEE Transactions on Terahertz Science and Technology* (TST), 2017

Design and Measurement of a Waveguide Probe Based WR3.4 Optically Controlled Modulator

Jake A. Connors[1*], C.-Y. E. Tong[2], Paul K. Grimes[2], A. Gebreyohannes[3] and Lei Liu[4]
[1]*Physics Department, Harvard University, Cambridge, MA 02138*
[2]*Smithsonian Astrophysical Observatory, Cambridge, MA 02138*
[3]*Center for Nanoscale Systems, Harvard University, Cambridge, MA 02138*
[4]*Electrical Engineering Department, Notre Dame University, Notre Dame, IN 46556*
*Contact: jconnors@physics.harvard.edu

Abstract— Presented is the design and measurement of an optically controlled modulator implemented in waveguide at WR3.4. The symmetric device consists of offset input and output waveguides bridged by a microstrip transmission line with radial waveguide probes patterned onto a thin silicon chip. Optical illumination of the chip at photon energies larger than the silicon bandgap generates free charge carriers in the silicon substrate, modulating its conductivity and dielectric constant. This photoconductive effect allows for the modulation of the millimeter-wave loss along the microstrip transmission line. Initial testing of the device has demonstrated up to 40dB of attenuation, with less than 1.5dB insertion loss in the off-state. A modulation bandwidth of 2MHz has been demonstrated, indicating a sub-100ns fall time when operated as a fast switch.

Introduction

The control and modulation of millimeter-wave source power is crucial to many applications from radar to antenna mapping to network analysis. Mechanically actuated variable attenuators, ferrite modulators or pin-diode attenuators can provide such control, however it comes at the cost of slow modulation speeds, narrow operating bandwidth or poor off-state insertion loss. Optically controlled modulators, however, can provide fast wideband modulation with wide attenuation range and low off-state insertion loss. The implementation of optically controlled devices has varied, from modulators built in dielectric waveguide[1,2] and dielectric loaded metal waveguide [3-5], quasi-optical modulators and beam-formers[6-9], ultrafast switches and programmable attenuators in planar transmission line[10-13], and photonic-bandgap switches[14]. We implement an optically controlled modulator/continuously variable attenuator in WR3.4 (220-330GHz) as a short section of microstrip transmission line bridging offset symmetric input/output waveguides. The microstrip line is terminated on both ends with a radial stub transition to WR3.4 waveguide, allowing modulation of the device insertion loss through optical illumination of the microstrip line where the electric field is concentrated and less frequency dependent than in waveguide.

Fabrication and Block Design

As shown in Figure 1, the device consists of a 25μm thick silicon chip bridging two offset rectangular waveguides. The chip is patterned with a 50Ω microstrip transmission line terminated on either end by a single-ended radial stub transition to WR3.4 waveguide[15,16]. The waveguide block, shown in Fig. 2b, was fabricated from Tellurium-Copper as an E-plane split block. A shallow trench in the lower waveguide block contains the probe chip, as seen in Fig. 2a. In the upper waveguide block, an air cavity sits above the microstrip line and constrains the silicon chip to sit within its trench when assembled. The probe chip was fabricated with simple photolithography to define deposition and etch masks, thermal evaporation for Cr/Au deposition and Bosch-process[17] deep reactive ion etching to free the barbell shaped chip from the parent wafer. The dimensions of the probe chip and the cavity above were chosen to make all higher-order and waveguide-like modes in this transmission line evanescent below 330GHz and thus not contribute to the coupling between waveguides. The radial stub transition with a single capacitive tuning step was designed and optimized in ANSYS HFSS[18] to achieve a better than -26 dB return loss across the entire WR3.4 band. With back to back transitions, this results in a better than -20dB device input return loss, which can be seen in the black dotted line of Fig. 3b.

Fig. 1 A view of the microstrip-to-waveguide probe chip as installed in the E-plane split block. Also visible are the WR3.4 waveguides each with a single capacitive tuning step.

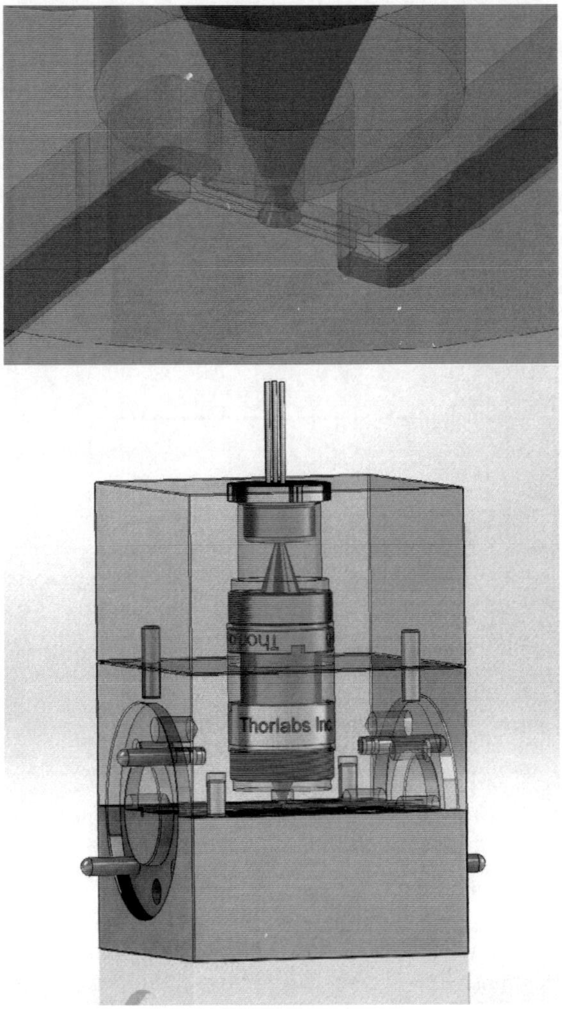

Fig. 2 (Top) View of probe chip as assembled in the waveguide block. Light passes through a small hole in the upper cavity onto the chip. (Bottom) Completed assembly of waveguide split-block and laser diode block. Shown in red is the beam of the 808nm laser diode as collimated and focused by the two-lens system.

Illumination of the probe chip is achieved using a laser diode and two-lens focusing system built into the waveguide and upper laser diode block, as shown in Fig. 2b. An 808nm multimode laser diode with up to 1W continuous output power was used for initial testing. The laser diode wavelength was chosen such that the absorption depth of the light impingent on the silicon chip was approximately half of the chip's thickness. This allows for carrier generation distributed throughout the thickness of the chip while keeping total absorption near 99%, discounting reflection from the air/silicon interface. We operate the laser diode using a current source to provide stable output power and to avoid the effects of thermal runaway in the diode. The laser diode is clamped into place in the upper block, allowing for sufficient alignment and heatsinking. A thread mounted aspheric lens was built into the same block as the laser diode and serves to collimate the laser diode output. The second aspheric lens, built into the waveguide block, focuses the light through a small aperture leading to the microstrip cavity, where the light can illuminate the center of the probe chip as shown in Fig. 2a. The two blocks are aligned with precision dowels to roughly center the optical beam. Because the optical beam is collimated between the two blocks, the upper block can be replaced with any collimated light source. The position of the lenses were tuned to maximize optical throughput as measured by an optical power meter, achieving nearly 50% coupling of the available optical power to the probe cavity.

Device Operating Principle

Optical illumination in a concentrated area of the microstrip transmission line modulates the bulk silicon substrate conductivity, leading to loss in the microstrip line. In this way the attenuation of the device can be controlled using the intensity of absorbed optical power. Simulations of the full device insertion loss and return loss are shown in Figure 3, as modelled by HFSS, for varying substrate conductivities. In these models, it is assumed that a 200µm long section of the microstrip line is illuminated and becomes conductive uniformly through the thickness of the chip. This model predicts that the insertion loss in dB increases nearly

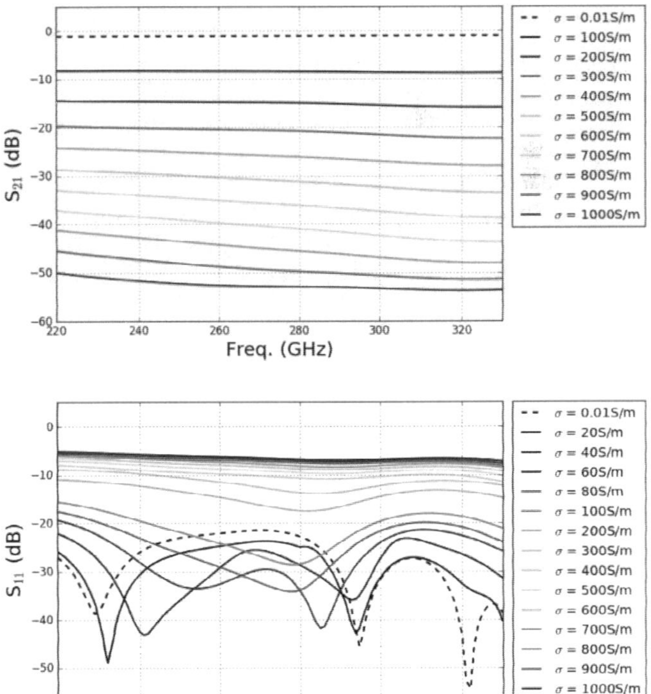

Fig. 3 Insertion loss (top) and return loss (bottom) of the full device as modelled by ANSYS HFSS for varying levels of substrate conductivity. It is assumed that illumination only changes the conductivity over a 200µm long section of the silicon probe chip.

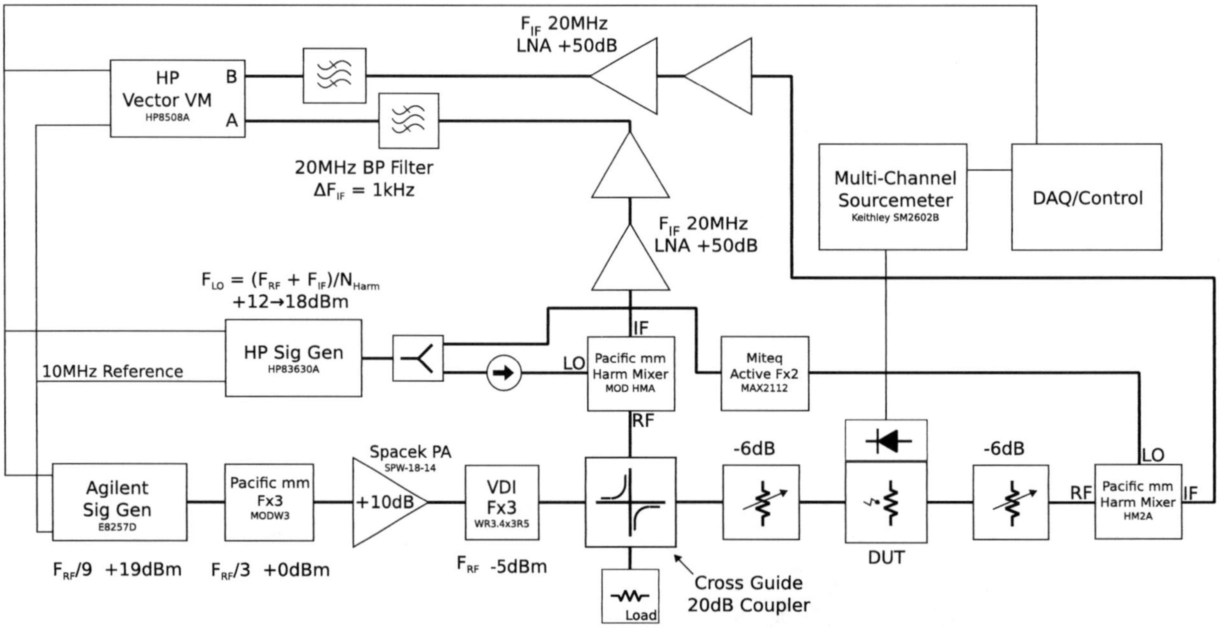

Fig. 4 Schematic of transmission measurement scheme. A coupled portion of the RF signal and the remainder after passing through the DUT are sent through a down-converting and amplification chain to an HP8508A Vector Voltmeter for measurement of the relative amplitude and phase of the transmitted signal.

linearly with substrate conductivity. The input is well matched at low substrate conductivity, but becomes increasingly reflective as the substrate conductivity increases beyond 100S/m.

Photons impingent on a semiconductor with energy larger than the bandgap energy can be absorbed by electrons in the semiconductor valence band, elevating them to the conduction band. The resulting electron-hole pairs are free to move under the influence of electric fields and thus

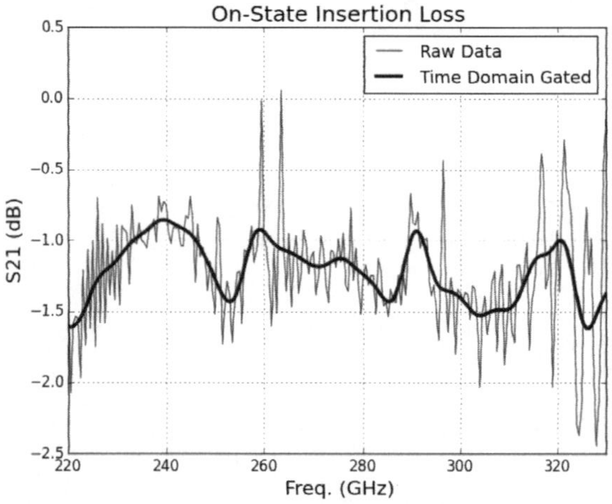

Fig. 5 Measurement of the insertion loss of the device in the no illumination or on-state. Time domain gating was applied to systematically smooth the data.

contribute to the bulk conductivity of the semiconductor, known as the photoconductive effect. In the low frequency Drude limit, the bulk conductivity is given by:

$$\sigma = e(n_e\mu_e + n_h\mu_h) \qquad [1]$$

where n_i is the carrier density and μ_i is the mobility for holes or electrons. The equilibrium carrier density under optical illumination is given by:

$$n_i = P_{inc}\frac{\lambda}{hc}\epsilon_{opt}\tau_i \qquad [2]$$

where P_{inc} is the incident optical power per unit volume, ϵ_{opt} is the optical coupling efficiency and τ_i is the hole or electron lifetime. In order to maintain a low off-state insertion loss, high-resistivity floatzone silicon was used having a bulk resistivity > 10^4ohm-cm, implying that the native dopant density and thus the native carrier density is very small. The carrier lifetime is a function of the various recombination mechanisms taking place within the wafer. The primary bulk recombination process at low carrier density, Shockley-Read-Hall recombination[19], occurs on timescales of ~1ms in very pure silicon, such as was used here. Other bulk recombination processes, such as radiative or Auger recombination only contribute at very high carrier densities, which were likely not seen in our testing. Due to the small thickness of chip used, the dominant recombination pathway is surface recombination, which occurs due to the large multiplicity of defect sites at the termination of the

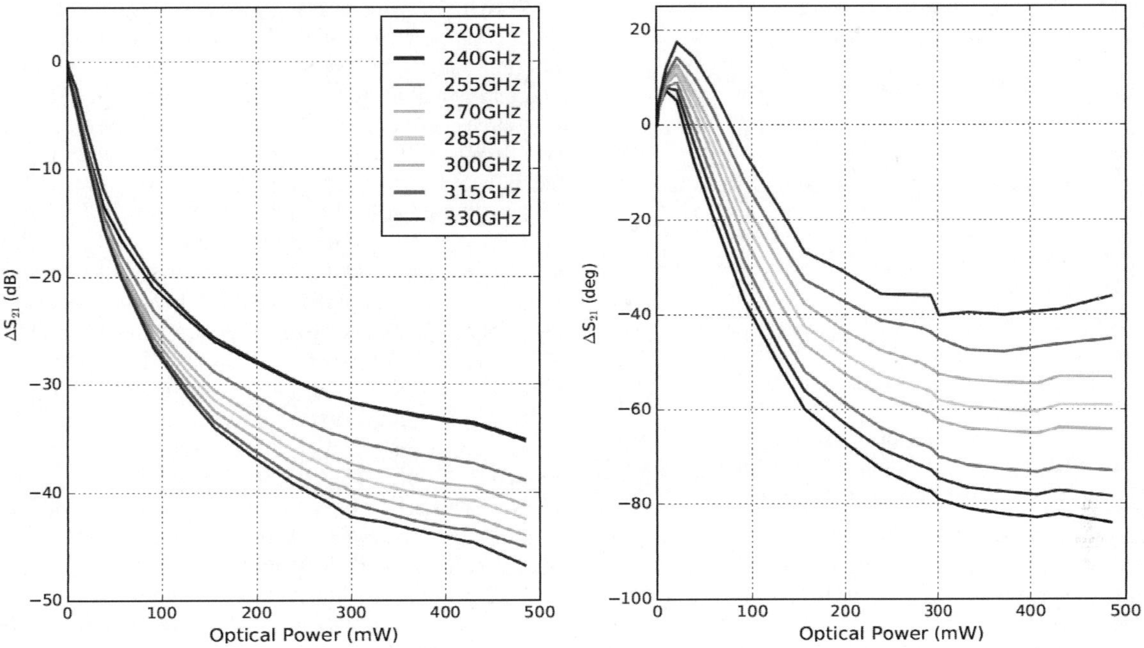

Fig. 6 Measurements of device insertion loss (left) and phase shift (right) versus incident optical power at 808nm for several frequencies across the WR3.4 band. HFSS modelling predicts a linear relationship between conductivity and insertion loss, which is seen for low incident power.

crystal lattice at the surface. The surface recombination time for a wafer with similar surfaces is given by [20,21]:

$$\tau_s = h/_{2s_v} + h^2/_{\pi^2 D} \qquad [3]$$

where h is the thickness of the wafer. The first term in Eqn. 3 represents the time necessary for a charge carrier to recombine once it has reached the surface, parameterized by s_v, the surface recombination velocity. This parameter is an empirical measure of the density of surface defects which is dependent upon the exact details of surface preparation and passivation. The second term represents the average time required for a charge carrier to diffuse to the surface, parameterized by D, the carrier diffusion constant. Using the known value for D in silicon at room temperature of $27cm^2/s$, the surface lifetime of a 25μm thick wafer can be estimated to be at least 24ns, but is ultimately determined by the surface recombination velocity. In the following sections, we therefore do not assume a carrier lifetime, but attempt to estimate it through various measurement techniques.

TRANSMISSION MEASUREMENTS

Measurements of the device were performed in waveguide over the WR3.4 band (220GHz-330GHz) using a coherent source/ receiver setup as shown in Fig. 4. The RF source is an amplifier-multiplier chain driven by a signal at $f_{RF}/9$ producing upwards of -5dBm of power across the WR3.4 band. A cross-guide coupler/harmonic mixer at the output of the RF source allows the amplitude and phase of the RF signal to be monitored. The signal then passes through the device under test and is detected in a final harmonic mixer. Local oscillator (LO) signals are provided via a power splitter to both harmonic mixers by a common signal generator. The LO harmonic number and power were optimized for each frequency point to maximize the dynamic range of the system. The 20MHz IF signal from both the reference and thru harmonic mixers are then amplified and fed into an HP Vector Voltmeter (VVM) which measures the relative amplitude and phase of the two signals. In the operating regime of our measurement setup, the harmonic mixers, amplifiers and VVM are linear with respect to the RF signal, so the IF signals are both linearly related to their RF counterparts allowing a relative amplitude and phase measurement of the device under test. In the limit that the device under test and harmonic mixers are perfectly matched, the resultant measurement is exactly calibrated by taking a set of reference measurements without the device under test. This is not true in practice, so a set of adjustable-vane attenuators are placed before and after the device under test to improve the system match.

Measurements of the insertion loss of the device without optical illumination are shown in Fig. 5. This measurement was calibrated by taking the ratio of measurements performed with and without the device in the measurement system. The insertion loss varies between 1 and 1.5 dB across the WR3.4 band. Time domain gating was performed using a gate the electrical length of the entire waveguide block to systematically smooth the data.

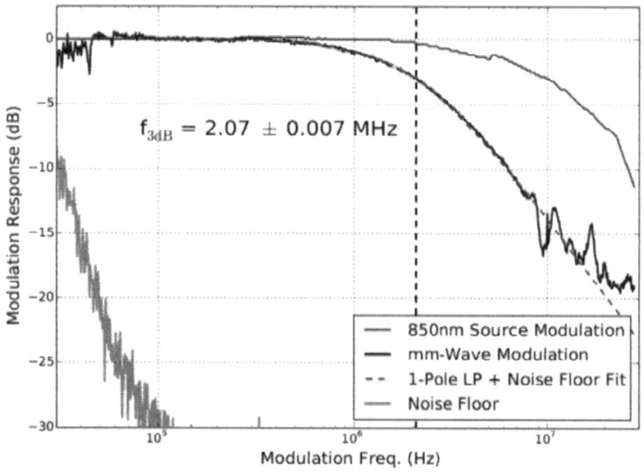

Fig. 7 Measurement of the modulation bandwidth of the device using a 20MHz 5mW IR LED source. A 3dB bandwidth of more than 2MHz is measured corresponding to a ~77ns carrier lifetime.

To measure the insertion loss of the device at varying levels of absorbed optical power, an 808nm laser diode was used to illuminate the device. By varying the current driven into the laser diode, the optical power can be controlled. The incident optical power through the upper portion of the waveguide split block was measured using an optical power meter for varying laser diode currents. Shown in Fig. 6 are measurements of the insertion loss of the device versus incident optical power for various frequencies across the WR3.4 band. The insertion loss is smoothly varying with frequency and monotonically increases with incident optical power. Up to 10dB of attenuation is achieved with incident optical powers of only 20mW, while up to 45dB of attenuation is achieved with an incident optical power of 500mW. While HFSS simulations imply that the insertion loss is a nearly linear function of substrate conductivity, the experimentally realized insertion loss shows diminishing returns at higher incident optical power. This is likely due to significant heating of the substrate by the incident light, causing an increase in temperature and thus lower carrier mobilities and conductivity.

For low incident optical power, where insertion loss is nearly linear with incident power and heating is unimportant, these measurements and the HFSS predictions for the insertion loss can be used to make an estimate of the effective ambipolar carrier lifetime. In making this calculation, the simple relations for Si substrate conductivity in Eqn. 1 and 2 are used to convert from delivered optical power to effective substrate conductivity. In doing so, we assume known carrier mobilities in silicon of $\mu_e \approx 1400\, cm^2/V \cdot s$ and $\mu_h \approx 450\, cm^2/V \cdot s$, an illuminated area of 200μm x 100μm, and an optical efficiency of 0.6 due to reflection from the air-Si interface. The illuminated area was chosen to match the calculated optical beam size at the substrate. Using these rough assumptions, a carrier lifetime of approximately 60ns was estimated.

Modulation Bandwidth Measurement

When used with a fast optical source, our device can operate as a millimeter-wave amplitude modulator. The switching speed or modulation bandwidth of our device is inherently limited by the carrier lifetime in Si. An identical RF source as used in the transmission measurement was used for this measurement. An 850nm IR LED with a 10ns rise and fall time was used as a fast optical source. The LED produced at most 5mW of optical power, of which only a small fraction was coupled onto the device. The millimeter-wave signal transmitted through the device was detected in a harmonic mixer pumped with an LO to produce an IF frequency of 200MHz. Using a spectrum analyzer, the carrier amplitude at 200MHz was compared to the sideband amplitude created by the optical modulation. Shown in Fig. 7 are results of the modulation bandwidth measurement. Results are normalized to the response at low modulation frequency.

As measured, the device exhibits a modulation bandwidth of approximately 2MHz. The modulation bandwidth of the optical source itself was measured to be approximately 10MHz, far enough above the measured bandwidth to be neglected in the estimate of the device bandwidth. Under the assumption that the carrier distribution is near equilibrium at all times, the modulation bandwidth can be calculated in the low-attenuation regime from the effective carrier lifetime as:

$$f_{3dB} = \frac{1}{2\pi\tau_{eff}} \quad [4]$$

where τ_{eff} is the effective carrier lifetime owing to contributions from all recombination pathways. Using this, we measure the effective carrier lifetime to be 76.9±0.3ns, in agreement with the rough estimate from transmission measurements.

Conclusions

We have designed and measured a new implementation of an optically controlled waveguide attenuator. Through the use of a back-to-back waveguide probe chip fabricated on thin silicon, up to 40dB of millimeter-wave attenuation was achieved over the entire WR3.4 band, with an off-state insertion loss less than 1.5dB. When operated as a fast modulator, the bandwidth of the device has been measured to be approximately 2MHz, limited by the effective photo-generated carrier lifetime. The device shows promising performance in both dynamic range and modulation bandwidth, suggesting its possible use in millimeter-wave test and measurement systems. Further improvements will be made in two directions. For use as a variable attenuator, we seek to decrease the required optical power per unit attenuation, which can be achieved through anti-reflection coating, surface passivation and heat sinking of the probe chip. For use as a fast modulator, we plan to fabricate similar probe chips out of GaAs, a direct bandgap semiconductor which has a much shorter carrier lifetime, allowing for a vastly larger modulation bandwidth.

REFERENCES

[1] Chi H. Lee, Paul S. Mak, and A.P. DeFonzo, "Optical Control of Millimeter-Wave Propagation in Dielectric Waveguides," *IEEE Journal of Quantum Electronics*, Vol. QE-16, No. 3, March 1980.

[2] M. Tsutsumi and A. Alphones, "Optical Control of Millimeter Waves in the Semiconductor Waveguide," *IEICE Trans. On Electronics*, Vol. E76-C, No. 2, pp. 175-182, 1993.

[3] B. Kazan, "Light-controlled Waveguide Attenuator," US Patent 2856589 A, 14 Oct 1958.

[4] E. Episkopou, "Reconfigurable optically-controlled waveguide for terahertz application," Ph.D. Thesis, Imperial College London, May 2014.

[5] Z. Jiang et. al, "Investigation and Demonstration of a WR-4.3 Optically Controlled Waveguide Attenuator," *IEEE Trans. on THz Sci. and Tech,* Vol. 7, No. 1, Jan. 2017.

[6] S. Biber, D. Schneiderbanger and L.-P. Schmidt, "Design of a Controllable Attenuator with High Dynamic Range for THz-Frequencies based on Optically Stimulated Free Carriers in High-Resistivity Silicon." *Frequenz,* Vol. 59, Iss. 5-6, June 2005.

[7] S. Lee, Y. Kuga and R.A. Mullen, "Optically Tunable Millimeter-Wave Attenuator Based on Layered Structures," *Microwave and Optical Tech. Lett.,* Vol. 27, No. 1, Oct. 2000.

[8] L.-J. Cheng and L. Liu, "Optical modulation of continuous terahertz waves towards cost-effective reconfigurable quasi-optical terahertz components," *Optics Express*, Vol. 21, Iss. 23, Nov. 2013.

[9] T. Gallacher et. al, "Optical Modulation of Millimeter-Wave Beams Using a Semiconductor Substrate," *IEEE Trans. MTT,* Vol. 60, No. 7, July 2012.

[10] A. Johnson and D. Auston, "Microwave switching by picosecond photoconductivity," *IEEE Journal of Quantum Electronics,* Vol. 11, Iss. 6, June 1975.

[11] W. Platte, "Optoelectronic Microwave Switching via Laser Induced Plasma Tapers in GaAs Microstrip Sections," *IEEE Trans. MTT,* Vol MTT-29, No, 10, Oct. 1981.

[12] R. Kremer et. al, "Optically Controlled Coplanar Transmission Lines for Microwave Signal Processing," *IEEE Trans. MTT,* Vol. 43, No. 9, Sept. 1995.

[13] S. Lee et. al, "Optically CW-Mode Controlled Microwave Switches with Carrier-Confinement on a Coplanar Waveguide," *IEEE Antennas and Propagation Soceity Int. Symposium,* July 2001.

[14] W. Otter et. al, "Photoconductive Photonic Crystal Switch," 2013 38th International Conference on *IRMMW-THz,* Mainz, 2013, pp. 1-2.

[15] S.Withington, G. Yassin, J. Leech, K.G. Isaak, " An accurate expression for the input impedance of one-sided microstrip probes in waveguide," *Tenth International Symposium on Space Terahertz Technology,* Charlottesville, March 1999.

[16] J.W. Kooi et. al, "A Full-Height Waveguide to Thin-Film Microstrip Transition with Exceptional RF Bandwidth and Coupling Efficiency," *International Journal of Infrared and Millimeter Waves,* Vol. 24, Iss. 3, pp 261-284, March 2003.

[17] F. Laermer and A. Schilp of Robert Bosch GmbH, "Method of Anisotropically Etching Silicon," US Patent No. 5501893, 1999.

[18] Ansoft Corporation, Four Station Square, Suite 200, Pittsburgh, PA 15219-1119, USA

[19] W. Shockley and W.T. Read, Jr., "Statistics of the Recombinations of Holes and Electrons," *Physical Review,* Vol. 87, No. 5, Sept 1952.

[20] K.L Luke and L.-J. Cheng, "Analysis of the interaction of a laser pulse with a silicon wafer: Determination of the bulk lifetime and surface recombination velocity," *Journal of Applied Physics,* Vol. 61, No. 6, pp. 2282-2293, March 1987.

[21] A.B. Sproul, "Dimensionless solution of the equation describing the effect of surface recombination on carrier decay in semiconductors," *Journal of Applied Physics,* Vol. 76, No. 5, pp. 2851-2854, Sept. 1994.

Ultra-Compact THz Multi-Pixel Local Oscillator Systems for Balloon-borne, Airborne and Space Instruments

Jose V. Siles[1], Jonathan Kawamura[1], Robert Lin[1], Choonsup Lee[1] and Imran Mehdi[1]

[1] *Jet Propulsion Laboratory, California Institute of Technology, Pasadena, CA 91109*
*Contact: Jose.V.Siles@jpl.nasa.gov

Abstract—Multi-pixel heterodyne receivers with dramatically increased mapping speeds are required to provide astronomers with the capability of obtaining large-scale high-angular resolution spectrally-resolved images of molecular clouds, and carrying out surveys of nearby galaxies. Since most important tracers of star forming regions lie in this frequency range ([CII] at 1.9 THz, [OI] at 2.06 THz, [NII] at 1.46 THz, etc.), the development of these receivers is key to understanding the processes governing the formation of interstellar clouds and stars, which is crucial for unraveling the evolution of galaxies. High mapping speeds are indeed extremely important for missions with limited observation times, such as balloon-borne and airborne astrophysics missions. These kinds of missions have lower costs and faster implementation plans than space missions. Therefore, they have become the primary driving force to enhance the capabilities of terahertz heterodyne receiver technology, which will greatly benefit future space instruments. The development of array terahertz local oscillator (LO) sources is key to enabling these systems. These sources need to be multi-pixel, compact, broadband, able to operate at room-temperature, have low dc power consumption, and exhibit state-of-the-art performance.

Recently, two 4-pixel local oscillator systems at 1.46 THz and 1.9 THz have been successfully flown on board the Stratospheric Terahertz Observatory (STO-2). STO-2, a balloon-born terahertz heterodyne instrument, was launched from Antarctica on December 2016 to look for carbon, oxygen and nitrogen in star forming regions of our galaxy. The sources operated in a temperature range of 0-45C with output power levels per pixel of > 20 μW. This is the first time that integrated multi-pixel LO sources have been successfully flown. Building upon the successful STO-2's LO designs, we have recently demonstrated a high-efficient ultra-compact 16-pixel source that can be easily extended up to 64-pixels. This new module will be the baseline for GUSTO, another balloon-borne mission currently under Phase A studies. GUSTO, a NASA Explorer Mission of Opportunity, will feature 8-pixel receivers at 1.56 THz, 1.9 THz and 4.7 THz and is planned to launch from Antarctica in 2019.

In this paper, we will present the development, integration and test of the STO-2 local oscillator subsystem. We will focus on some key operational aspects and lessons learned that were taken into account in the design of our new 16-pixel 1.9 THz-2.016 THz LO module, which is the largest pixel count local oscillator source demonstrated so far in the terahertz range. The new 16-pixel module is also approximately five times smaller in size than the 4-pixel LO module delivered for STO-2. Its power consumption per pixel is around 10 times lower and each pixel delivers around 30 μW output power that can be individually adjusted, The performance of this new LO module will also be presented together with a roadmap to extend this scheme to 64- 128- or even 256-pixel local oscillator systems, taking advantage of the recent progress on backend receiver technology and low power low noise amplifiers. This will be a major step forward for future airborne instruments on-board SOFIA, as well as future NASA missions under concept study, such as the Cosmic Origins Space Telescope (CST) and the Terahertz Space Telescope (TST).

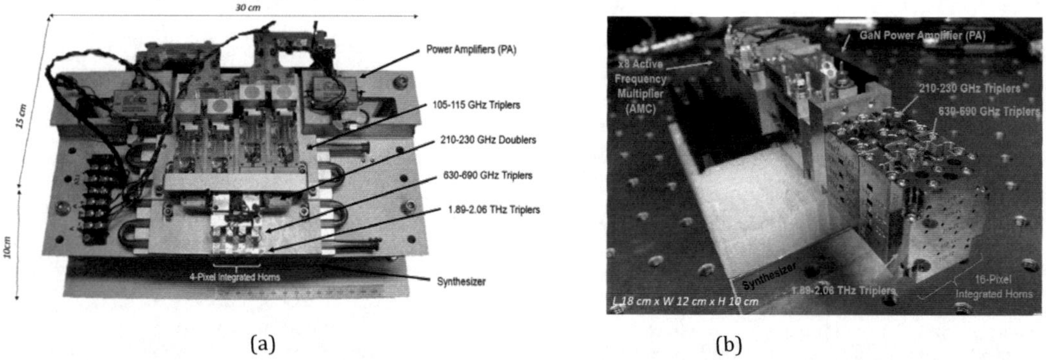

(a) (b)

Fig. 1. (a) JPL 4-pixel 1.9 THz local oscillator source flown on-board STO-2 (dc power consumption is 28 Watt/pixel); (b) New generation ultra-compact 16-pixel 1.9 THz local oscillator source recently demonstrated at JPL (power consumption is only 2.3 Watt/pixel).

This work was carried out at the Jet Propulsion Laboratory, California Institute of Technology, Pasadena, CA under a contract with National Aeronautics and Space Administration (NASA).

A continuous wave terahertz molecular laser pumped by a quantum cascade laser

Antoine Pagies[1], Guillaume Ducournau[1], and Jean-François Lampin[1*]

[1]*Terahertz Photonics Group, Institut d'Electronique de Microélectronique et de Nanotechnologie, Univ. Lille, CNRS, Centrale Lille, ISEN, Univ. Valenciennes, UMR 8520 –IEMN, F-59652 Villeneuve d'Ascq, France*
[*]Contact: jean-francois.lampin@univ-lille1.fr

Abstract— We demonstrate a new approach to realize a continuous wave (CW) THz laser source that can be used as a local oscillator. It is based on a molecular gas optically pumped by a mid-infrared beam. Generally optically pumped terahertz lasers (OPTL) are pumped by CO_2 discharge lasers. They are bulky and have a low efficiency. Here we demonstrate for the first time the use of a solid-state mid-IR quantum cascade laser (QCL) as an OPTL pump laser [1]. The main advantage of the QCL is its continuous tunability compared to CW CO_2 lasers which are only tunable on discrete lines. It allows a larger degree of freedom in the choice of the molecule and in the transitions. Small molecules with large electric dipoles are good candidates to realize high efficiency and compact OPTL. Here, the active medium of the laser is made of low-pressure NH_3 gas enclosed in a cylindrical metallic waveguide closed by two flat mirrors. The mid-infrared QCL work at room temperature and its beam is focused in the cylindrical cavity through an input coupler. It is tuned to a transition between a ground state level and a $v_2 = 1$ excited state level of the NH_3 molecule. Population inversion is achieved between excited state levels. The molecules de-excite by stimulated emission on pure inversion "umbrella-mode" quantum transitions. These transitions are allowed by the tunnel effect and their frequencies are close to 1 THz [1,2]. We have already demonstrated a CW output power of 0.35 mW at 1073 GHz. More than ten other laser lines can be also obtained around 1 THz. The generated power is sufficient to pump HEB or Schottky diode mixers. We believe that this source can be used as a local oscillator for heterodyne receiver applications.

[1] A. Pagies, G. Ducournau, and J.-F. Lampin, *Low-threshold terahertz molecular laser optically pumped by a quantum cascade laser*, APL Photonics **1**, 031302 (2016).

[2] A. Pagies, G. Ducournau, and J.-F. Lampin, *Continuous wave terahertz molecular laser optically pumped by a quantum cascade laser*, 41st International Conference on Infrared, Millimeter, and Terahertz Waves (2016).

Local Oscillator for a 4.7-THz Multi-Pixel Heterodyne Receiver Based on a Quantum-Cascade Laser

H. Richter[1,*], N. Rothbart[1], M. Wienold[1,2], L. Schrottke[3], K. Biermann[3], H. T. Grahn[3], and H.-W. Hübers[1,2]

[1]*German Aerospace Center (DLR), Institute of Optical Sensor Systems, Rutherfordstr. 2, 12489 Berlin, Germany*
[2]*Humboldt-Universität zu Berlin, Department of Physics, Newtonstr. 15, 12489 Berlin, Germany*
[3]*Paul-Drude-Institut für Festkörperelektronik, Leibniz-Institut im Forschungsverbund Berlin e. V., Hausvogteiplatz 5−7, 10117 Berlin, Germany*
*Contact: Heiko.Richter@dlr.de

Abstract— We report on the design and performance of a 4.7-THz local oscillator (LO) for the seven-pixel heterodyne spectrometer upGREAT, which is the upgrade of the German Receiver for Astronomy at Terahertz Frequencies (GREAT). The upGREAT instrument has been developed for SOFIA, the Stratospheric Observatory for Infrared Astronomy. The LO is based on a quantum-cascade laser (QCL). The first operation took place in October 2016. During this measurement campaign the fine structure line of atomic oxygen has been observed in a variety of astronomical objects.

The LO combines a QCL with a compact, low-input-power Stirling cooler. The 4.7-THz QCL is based on a hybrid design and has been developed for continuous-wave operation, high output powers, and low electrical pump powers [1]. Efficient carrier injection is achieved by resonant longitudinal optical phonon scattering. This design allows for an operating voltage below 6 V. The amount of generated heat complies with the cooling capacity of the Stirling cooler of 7 W at 65 K with 240 W of electrical input power [2]. The QCL has a single-plasmon waveguide with a lateral distributed feedback (DFB) grating, which is optimized for 4.745 THz. This yields single mode emission over most of the driving current range of the laser through one of the end facets of the waveguide. The beam of the QCL is formed with dedicated optics into an almost Gaussian profile. The peak output power of the QCL is 2.5 mW. The frequency tunability ranges from about −1.5 GHz to +5 GHz around the OI rest frequency. The LO is a significant improvement over its predecessor, which has been in routine operation in the GREAT heterodyne spectrometer on SOFIA since 2014 [3].The design of the LO and its performance in terms of output power, frequency accuracy, frequency stability, and beam profile as well as its implementation in upGREAT will be presented.

[1] L. Schrottke, M. Wienold, R. Sharma, X. Lü, K. Biermann, R. Hey, A. Tahraoui, H. Richter, H.-W. Hübers, and H. T. Grahn, Semicond. Sci. Technol. vol. 28, 035011, 5 pages (2013).
[2] H. Richter, M. Greiner-Bär, S. G. Pavlov, A. D. Semenov, M. Wienold, L. Schrottke, M. Giehler, R. Hey, H. T. Grahn, and H.-W. Hübers, Opt. Express vol. 18, pp. 10177–10187 (2010).
[3] H. Richter, M. Wienold, L. Schrottke, K. Biermann, H. T. Grahn, and H.-W. Hübers, IEEE Trans. Terahertz Sci. Technol. vol. 5, pp. 539–545 (2015).

A single-mode BCB-embedded antenna-integrated continuous wave quantum cascade laser for heterodyne measurement at 4.745 THz

Lorenzo Bosco[1,*], C. Bonzon[1], K. Otani[1], M. Justen[2], M. Beck[1], and J. Faist[1]

[1]*Eidgenössische Technische Hochschule Zürich, 8093, Zurich, Switzerland*
[2]*Universität zu Koln, 50923, Köln, Germany*
[*]Contact: lbosco@phys.ethz.ch

Abstract**—**Nowadays, the demanding of an efficient source for the terahertz (THz) region is increasing. Such a source is particularly important for astrophysical application. Indeed, the THz spectrum includes important atomic cooling lines which detection gives information about interstellar mediums. This work purpose is to develop a monochromatic source for detecting [OI] cooling line at 4.745 THz thanks to a heterodyne measurement setup. The source is a THz Quantum Cascade Laser (QCL) based on a four quantum well structure. The single modality of the laser is achieved engineering a double metal cavity, which is composed by a Distributed Bragg Reflector (DBR), a main region with a second order lateral grating, a multi-section tapering and a first order lateral grating used as a front reflector. Moreover, the whole structure is embedded in Benzocyclobutene, also known as BCB, an insulating THz-transparent organic polymer, which is essential for the deposition of a patch-array antenna together with the top metal contact. The antenna function is to out-couple the lasing mode in a single lobe vertically emitted beam. The device is also able to operate up to 55 K in continuous wave and it is tunable up to 3.5 GHz with a peak power close to 2.0 mW.

Solid State Terahertz Sources

Steven A. Retzloff, Jeffrey L. Hesler, Thomas W. Crowe*

Virginia Diodes, Inc. (VDI), Charlottesville, VA, USA
*Contact: crowe@vadiodes.com

Abstract— Solid-state sources using diode based frequency multipliers are critical for radio astronomy and atmospheric studies, as well as many other important scientific applications. VDI shipped its first multiplier based source operating above 1 THz in 2005, generating about 5 uW in the band from 1.3 to 1.35 THz. Since then, VDI and many other groups have worked to improve this technology with great success. However, the needs of the scientific community continue to increase and there is continuous demand for improved power, tuning bandwidth, noise and system reliability. An important example is the requirement for greatly increased power to allow the development of imaging array for radio astronomy; the more receiver pixels that can be pumped, the more rapidly scientific data can be collected. This talk will describe many of the technological innovations that are now used to achieve these goals and two specific recent examples that demonstrate the current capability of this technology.

VDI's THz sources rely on two types of multiplier circuits. Varactor multipliers are used at lower frequency because of their very high efficiency and power handling. For the varactor multipliers, the primary goal is to achieve the power needed to effectively pump the terahertz varistors with the greatest possible efficiency and reliability. This has required significant development of the thermal design of the complete circuit, including the use of diamond heat spreaders, optimal geometric layout of the assembly and more thermally robust materials. In-phase power combining is also used to achieve a roughly 3dB power handling improvement. However, we have found that for very high efficiency multipliers excessive AM noise can be generated due to complex interactions between the combined multiplier circuits if necessary counter measures are not employed.

Above about 500 GHz, varistor multipliers are generally used due to their inherently broad bandwidth and relative simplicity. Also, the efficiency benefit of varactor multipliers is not as significant at these higher frequencies. For these circuits, the primary improvements have been increased operating bandwidth and appropriate scaling to higher frequency. Also, it has been necessary to optimize the designs to achieve good performance at input power levels equal to those available from the lower frequency varactor multipliers. In fact, matching the input power requirement of each multiplier with the power available from the preceding multiplier is critical for the success of the system as a whole.

To demonstrate the success of this effort, two recent results will be briefly reviewed. The first is a 215 GHz varactor source that generates greater than 300 mW. The second is a 2.5 THz source that generates about 5uW in the band from 2.48 to 2.7 THz. This power level can be increased by several dB through the cooling of the final varistor multiplier stages. Prospects for further improvements and technical challenges and opportunities will be discussed.

Local oscillator requirements and noise performance of a cryogenic 360 GHz Schottky diode subharmonic mixer

D. Pardo[1*], H. Wang[1], B. Ellison[1], P. Huggard[1], B. Alderman[1], and J. Grajal[2]
[1]*STFC-Rutherford Appleton Laboratory, OX11 0QX, Oxfordshire, UK*
[2]*Universidad Politécnica de Madrid, E.T.S.I. Telecomunicación, 28040, Madrid, Spain*
*Contact: diego.pardo@stfc.ac.uk

Abstract— Improving the cryogenic performance of subharmonic Schottky mixers in terms of minimum noise and LO power is of high relevance for a large number of space missions. In this work we carry out a theoretical and experimental analysis of a 360 GHz subharmonic mixer at 77 K in order to improve our understanding of cryogenic mixers and provide a procedure to reduce the LO power required by these receivers. By using a physics based model of the diode it will be shown that the increase on the LO power reported in the literature for cryogenic subharmonic mixers is related to the increase of the mismatch of the LO source, and that this effect can be taken into account in the design stage by using proper physics based diode models.

I. Introduction

The majority of heterodyne receivers used in spaceborne Earth observation and planetary science missions use GaAs Schottky barrier diodes as the first frequency down-conversion (mixer) element. This is primarily due to the maturity of the technology, its ability to be configured as different mixer topologies, e.g. as multiple diode subharmonic mixers, and to operate at both room and cryogenic temperatures. A brief review of related planned European space missions reveals the continued importance of this technology with mission examples including: MicroWave Sounder (MWS), Microwave Imager (MWI) and Ice Cloud Imager (ICI) on-board the European Space Agency's (ESA) MetOp-SG satellites with heterodyne receivers operating at frequencies between 183 GHz and 664 GHz [1]; the Submillimetre Wave Instrument (SWI) of the Jupiter ICy Moons Explorer (JUICE) mission of the ESA consisting of two heterodyne receivers working at approximately 600 GHz and 1.2 THz [2]; and the LOw Cost Upper atmosphere Sounder (LOCUS) concept mission [3] that aims to use Schottkky mixers in four discrete frequency bands of 0.8 THz, 1.15THz, 3.5 THz and 4.7 THz.

In the above examples, Schottky diode receivers operate within either a room or cryogenic temperature environment. Under cryogenic conditions, a well-known enhancement in receiver system sensitivity is gained through a corresponding reduction in mixer and intermediate frequency amplifier noise, with the former the dominating contributor. Less well understood is the change in local oscillator (LO) power required to pump the cryogenic mixer, particularly with respect to a subharmonic (SHP) mixer configuration which can be difficult to dc bias for optimum performance. Results from the former SWI mission [2], for instance, indicated a reduction by nearly a factor 2 in receiver noise temperature at 560 GHz when cooled to 140 K, but at the expense of a doubling in LO power. Similar performance for the noise and the LO power of a subharmonic 1.46 THz Schottky receiver have been reported by VDI when the receiver is cooled to 75 K [4]. Additionally, modern mixer design techniques have obviated the need for mechanical backshort tuning leading to compact device structures exhibiting excellent room temperature performance, but that lack a means of optimisation when cooled. Thus, despite the improvement in the noise temperature achieved via cryogenic receiver operation, the correlated increase in LO power is a significant disadvantage and has impact upon the space payload design and can ultimately limit science return through constrained operational frequency, e.g. tuneable bandwidth, and a restriction in the number of reception channels. It is therefore essential to understand the theoretical attributes of the mixer diode, including circuit requirements, in terms of achieving optimum LO power and noise in a cryogenic context. Doing so allows the development of device design models that are optimised for low temperature performance.

In this work, we report on a theoretical analysis and experimental investigation of the performance of a 360 GHz Schottky diode subharmonic mixer operated at an ambient temperature of ~77K. By using a physics-based drift diffusion (DD) model of the diode coupled to a multi-tone harmonic balance (HB) circuit simulator [5], it will be shown that it is possible to gain a significant improvement in the LO power required to achieve optimum noise performance. We compare results obtained for the same device at both low and room temperatures, and also with results obtained via other researchers using similar mixer configurations.

II. Cryogenic Tests of Schottky Diodes under DC Conditions

GaAs Schottky diodes with anode diameter 5.62 μm and 0.95 μm fabricated by *Teratech Components Ltd* were used in this test. On block current versus voltage measurements of the diodes for temperatures between 77 K and 295 K (IVT) were carried out. The temperature of the diode block was controlled with a liquid nitrogen cryostat. Fig. 1 shows measured results for the diodes with smaller anode diameter.

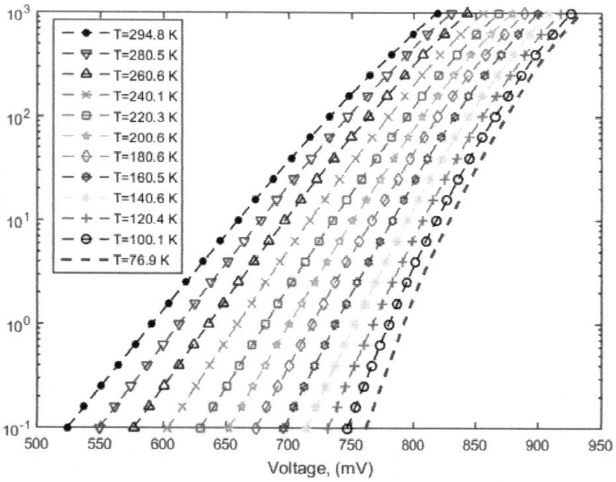

Fig. 1 Current versus voltage measurements as a function of the temperature for diodes with anode diameter 0.95 μm.

From the IVT measurements, the series resistance of the diodes was extracted using the technique described in [6], see Fig. 3 were a comparison with DD simulation results is included. For the diodes with anode diameter 5.62 μm, the series resistance decreases when the temperatures decreases, as expected from the DD simulations. Such a decrease is related with the increase of the electron mobility with the decrease of the temperature. However, the measurements for the diodes with anode diameter 0.95 μm show an increase of the series resistance related to self-heating effects in the diode. Such effects have a significant impact on submicron diodes as described in [7].

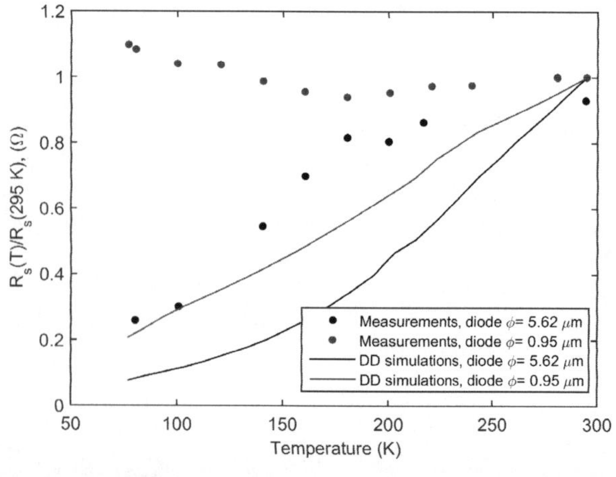

Fig. 2 Series resistance normalized by the values at 295 K as a function of the temperature for the diodes with anode diameter 5.62 μm and 0.95 μm.

The temperature dependence of the ideality factor and fundamental barrier of these diodes were also evaluated by fitting to the standard exponential performance of thermionic emission, as is shown in Figs. 3 and 4. The ideality factor increases as the temperature decreases due to the increase of the thermionic field emission current component [8]. The differences between the measurements and the prediction of the thermionic emission current component are generally attributed to density of states on the surface of the semiconductor, crystal defects, surface roughness or the presence of insulating interfacial layers.

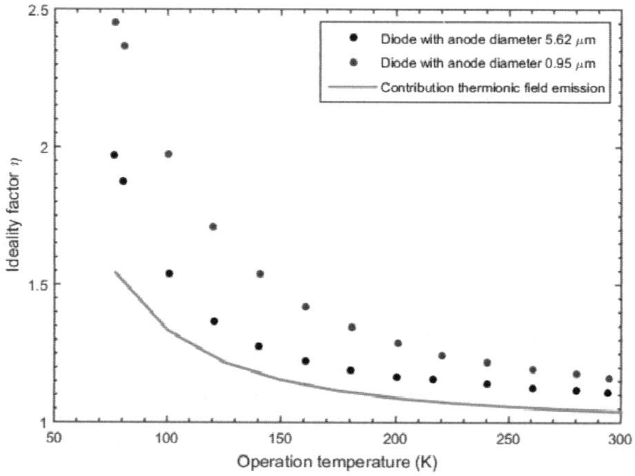

Fig. 3 Ideality factor as a function of the temperature for the Schottky diodes with anode diameter 5.62 μm and 0.95 μm.

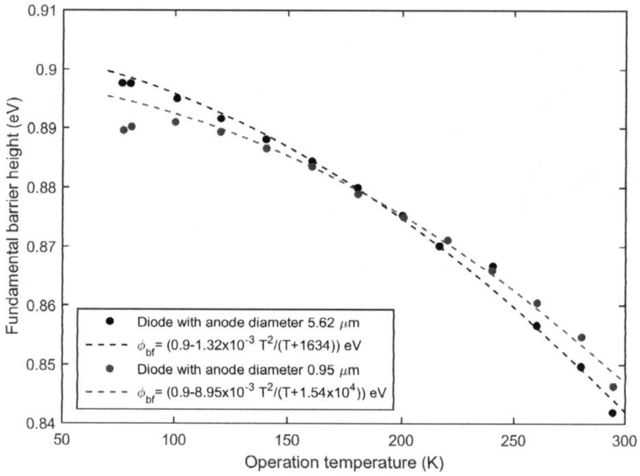

Fig. 4 Fundamental barrier height as a function of the temperature for the Schottky diodes with anode diameter 5.62 μm and 0.95 μm.

The fundamental barrier height is defined by the following equation [9]:

$$\emptyset_{bf} = \eta\emptyset_{b0} + (\eta - 1)\frac{k_B T}{q} ln\left(\frac{N_c}{N_d}\right) \quad (1)$$

where η is the ideality factor, $\emptyset_{b0}$ is zero bias barrier height, N_c is the effective density of states in the conduction band, N_d is the doping density, and k_B is Boltzmann's constant. The measured dependence of $\emptyset_{bf}$ on the temperature, see Fig. 4, is in agreement with the assumption that it is entirely due to the change of the energy gap of the semiconductor (E_g) with the temperature as suggested by some authors [10]. According to [10], the dependence of E_g on the temperature for GaAs is given by:

$$E_g(T) = \alpha - \beta T^2/(T + \gamma) \quad (2)$$

with α=1.522, β= 5.8x10^{-4}, and γ=300.

III. Analysis of a 360 GHz Subharmonic Mixer at Cryogenic Temperatures

As was mentioned in the introduction, results published by different groups [2, 4] have shown that the operation of subharmonic Schottky mixers without tuning elements at cryogenic temperatures is accompanied by a reduction of the noise temperature but at the cost of considerably higher levels of LO power compared with room temperature operation. In this section, we present experimental and theoretical analyses of a 360 GHz subharmonic mixer in order to improve our understanding of the cryogenic performance of Schottky receivers. The subharmonic mixer used in the tests was designed to operate at room temperature, and it uses an antiparallel diode pair of anode diameter 0.95 µm, whose experimental characterization under dc conditions was presented in the previous section.

The two experimental setups in Fig. 5 have been analysed for the 360 GHz SHP mixer. In setup 1, the Schottky based frequency doubler pumping the mixer is outside the cryostat and, hence it is not cooled at 77 K. In the setup 2, the doubler is inside the cryostat and it is cooled with the mixer. The analysis of these setup will provide information about the coupling of the LO power to the mixer at 77 K.

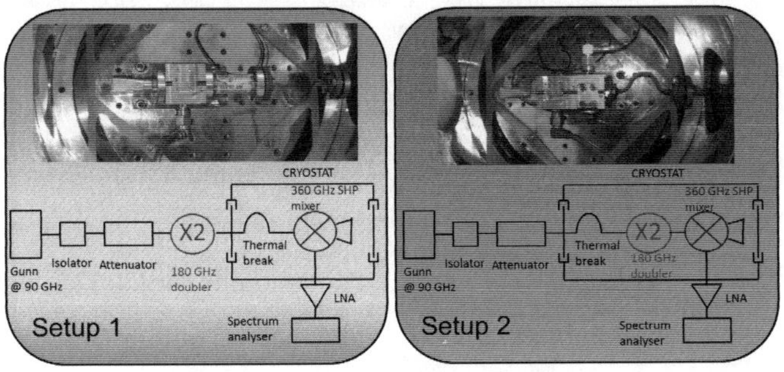

Fig. 5 Sketches of the experimental setups analysed for the 360 GHz SHP mixer. Pictures of the receiver in the cryostat for each setup are also included.

A. Setup 1: doubler outside the cryostat

Figs. 6 and 7 show the DSB mixer noise temperature and conversion loss measured for the 360 GHz SHP mixer in setup 1 at 295 K and 77 K. When the mixer is cooled at 77 K, there is a reduction of the noise temperature by nearly a factor 2 and an improvement of 1 dB in the conversion losses. Contrary to published results for SHP mixers [2, 4], where the LO power required to achieved minimum noise increases when the mixer is cooled, these figures indicate that it is practically the same at 295 K and 77 K.

The DDHB simulations of the conversion losses of the mixer are included in Fig. 7, showing a good agreement with the measurements at both operation temperatures. The DD model takes into account the change of the diode properties when it is cooled, i.e. the reduction of the series resistance, increase of the ideality factor, increase of the fundamental barrier height, and a reduction of the zero-bias junction capacitance associated to the increase of the built-in potential , see section II. Despite the variation of those parameters for the cryogenic mixer diodes, the simulations indicate that there is not a significant change of the coupling of the LO power to the diode pair when operated at 77 K.

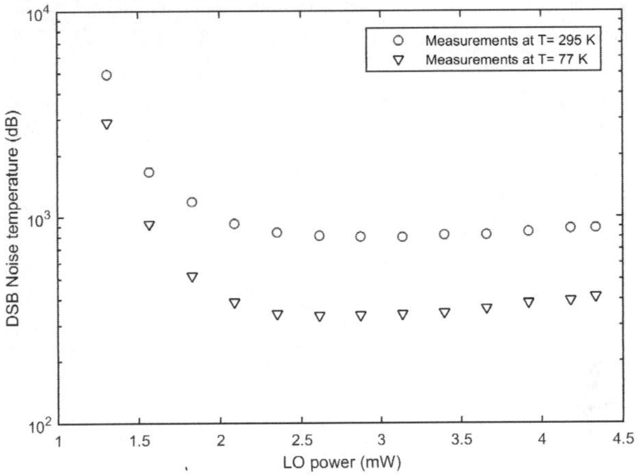

Fig. 6 Measured DSB equivalent input noise temperature of the 360 GHz SHP mixer in setup 1 operating at 295 K and 77 K.

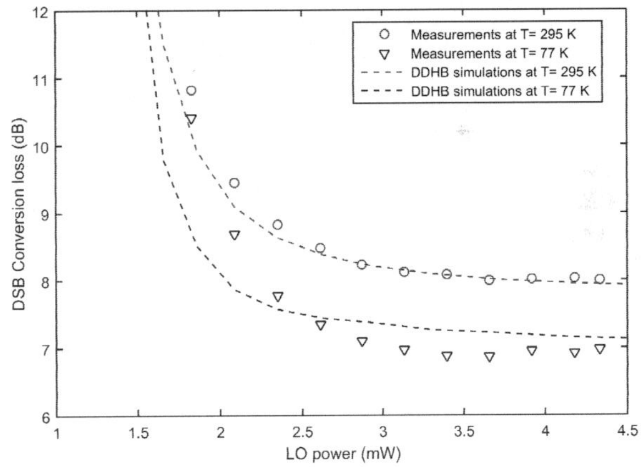

Fig. 7 Measured DSB conversion loss of the 360 GHz SHP mixer in setup 1 operating at 295 K and 77 K. DDHB simulation results are also shown.

A. Setup 2: doubler inside the cryostat

The measured results of the 360 GHz mixer in the setup 2 at 295 K and 77 K are shown in Figs. 8 and 9. When the setup 2 is cooled at 77 K, there is an improvement on the noise temperature and conversion loss similar to the improvement observed in setup 1. However, in this case the LO power required to achieved the minimum noise increases by around 0.5 mW when the receiver is cooled at 77 K.

The conversion losses of the mixer were simulated with the DDHB tool, see Fig. 9. A good agreement with measurements is observed at room temperature, but not at 77 K. In this simulations, the LO circuit impedance at 295 K was also used in the simulations at 77 K (blue dashed line in Fig. 9). In order to reproduce the measurements at 77 K in setup 2, the LO circuit impedance has to be increased by $j35\ \Omega$ (black solid line in Fig. 9). This result indicates that there is a degradation of the coupling of the LO power to the mixer when the LO source is cooled, which causes an increase of the LO power required to achieved minimum noise.

Accounting for this degradation of the LO power coupling at cryogenic temperatures in the design stage will avoid the increase of the LO power detected in this analysis.

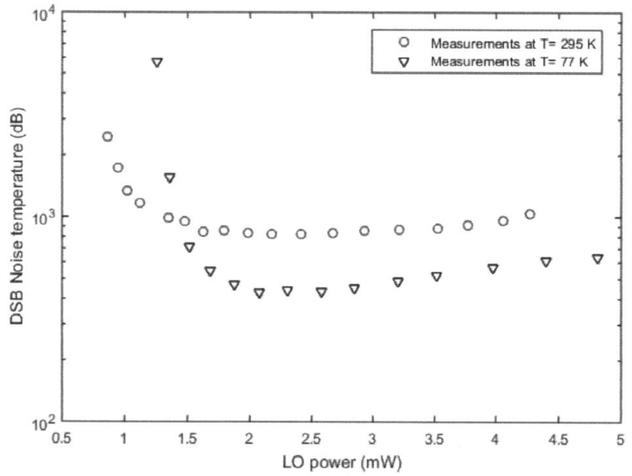

Fig. 8 Measured DSB equivalent input noise temperature of the 360 GHz SHP mixer in setup 2 operating at 295 K and 77 K.

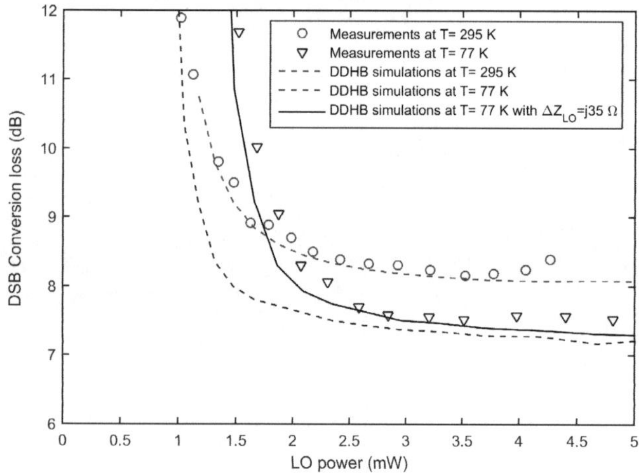

Fig. 9 Measured DSB conversion loss of the 360 GHz SHP mixer in setup 2 operating at 295 K and 77 K. DDHB simulation results are also shown.

IV. Conclusion

The cryogenic performance of a 360 GHz subharmonic Schottky mixer has been analysed experimentally and theoretically by means of a physics based drift diffusion diode model. Measurements of the receiver show an improvement of the conversion loss by 1 dB and the noise temperature by a factor 2 when the receiver is cooled at 77 K. The experimental tests carried out have shown an increase of the LO power required to reach minimum noise when the LO source also operates cryogenically, in agreement with published results. The simulation of the experimental setup with a DD model has shown that such increase of the LO power is due to a degradation of the coupling of the LO power to the mixer diode when the LO source operates cryogenically. Accounting for this LO mismatch during the design stage with an accurate diode model like the DD model used in this work, will lead to cryogenic receivers with improved LO power requirements.

Acknowledgment

This project has received funding from the European Union's Horizon 2020 research and innovation programme under the Marie Skłodowska-Curie grant agreement No 665593 awarded to the Science and Technology Facilities Council, the Spanish National Research and Development Program under project TEC2014-53815-R, and the Madrid Regional Government under project S2013/ICE-3000 (SPADERADAR-CM).

References

[1] M. Henry et al., “Design and development of Schottky diode frequency multipliers for the MetOp-SG satellite instruments,” in *Proc. 41st Int. Conf. Infrared, Millimeter and Terahertz Waves*, 2016.

[2] K. Jacob et al., “Characterization of the 530 GHz to 625 GHz SWI receiver unit for the Jupiter mission JUICE,” in *36th ESA Antenna Workshop on Antennas and RF Systems for Space Science*, Sep. 2015.

[3] D. Gerber et al., “LOCUS: Low cost upper atmosphere sounder,” in *Proc. SPIE*, vol 8889, 2013.

[4] S.A. Retzloff, A. Young, and J.L. Hesler, “A 1.46 THz Schottky receiver at cryogenic temperatures,” in *39th Int. Conf. Infrared, Millimeter and Terahertz Waves*, pp 1–2, Sept 2014..

[5] J. Grajal, V. Krozer, E. Gonzalez, F. Maldonado, and J. Gismero, “Modeling and design aspects of millimeter-wave and submillimeter-wave Schottky diode varactor frequency multipliers,” *IEEE Trans. Microw. Theory Techn.*, vol. 48, no. 4, pp. 700–711, Apr. 2000.

[5] J. Grajal, V. Krozer, E. Gonzalez, F. Maldonado, and J. Gismero, “Modeling and design aspects of millimeter-wave and submillimeter-wave Schottky diode varactor frequency multipliers,” *IEEE Trans. Microw. Theory Techn.*, vol. 48, no. 4, pp. 700–711, Apr. 2000.

[5] J. Grajal, V. Krozer, E. Gonzalez, F. Maldonado, and J. Gismero, “Modeling and design aspects of millimeter-wave and submillimeter-wave Schottky diode varactor frequency multipliers,” *IEEE Trans. Microw. Theory Techn.*, vol. 48, no. 4, pp. 700–711, Apr. 2000.

[5] J. Grajal, V. Krozer, E. Gonzalez, F. Maldonado, and J. Gismero, “Modeling and design aspects of millimeter-wave and submillimeter-wave Schottky diode varactor frequency multipliers,” *IEEE Trans. Microw. Theory Techn.*, vol. 48, no. 4, pp. 700–711, Apr. 2000.

[6] A. Jelenski, E. Kollberg, M. V. Schneider, and H. Zirath, “Accurate determination of the series resistance of mm-wave Schottky-diodes,” *in 15th European Microwave Conference*, pp. 279–284, Sept 1985.

[7] T. Kiuru, J. Mallat, A. V. Raisanen and T. Narhi, "Schottky Diode Series Resistance and Thermal Resistance Extraction From S -Parameter and Temperature Controlled I–V Measurements," *IEEE Trans. Microw. Theory Techn.*, vol. 59, no. 8, pp. 2108-2116, Aug. 2011.

[8] F. Padovani and R. Stratton, “Field and thermionic-field emission in schottky barriers,” *Solid-State Electronics*, vol. 9, no. 7, pp. 695 – 707, 1966.

[9] L. F. Wagner, R. W. Young, and A. Sugerman, “A note on the correlation between the schottky-diode barrier height and the ideality factor as determined from i-v measurements,” *IEEE Electron Device Letters*, vol. 4, no. 9, pp. 320–322, Sep 1983.

[10] R. Hackam and P. Harrop, “Electrical properties of nickel-low-doped n-type gallium arsenide Schottky-barrier diodes,” *IEEE Transactions on Electron Devices*, vol. 19, no. 12, pp. 1231–1238, Dec 1972.

Spectroscopy around 245 GHz based on a SiGe Transmitter and Heterodyne Receiver

Nick Rothbart[1,2*], Klaus Schmalz[3], Johannes Borngräber[3], Dietmar Kissinger[3,4], and Heinz-Wilhelm Hübers[1,2]
[1]*German Aerospace Center (DLR), Institute of Optical Sensor Systems, 12489, Berlin, Germany*
[2]*Humboldt-Universität zu Berlin, Department of Physics, 12489, Berlin, Germany*
[3]*IHP, 15236, Frankfurt (Oder), Germany*
[4]*Technische Universität Berlin, 10623 Berlin, Germany*
*Contact: nick.rothbart@dlr.de

Abstract— Terahertz/millimeter wave gas spectroscopy is an important tool for space exploration due to strong rotational transitions of many gases. We report on a tunable 245 GHz SiGe transmit/receive system based on SiGe technology. The system performance is demonstrated by high resolution gas spectroscopy.

The transmitter (TX) and receiver (RX) chips are fabricated in 0.13 µm BiCMOS SiGe technology at IHP [1]. Both, TX and RX have integrated antennas fabricated by localized backside etching (LBE). The heterodyne RX consists of a 120 GHz push-push voltage controlled oscillator (VCO), a 1/64 frequency divider for the fundamental frequency, a Gilbert-cell based subharmonic mixer, and a five-stage low noise amplifier. The TX is also based on a 120 GHz VCO. The tuning ranges of the TX and RX are 234 to 250 GHz and 237 to 254 GHz, respectively. The emitted output power of the TX is 1 dBm with an equivalent isotropically radiated power of 7 dBi.

For gas spectroscopy experiments, TX and RX are integrated in compact modules that provide e.g. the drive voltages, divider outputs, and input ports for the frequency control voltage. The modules are attached to a gas cell with a 1.9 m path length and plano-convex HDPE lenses as entrance windows for focusing of the radiation. The setup including TX, RX, the folded gas cell, and vacuum pumps is installed on a 45 cm by 75 cm ground pate [2].

The frequency tuning can be realized either by a phase-locked loop (PLL) or by directly applying a voltage to the VCO. For the latter setup, a voltage-frequency calibration was made by tuning the frequency with a PLL and reading out the corresponding voltage applied to the VCO. With respect to the calibration, a linear frequency scan can be performed by a nonlinear voltage ramp. By direct voltage frequency tuning a 2f spectrum of 12 Pa methanol from 238 to 251 GHz was measured. It exhibits a high SNR of 560 at an absorption line with an integrated absorption coefficient of $S = 4.8\times10^{-23}$ cm. The intermediate frequency (IF) was set to a constant value of 400 MHz and the power was detected by rectification with a Schottky diode. Measurements with higher IFs of up to 10 GHz were performed showing the potential of the RX for broadband heterodyne detection.

Acknowledgment

This work was supported by Deutsche Forschungsgemeinschaft (DFG) through the DFG project ESSENCE under Grant SPP 1857.

References

[1] K. Schmalz, J. Borngräber, P. Neumaier, N. Rothbart, D. Kissinger, and H.-W. Hübers, "Gas spectroscopy system at 245 and 500 GHz using transmitters and receivers in SiGe BiCMOS," in Millimeter Waves (GSMM) & ESA Workshop on Millimetre-Wave Technology and Applications, 2016 Global Symposium on, 2016, pp. 1–4.

[2] K. Schmalz, N. Rothbart, P. F.-X. Neumaier, J. Borngräber, H.-W. Hübers, and D. Kissinger, "Gas Spectroscopy System for Breath Analysis at mm-wave/THz Using SiGe BiCMOS Circuits," IEEE Transactions on Microwave Theory and Techniques, vol. 65, no. 5, pp. 1807–1818, May 2017.

The diode heterostructures for THz devices

D. G. Pavelyev*, A.P.Vasilev[2], V.A. Kozlov [1,3], E.S.Obolensky[1], S.V.Obolensky [1], V.M. Ustinov [4]
[1]*Lobachevsky State University, Nizhny Novgorod, Russia*
[2] *SHM R&E Center, RAS, Saint Petersburg, Russia*
[3]*IPM RAS, Nizhny Novgorod, Russia*
[4]*Ioffe Institute, Saint Petersburg, Russia*
*Contact: pavelev@rf.unn.ru

Abstract. – **Planar diode structures with small area active region (~ 1 μm²) based on highly doped GaAs / AlAs superlattices are investigated. The possibility of effective application of such diodes in the terahertz (THz) frequency range is discussed. Monte Carlo simulation of electron transport in 6-period superlattices was carried out. The possibility of increasing the operating frequencies of devices on superlattices by optimizing their parameters and selecting the appropriate diode design is discussed.**

At present time, big interests to high resolution gas spectroscopy in the terahertz (THz) frequency range were observed. Nowadays, the quantum-cascade laser (QCL) is very promising source for spectroscopy. For stabilization of QCL frequency, uncooled harmonic mixers with Schottky planar diodes [1] and superlattice (SL) diodes are currently used [2, 3]. The usage of highly doped SL diodes gives the possibility to stabilize the operating frequencies up to several terahertz [3]. It was shown [4] the possibility of such diodes allow receiving a signal at harmonics up to 8 THz. In [2], a theoretical and experimental comparison of the parameters of the 18- and 6-period superlattices was made. The advantages of usage of small number of periods in terahertz SL diodes were demonstrated.

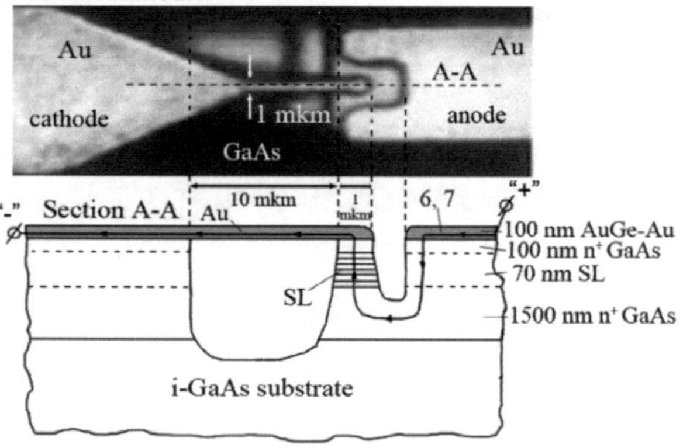

Fig. 1. Design of the planar diode heterostructure. Label "SL" and hatching marked the working area of the diode. Diode working area means the place of high current density flow. This area determines the external parameters of the diode. The heterostructure parts with low current density do not play a fundamental role in the operation of the diode. They are not hatched and are not marked with the label "SL".

In this paper, the transport of electrons in SL is studied by the simulation of electrons motion in the quasihydrodynamic approximation [5] and by the Monte Carlo method [5, 6]. The results of calculations point to the possibility of increasing the operating frequencies of small period SL diodes using the optimization of their parameters and design. For investigation the SL consisted of heterostructure with 6-periods GaAs/AlAs was prepared (Fig. 1). The heterostructure were grown by molecular-beam epitaxy in a Riber 32P unit on semi-insulating GaAs substrates with the (100) orientation. The growth rates of the binary components AlAs and GaAs were calibrated by the observation of x-ray sweep curves at a wide angle near the reflex (004) from the GaAs / AlAs test superlattices. The growth rates were about 1 monolayer / s for GaAs and 0.5 monolayer / s for AlAs. Si was used as the doping impurity.

Figure 1 shows the picture of the planar diode structure. The upper part of the picture is the top view photo. The lower part of the picture is the section A-A. The label «SL» indicates the diode working area, formed by liquid etching. The arrows in the figure show the channel of current flow in the diode: through the metal contact of the cathode, through the diode working area, and then into the metal contact of the anode.

To simulate electron transport in small-period SL, the quasi-hydrodynamic approximation and the Monte Carlo method [6] were applied, and the band diagram of the superlattice was taken into account.

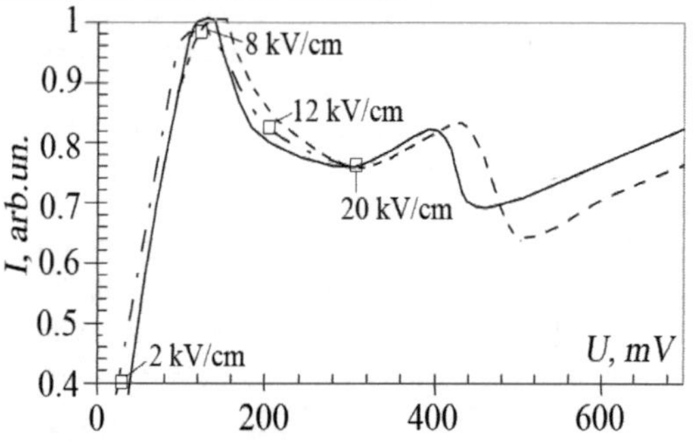

Fig. 2. Experimental line (solid line), inverse (dotted line) and calculated (dash-dotted) volt-ampere characteristics for 6-period SL. The abscissa represents the total voltage applied to the diode. The ratio of the field strength in the superlattice and the external supply voltage was determined by taking into account the resistance of the contacts. The figure shows the applied field to the diode.

Fig. 2 shows experimental and calculated volt-ampere curve for a 6-period superlattice. Two areas are observed for each polarity of voltage: the first area - the ohmic type current flow, the second area – is negative differential conductivity (NDC) which consists of two characteristic areas: first voltage «drop» and second voltage «drop». One can be seen that the values of the corresponding voltage «drops» for forward and reverse bias are different.

It is necessary to note, that ballistic electrons may produce additional dynamic negative differential conductivity in the THz frequency range [7, 8, 9]. Structures on superlattices consisting of 14, 16, 18, 20 monolayers of GaAs and 2, 4, and 6 monolayers of AlAs were investigated theoretically.

The produced 6-period (18x4 monolayers) diode structures were studied experimentally in harmonic mixers of the frequency range 0.2-4.76 THz. Harmonic mixers were used in a precision terahertz spectroscopy apparatus in which a frequency comb was used as the source, excited by femtosecond laser pulses [10, 11], as well as gas and quantum-cascade lasers. The conversion losses for the second (300 GHz, 12 dB) and fourth (450 GHz, 20 dB) LO harmonics on the backward wave tube were determined.

Figure 3 shows a block diagram of a heterodyne receiver with a harmonic mixer based on the investigated diode, where 1 is the pump waveguide (WR-06) with a comb line, 2 - the place where the diode is connected to the output waveguide with a cut-off frequency more than 800 GHz , 3 – the diagonal horn. The investigated beat signal between the harmonics arising in the diode under the influence of pumping and the signal from the terahertz source are fed to the intermediate frequency amplifier and then to the spectro-analyzer.

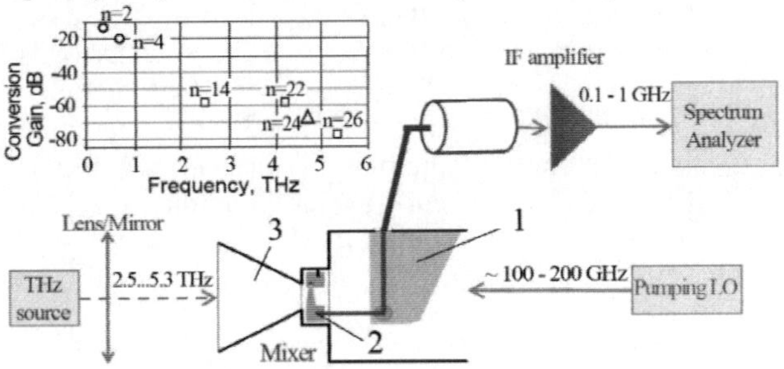

Fig. 3 Block diagram of a heterodyne receiver with a harmonic mixer based on the investigated diode, where 1 is the pump waveguide (WR-06) with a comb line, 2 - the place where the diode is connected to the output waveguide with a cut-off frequency more than 800 GHz , 3 - diagonal horn. The inset shows the dependence of the conversion coefficient on the frequency of the signal for the diode from 6-period superlattice. This dependence was measured using several sources of a terahertz signal: frequency comb excited by a femtosecond laser (n = 2.4); gas laser on CH_3OH (n = 14, 22, 26); and QCL (n = 24). The digits (2,4, ... 26) on the sidebar indicate the numbers of harmonics used by the harmonic mixer pumped from the local oscillator at a frequency of 0.1 - 0.2 THz.

The conversion losses were compared for mixers made on planar diode structures containing superlattices with 18 and 6 periods. A significant difference in the conversion parameters of the fabricated structures was noticed during usage them in phase-stabilization systems of the frequency of quantum cascade lasers at frequencies above 3 THz. Structures with six periods have a conversion loss of 20 dB less, and were successfully applied to stabilize the frequency of such lasers at 4.7 THz [12].

Application of external DC bias leads to arising of self-oscillation in the investigated planar structures. The frequencies of self-oscillation in the SL structures were measured. For the measurements, a similar block diagram was used (Fig. 3), in which DC bias was applied to the mixing diode. The self-oscillation power was measured from the back output of the pump waveguide using a Fourier transform spectrometer with a cryogenic Bolometer as a receiver. Measured frequencies of self-oscillation were 0.25 THz for positive bias and 0.18 THz for negative bias. The maximum power was observed when the operating point was selected in the region of the second "drop" at a bias voltage of ± 500-600 mV.

The difference in the generated frequencies for opposite bias voltages is explained by the fact that the equivalent circuit for the planar diode structure contains capacitance and inductance, which form a serial low-Q circuit. The frequency of the serial resonance is given by: $f = \frac{1}{2\pi\sqrt{LC}}$, where L is the inductance of the supply conductor, and C is the capacitance of the SL. Since the inductance L of the input does not change for opposite polarities of bias voltage, then, the change in frequency is explained by a change of the capacitance by a factor of two, which is due to the difference of the conical shape working area (see Figure 1). Conical shape appears during the formation procedure of diodes by liquid etching.

The work is supported by RFFI grant №15-02-07935.

References

[1] A. Danilov, N. Erickson, A. Light, and J. Waldman, "Phase locking of 2.324 and 2.959 terahertz quantum cascade lasers using a Schottky diode harmonic mixer", *Opt. Lett.*, Vol. 40, p.5090, 2015.

[2] D.G. Pavelyev, A.P. Vasilev, V.A. Kozlov, Yu.I. Koshurinov, E.S. Obolenskaya, S.V. Obolensky, and V.M. Ustinov, "Diode heterostructures for terahertz frequency range devices", *Zh. Radioelektron.*, No. 1, p. 1, 2016.

[3] D.G. Pavelyev, N.V. Demarina, Yu.I. Koshurinov, A.P. Vasilev, E.S. Semenova, A.E. Zhukov, and V.M. Ustinov, "Characteristics of planar diodes based on heavily doped GaAs/AlAs superlattices in the terahertz frequency region", *Semiconductors*, vol. 38, p.1105, 2004.

[4] D.G. Pavelyev, Yu.I. Koshurinov, A.S. Ivanov, A.N. Panin, V.L. Vaks, V.I. Gavrilenko, A.V. Antonov, V.M. Ustinov, and A.E. Zhukov, "Experimental study of frequency multipliers based on a GaAs/AlAs semiconductor superlattices in the terahertz frequency range", *Semiconductors*, vol. 46, p.121, 2012.

[5] E.S. Obolenskaya, A.Yu. Churin, S.V. Obolensky, V.I. Shashkin, and A.V. Murel', "On the radiation resistance of planar Gunn diodes with delta-doped layers", *Semiconductors*, vol. 49, p. 1459, 2015.

[6] S.V. Obolensky, A.V. Murel, N.V. Vostokov and V.I. Shashkin, "Simulation of the Electron Transport in a Mott Diode by the Monte Carlo Method", *IEEE Trans. on Electron Devices*, vol. 58, №8, p.2507, 2011.

[7] V.A. Kozlov, A.B. Kozyrev, A.V. Samokhvalov. "Population inversion and negative differential conductivity in GaN based heterostructures at terahertz frequencies", *XXVI Workshop on Compound Semiconductor Devices and Integrated Circuits held in Europe*, Chernogolovka, Russia, 21-25 May, p.157, 2002.

[8] V.A. Kozlov, A.B. Kozyrev, A.V. Samokhvalov. "Inverse population and the possibility of radiation generation in nano method structures based on GaN", *Bulletin of the Russian Academy of Sciences. Physics*, vol. 67, №2, p. 265, 2003.

[9] V.A. Kozlov, A.V. Nikolaev, A.V. Samokhvalov. *Semiconductor Science and Technology*, "The population inversion and the terahertz negative differential conductivity induced by hot carriers transit time effects in coordinate and momentum spaces", vol. 19, S99, 2004.

[10] A.V. Khudchenko, D.J. Hayton, D.G. Pavelyev, A.M. Baryshev, J.R. Gao, T.-Y. Kao, Q. Hu, J. L. Reno, and V.L. Vaks, "Phase Locking a 4.7 THz Quantum Cascade Laser using a Super-Lattice Diode as Harmonic Mixer", in *Proceedings of the 39th International Conference on Infrared*, Millimeter, and Terahertz Waves IRMMWTHz, Tucson, Arisona, Sept. 14–19, 2014.

[11] D.G. Pavelyev, A.S. Skryl, and M.I. Bakunov, "High-resolution broadband terahertz spectroscopy via electronic heterodyne detection of photonically generated terahertz frequency comb" *Opt. Lett.*, p. 5669, 2014

[12] A.S. Skryl, D.G. Pavelyev, M.Y. Tretyakov, and M.I. Bakunov, "High-resolution terahertz spectroscopy with single tunable frequency comb" *Optics Express 22*, p. 22537, 2014.

Qualification of Direct Detection Technology for ESA MetOp-SG Space Mission

Matthias Hoefle[1,*], Martin Rickes[1], Ion Oprea[1], and Oleg Cojocari[1]
[1]*ACST GmbH, 63457 Hanau, Germany*
*Contact: matthias.hoefle@acst.de

Abstract— This contribution presents the ACST 89 GHz Schottky detector module performance and pre-qualification results towards the second European meteorological operational satellite program MetOp-SG. The detector flight module prototype has been assembled, characterized, and successfully implemented into the 89 GHz bread board receiver within the procurement activity for MetOp-SG. Performances of the detector and overall receiver are discussed.

INTRODUCTION

ACST GmbH is a leading European commercial supplier of Schottky diodes for millimeter wave and sub millimeter wave applications. With core business in the Schottky diode fabrication technology, ACST has recently extended its activity towards assembly and environmental testing of THz components and modules for Space missions. This brings gathering experience and know-how in different critical processes in house and provides opportunity for linkage between different stages within the development process of Space THz-electronics. Being involved in several space-related activities, ACST has become a strong partner in the area of development and qualification of THz electronics for space applications.

Millimeter wave space borne imaging and sounding radiometer instruments accommodate an increasing number of Direct Detection Radiometer (DDR) channels. With this rather compact installation, more hardware can be implemented in the focal plane of the instrument antenna. Medium power local oscillator sources are omitted with the direct detection approach, reducing mass, volume and power consumption of mm-wave receivers [1]. Still challenging is the design and implementation of reliable low noise amplifiers and low noise detectors at the radiometer RF input frequency. ACST has demonstrated the world wide best detector performance regarding signal to noise ratio with excellent linearity, based on matched Schottky diode parameters, detector design, and video amplifier design.

For the highest frequency channel in MetOp-SG with direct detection, ACST is carrying out the full Schottky diode and 89 GHz RF detector module development and qualification process. Starting from the fundamental technology process development of low noise zero-bias Schottky diodes [2], the suitability of the diode technology has been proven in an ESA study with preliminary reliability testing [3][4]. The RF module has been designed to meet the stringent MetOp-SG requirements and will be implemented as a core element of the 89 GHz receiver onboard of the MicroWave Sounder (MWS) and MicroWave Imager (MWI) instrument. The majority of RF module reliability testing was finished at the beginning of 2017, covering strength (bond pull, die shear), endurance (DC life, RF life) and environmental testing (humidity, mechanical shock/vibration, thermal cycling). Most of these tests are performed at ACST measurement and cleanroom facilities.

DIRECT DETECTION MODULE TOWARDS METOP-SG

The original module applied for reliability testing had to be adapted to meet the stringent MetOp-SG requirements [4]. The overall volume is reduced by a factor of 3.5, the interface is reworked and the video amplifier section is fine tuned to meet the deviating MWI and MWS requirements. The main technology processes has proved to be very reliable and remain unchanged for actual flight hardware.

PERFORMANCE AT 89 GHZ

Before being implemented into the 89 GHz receiver, the detector module (Fig. 1) undergoes separate full functional tests at ACST.

Detector module level

Main tests analyze the performance regarding RF voltage responsivity, linearity, input matching, noise performance, all at ambient temperature and over the mission temperature range.

Fig. 1 89 GHz detector module.

The RF voltage responsivity is in the range of 300 V/mW with a considered amplification factor of Av = 100. The required linearity of a maximum deviation of 0.04% within 3 dB dynamic is achieved for power levels up to -25 dBm. The minimum power level to meet the required SNR = 30 dB over the temperature range is -38 dBm. The target input power at detector level is -33 dBm to -30 dBm. The temperature stabilized RF voltage responsivity shows a maximum drift of 0.022dB/°C.

Receiver level

The receiver (Fig. 2) contains a low noise amplifier chain with technology from Fraunhofer IAF, noise and channel filters, a coupler for test port access, the detector module with video filtering and amplification, and the general video section for power supply and other electronics.

The breadboard receiver shows excellent performance results. The RF to video response matches the bandwidth specification over the temperature range for MWS and MWI. The radiometric gain is above 4.5 mV/K over the full temperature range with specification of minimum 3.5 mV/K. The noise figure stays below 3.5 dB, as specified.

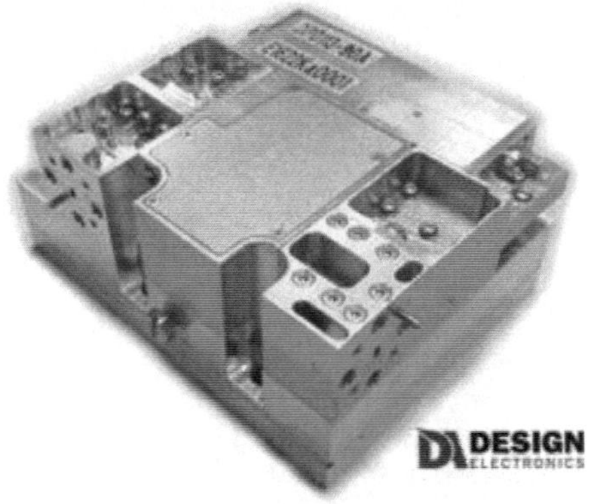

Fig. 2 89 GHz detector module (centre) implemented into the 89 GHz breadboard receiver by DA Design.

Reliability of Direct Detection Technology from ACST

In the framework of the ESA study "Preliminary Reliability Assessment of Millimeter-wave Detectors", several tests are performed on representative test samples, i.e. diodes on RF substrates. Therefore, not only the diode technology is investigated, but also the critical diode assembly on the RF substrates. The objective of the activity is to reveal and correct at an early stage possible degradation or failure mechanisms.

A reliable mounting process has been developed and verified with step stress tests, life tests, and environmental tests in Phase 1 of the activity. All tests have been finished successfully without a single failure of a diode or a diode assembly contact. In a second phase, the tests have been repeated with RF modules and extended test length. Additional tests are added with RF stress and life tests, as well as mechanical tests.

High humidity high temperature (HHHT – 85%/85°C) test with and without bias and thermal cycling (-55°C/+100°C) at ACST, as well as mechanical shock and vibration tests at MILLILAB did not reveal any weakness of the materials and processes within the required MetOp-SG test length and stress levels.

RF life tests with 2 dB margin on the operating power level of -30 dBm for 2000h, as well as 168h at -9 dBm (21 dB margin), did not show any sign of performance deviation.

The most demanding tests for the technology proved to be the DC life and storage tests, with its long test length >1500h, due to the technology limitation of the applied temperature test level. The test at 100°C showed an acceptable parameter drift. The test at the smaller high temperature level at 85°C is ongoing as the last preliminary reliability test.

Conclusions

Presented is the ACST 89 GHz detector module for direct detection radiometry. The detector is designed to meet the MetOp-SG specifications, and is intended to be used in the 89 GHz channel receivers on board MetOp-SG MWS and MWI. The concept is based on low-barrier Schottky diodes, developed for minimum white and 1/f noise contribution. From measurement results, all specifications are fulfilled at ambient temperature with excellent noise and linearity performance, only possible with detailed knowledge on noise reduction at diode, detector and amplifier level. The technology is verified with pre-evaluation tests, regarding long term operation on board the MetOp-SG satellites.

Acknowledgment

The 89 GHz detector module is developed and tested regarding performance and reliability within the ESA activity "Preliminary Reliability Assessment of Millimeter-wave Detectors" (ESTEC Contract Nr.: 4000109751/13/NL/RA).

DA-Design and MILLILAB are acknowledged for the excellent work and cooperation within the procurement phase and the valuable input for this publication.

References

[1] C. Goldstein, M. Trier, A. Maestrini, and J.C. Orlhac, "Present and future R&T development in CNES for Microwave radiometer", MicroRad, 2006.

[2] G. Carpintero, L.E. García-Muñoz, H.L. Hartnagel, S. Preu, and A. Räisänen, "Semiconductor THz Technology: devices and systems for room temperature operation", West Sussex, UK: John Wiley & Sons, 2015.

[3] T. Decoopman, M. Trier, N. Martin, D. Boisbunon, A. Lemasson, J. Tailhades, P. Frijlink, M. Hoefle, O. Cojocari, P. Piironen, M.G. Périchaud, "Millimetre-Wave Detectors for Direct Detection Radiometers", The 9th Global Symposium on Millimeter-Waves (GSMM 2016) and The 7th ESA Workshop on Millimetre-Wave Technology and Applications, June 6-8, 2016, Espoo, Finland

[4] M. Hoefle, O. Cojocari, M. Rickes, M. Sobornytskyy, J. Montero-de-Paz, T. Decoopman, M. Kantanen, M.G. Périchaud, P. Piironen, "89 GHz Schottky Detector Modules for MetOp-SG," The 9th Global Symposium on Millimeter-Waves (GSMM 2016) and The 7th ESA Workshop on Millimetre-Wave Technology and Applications, June 6-8, 2016, Espoo, Finland